AF358722

Acoustics, Soundscapes and Sounds as Intangible Heritage

Acoustics, Soundscapes and Sounds as Intangible Heritage

Editors

Lidia Alvarez-Morales
Margarita Díaz-Andreu

Basel • Beijing • Wuhan • Barcelona • Belgrade • Novi Sad • Cluj • Manchester

Editors
Lidia Alvarez-Morales
Universitat de Barcelona
Barcelona
Spain

Margarita Díaz-Andreu
ICREA, Universitat de
Barcelona
Barcelona
Spain

Editorial Office
MDPI AG
Grosspeteranlage 5
4052 Basel, Switzerland

This is a reprint of articles from the Special Issue published online in the open access journal *Acoustics* (ISSN 2624-599X) (available at: https://www.mdpi.com/journal/acoustics/special_issues/Acoustics_Soundscapes_Intangible_Heritage).

For citation purposes, cite each article independently as indicated on the article page online and as indicated below:

Lastname, A.A.; Lastname, B.B. Article Title. *Journal Name* **Year**, *Volume Number*, Page Range.

ISBN 978-3-7258-1503-6 (Hbk)
ISBN 978-3-7258-1504-3 (PDF)
doi.org/10.3390/books978-3-7258-1504-3

Cover image courtesy of Jerónimo Vida Manzano
Palace of Charles V in the Alhambra, Granada, Spain

Contents

About the Editors

Lidia Alvarez Morales

Lidia Alvarez Morales is an acoustic engineer specialising in the acoustics of historical and archaeological sites. During her professional career, she has had the opportunity to work on several international projects and carry out extensive research on the present and past acoustics of heritage sites located all around the world, with a special focus on cathedrals and rock art sites. Her pathway to acoustics began at the University of Malaga (Spain), where she obtained a degree in telecommunication engineering with specialisation in image and sound. During her master's studies at the University of Granada (Spain), she focused on architectural acoustics. Lidia's background in the field continued to grow through her participation for more than 7 years in three multidisciplinary research projects developed at the Architecture School of the University of Seville, where she completed her PhD on the acoustic study of large places of worship in 2016. In 2017, Lidia was awarded a Marie Sklodowska-Curie Individual Fellowship to develop a 2-year research project (CATHEDRAL ACOUSTICS, Oct 2018-Sept 2020) at the department of Theatre, Film, Television and Interactive Media of the University of York (UK). In March 2021, Lidia joined the ERC Artsoundscapes project as a senior postdoctoral researcher working on the acoustic characterisation of rock art sites and landscapes.

Margarita Díaz-Andreu

Margarita Díaz-Andreu is an ICREA Research Professor based at the University of Barcelona (2012-ongoing). She was a Lecturer and Reader in Archaeology at Durham University (1996-2011) and an Assistant Lecturer in the Department of Prehistory at the Complutense University of Madrid (1994-95). She is the author of many publications, including books, edited books and articles. She has worked on rock art for more than three decades, and in the last decade she has specialised in archaeoacoustics. Her pioneering work in this field has been acknowledged with an ERC Advanced award, the Artsoundscapes project ("The sound of special places: exploring rock art soundscapes and the sacred" ref. 787842, 2018-24). Together with her team of post-doctoral and pre-doctoral students and research assistants, she has undertaken several archaeoacoustic seasons in diverse areas of the world with the aim of analysing rock art landscapes through physical acoustics. The results have also been studied by members of the project working on psychoacoustics, neuroacoustics and ethnomusicology. She has conducted research in Siberia, Spain, Baja California and South Africa. During her academic career, Díaz-Andreu has supervised 17 PhD students through to completion and she is currently supervising 5. She has tutored 13 post-doctoral researchers and now she is tutoring 3 more; all working on archaeoacoustics. She is regularly part of evaluating panels for grants, projects and prizes at a national and international level. Margarita Díaz-Andreu was awarded the Menéndez Pidal National Prize for the Humanities in 2021. Her contribution to the study of archaeoacoustics was explicitly mentioned. In 2022, she was invited to give the prestigious Felix Neubergh lecture in Sweden.

Preface

Thanks to the Artsoundscapes ERC project, we, an acoustic engineer who first specialised in the acoustics of European cathedrals, and an archaeologist exploring the sonic features of rock art sites, found our research paths converging. Our shared experience in facing the challenges of empirically studying the past acoustics of these sites sparked the idea of compiling "Acoustics, Soundscapes, and Sounds as Intangible Heritage", now reprinted in this volume.

As editors, we were priviledged to feature the contributions of many well-known researchers in the fields of historical acoustics and archaeoacoustics. The archaeological and historical sites explored in their articles offer great variety regarding architectural features and cultural backgrounds, leading to diverse methodological approaches in assessing their acoustic properties. Such variety greatly enriched this Special Issue by providing new understanding of the acoustic heritage of these sites. Therefore, this reprint makes a significant contribution to future research in this discipline.

This volume was partially funded by the Artsoundscapes Advanced ERC project (Grant Agreement No. 787842, Principal Investigator: ICREA Research Professor Margarita Díaz-Andreu) funded by the European Research Council (ERC) under the European Union's Horizon 2020 research and innovation programme.

Lidia Alvarez-Morales and Margarita Díaz-Andreu
Editors

acoustics

MDPI

Editorial

Acoustics, Soundscapes and Sounds as Intangible Heritage

Lidia Alvarez-Morales [1,2,*] and Margarita Díaz-Andreu [1,2,3,*]

[1] Departament d'Història i Arqueologia, Universitat de Barcelona, 08001 Barcelona, Spain
[2] Institut d'Arqueologia, Universitat de Barcelona (IAUB), 08001 Barcelona, Spain
[3] Institució Catalana de Recerca i Estudis Avançats (ICREA), 08010 Barcelona, Spain
[*] Correspondence: lidiaalvarez@ub.edu (L.A.-M.); m.diaz-andreu@ub.edu (M.D.-A.)

Citation: Alvarez-Morales, L.;
Díaz-Andreu, M. Acoustics,
Soundscapes and Sounds as
Intangible Heritage. *Acoustics* 2024, 6,
408–412. https://doi.org/10.3390/
acoustics6020022

Received: 18 February 2024
Accepted: 24 April 2024
Published: 2 May 2024

Since UNESCO unveiled its declaration for an integrated approach to safeguarding tangible and intangible cultural heritage in 2003 [1], increased emphasis has been placed on the intangible and immaterial components linked to archaeological and historical sites. Their sonic aspects may be regarded as one of these intangible components, as they have played a crucial role in shaping individuals' perceptions about past spaces and buildings and the way people have interacted with such spaces over time [2,3]. Thus, acoustics, soundscapes, and sounds have been embraced as key subjects for a better understanding of our cultural heritage [4].

Adopting the broad definition of the term 'acoustical heritage' proposed by Zhu, Oberman, and Aletta [5], a thorough examination of acoustical heritage should encompass diverse approaches and disciplines. This entails not only delving into the realms of physical acoustics but also exploring the intersections with music, psychoacoustics, history, archaeology, and various other fields. However, there has been a marked emphasis on conducting comprehensive initial acoustic characterizations that have subsequently served as the cornerstone for further multidisciplinary approaches [6–9]. At archaeological and historical sites this type of acoustic research focuses on recovering and analyzing their physical and quantifiable acoustic features. This is carried out through a variety of experimental and simulation techniques, that are mainly based on room acoustics theories and methods [10,11]. Nonetheless, dealing with such sites requires a broad understanding of the term "room", as such spaces often exhibit complex forms and peculiar finishing materials whose acoustic performance is not well known. This is particularly evident in the context of archaeological sites in natural locations, where the morphology is entirely fortuitous and singular. In addition, uncertainties concerning the details of the sound-related events performed in these spaces typically arise due to the scarcity of available evidence.

Therefore, given the inherent complexity of applying standardized methodologies and procedures to archaeological and historical sites, specific considerations are required for a comprehensive analysis of each individual case study. In this regard, Aletta and Kang remarked on the considerable challenges present when dealing with these sites [12]. They emphasized not only the difficulties in conducting standardized impulse response (IR) measurements [13], but also in making plausible acoustic simulations. In this respect, simulating the past acoustics of sites that no longer exist or have undergone significant changes over time is particularly challenging. In such cases, conducting field measurements to follow the recommended validation process for a reliable simulation model is not possible [14]. Moreover, accurately delineating the lost geometry of these sites and realistically estimating the acoustic performance of their finishing materials in the period under study may be difficult or even impossible. In addition, in order to interpret the results, it is essential to take into account the socio-cultural component in which both the sites and the sonic events being investigated are framed.

This dedicated compilation of articles focuses on the acoustical analysis of archaeological and historical sites, placing special emphasis on the methodologies and procedures used. It aims to highlight the necessity of diverse and interdisciplinary approaches to a comprehensive exploration of past sonic experiences. With this purpose, it brings together a collection of original research articles that delve into the acoustics and sonic aspects of a diverse group of sites worldwide. It includes a considerable range of case studies that span various approaches and methodologies, such as acoustic simulations based on finite-difference time-domain (FDTD) methods and geometrical acoustics (GA), as well as spatial IR measurements. The authors have faced and successfully navigated several of the challenges mentioned above, thus contributing to more rigorous and significant research in the field, while thereby expanding our acoustical heritage.

The first article published as part of this Special Issue, *Sound Reflections in Indian Stepwells: Modelling Acoustically Retroreflective Architecture* by Cabrera et al., focuses on the acoustics of two Indian stepwells (or stepped ponds). It analyzes the retroreflective effect—a notable acoustic phenomenon consisting of a multitude of reflections returning to the source, translating into a support for the performer—potentially present in step-based architectures such as the Indian stepwells. The authors used FDTD simulations validated by image source-based calculations to assess the stepped ponds' retroreflective potential and audibility. This assessment depended on the structures' size and geometry, taking into account different positions of the sound source. In addition to introducing a new dimension to the exploration of Indian stepwells, the analysis conducted in these semi-enclosed ancient structures contributes to a deeper understanding of this complex acoustic phenomenon. These structures can be considered an exemplary case study despite not having been initially designed for acoustic purposes but for descending to the pond.

The research presented by Weber et al. delves into another complex acoustic phenomenon analyzed by applying a methodological approach based on the FDTD method. On this occasion, the FDTD simulation is used to investigate the *Sound Scattering by Gothic Piers and Columns of the Cathedrale Notre-Dame de Paris* (France). The simulation is validated through a meticulous comparison of the simulation results with experimental measurements taken on scale models of seven piers and columns of different designs, and analytical models. The simulated scattered fields are analyzed in terms of sound propagation and the results are discussed in terms of audibility under different listening scenarios. With this case study, the authors not only add new knowledge to the acoustics of the emblematic Gothic Cathedral of Notre-Dame de Paris, but also offer new insights into the diffraction and higher-order propagation effects caused by these architectural elements, typically present in other cathedrals and historical buildings.

Computation based on FDTD is also used by Martellotta et al. in their article entitled *Low-Frequency Response of a Rupestrian Church by Means of FDTD Simulation*. An FDTD modeling is chosen to simulate the low-frequency behavior of the 11th-century Rupestrian Church Saints Andrew and Procopius, in Monopoli (Apulia, Italy). They first analyze the acoustics of the church remains using information derived from a set of IRs measured on-site, focusing on the results of several room acoustical parameters, spectrograms, and modal reverberation time. These data also serve to calibrate the simulation model of the church in its current conservation state, whose finishing was reconstructed by adding plaster and frescoes in the late medieval period. In order to accurately define the model's boundary conditions, the authors consider it essential to characterize the acoustic performance of the tuffaceous surfaces of the church using an impedance tube. Seating audience is also incorporated in the reconstructed model to investigate whether the strong modal behavior caused by the morphology of the space may have impacted the liturgical celebrations held in the church during that period.

In their *Investigation of a Tuff Stone Church in Cappadocia* via *Acoustical Reconstruction*, Adeeb et al. use room acoustic simulations based on ray tracing and image source methods to acoustically reconstruct the Bell Church, a partially demolished Middle-Byzantine masonry church in Cappadocia (Turkey). Since acoustic measurements cannot be performed in the church in its current condition, the simulation models are tuned taken as a reference a set of acoustic measurements previously conducted in two structures in the nearby region built with similar materials. Material tests are again necessary during the modelling for determining the sound absorption characteristics of tuff rocks [15]. This acoustic study assesses the evolution of the acoustical performance of the church over time based on three different acoustic models of the space: Phase I (without frescoes), Phase I (with frescoes), and Phase II (with frescoes and narthex), in an attempt to understand the context of the use of a Middle-Byzantine church for different cultural practices.

An Acoustic Reconstruction of the House of Commons, c. 1820 and 1834 is presented by Foteinou et al. In this article, room acoustic simulations based on ray tracing and image source methods serve to evaluate how the lost Commons chamber (Palace of Westminster, London, UK) might have shaped political speeches and discussions during the period under investigation. Because the calibration of the acoustic model by means of IR measurements is deemed not to be possible, several versions of the model are assessed in order to ensure a feasible representation of its boundary conditions. The authors opt for a validation approach similar to that mentioned by Adeeb and Sü Gül [16]. Such an approach uses on-site measurements previously conducted in five spaces sharing geographical location, historical uses and/or architectural characteristics with the House of Commons, as a reference to support the modeling decisions made. The simulated IRs are used not only to describe the acoustics of the space through acoustic parameters but also to conduct historically-informedauralizations.

The article entitled *The Bacinete Main Shelter: A Prehistoric Theatre?* by Alvarez-Morales et al. assesses the potential influence of acoustics on the sociocultural practices of prehistoric societies in two shelters at the Bacinete rock art complex (Cádiz, Spain). The selection of these shelters is guided by both their spatial configuration and the presence of rock art, rendering plausible the hypothesis that they could have functioned as "performance" spaces. The archaeoacoustic study is based on the analysis of a comprehensive set of spatial IRs measured on-site using Ambisonics techniques. The authors examine the acoustic properties of the shelters under different usage hypotheses. These hypotheses are designed to partially compensate for the uncertainties related to, firstly, the lack of evidence regarding the particular sound events undertaken by post-Paleolithic hunter-gatherer societies and, secondly, the potential alterations in the site due to prolonged exposure to weathering agents. Additionally, Alvarez-Morales and her colleagues delve into considerations regarding the applicability and limitations of standardized protocols described in the ISO 3382-1 [13] for characterizing the acoustics of open-air spaces.

The Renaissance palace of Charles V, built inside the fortification of the Islamic Alhambra Palace (Granada, Spain), is the research object of *The Acoustics of the Palace of Charles V as a Cultural Heritage Concert Hall* by Almagro-Pastor et al. Despite its original design not catering for acoustic purposes, since 1883 this heritage site has been used as an unroofed concert hall. This motivated the authors to examine the acoustics of the venue on the basis of data derived from experimental measurements taken on-site. The methodology employed involves a substantial number of source and receiver positions designed to explore the impact of the palace's geometry on the acoustics experienced by the performers and audience. The acoustic parameters derived from the collected IRs and the spatial distribution of reflections are analyzed. The authors explain that an additional set of Impulse Responses (IRs) was recorded using a studio monitor as a sound source. This was intended for auralization purposes with the view of minimizing the impact of both the frequency response and the directivity of the dodecahedral source.

In their article *Measuring the Acoustical Properties of the BBC Maida Vale Recording Studios for Virtual Reality*, Kearney et al. implement a detailed and time-consuming optimized protocol for registering spatial IR sets for virtual reality (VR) applications at heritage sites. The proposed protocol requires a proper definition of the characterized location grid to ensure sufficient IRs for real-time interpolation. It involves different types of loudspeakers and microphones to enable reconstruction from the perspective of both the performers (3DoF) and the audience (6DoF). Moreover, multiple spatial IR measurements at each characterized location considering different loudspeaker rotations are included in order to achieve a realistic representation of the directional properties of the sound source. The proposed workflow is then applied in four iconic studio spaces at the British Broadcasting Corporation (BBC) Maida Vale studios (London, UK), which vary in size, geometry, and purpose. A crucial aspect of this analysis is the open accessibility of the captured impulse response database, making it available to a global audience.

Last but not least, this Special Issue also features an insightful review on the *Intangible Mosaic of Sacred Soundscapes in Medieval Serbia* by Đorđević et al. The initial part of this work provides an overview of previously published research concerning the indoor acoustics of medieval churches in Serbia. The surveyed works predominantly used IR measurements and virtual models to characterize their acoustic features, with a particular emphasis on their impact on speech and singing in medieval religious practices. The subsequent section delves into the limited body of literature addressing open-air soundscapes associated with religious activities in the area. Following a sociocultural analysis of the significance of outdoor soundscapes in Orthodox Christian practices during this historical period, the authors advocate the inclusion of significant sound-related elements such as the use of the semantra and bells, the sonic events integral to litany processions, and the state and church assemblies held in the immediate proximity of churches and monasteries. These, together with an examination of indoor church acoustics should be part of future archaeoacoustic enquiries. In their view, this holistic approach is essential for a comprehensive understanding of the aural environment of sacred soundscapes in medieval Serbia.

To conclude, we would like to highlight the richness of this Special Issue by acknowledging both the broad range of methodologies used and approaches taken by the authors, as well as the chronological and cultural diversity represented by the investigated case studies. We hope this Special Issue, in conjunction with the enduring topic collection on 'Historical Acoustics', will contribute to enriching the academic discourse on the world's acoustical heritage by fostering the latest and innovative research in the field. Finally, and equally important, we hope it will serve to reinforce the need for a multidisciplinary approach that encompasses acoustics, soundscapes, and sounds as integral components of our intangible heritage.

Author Contributions: Conceptualization, L.A.-M. and M.D.-A.; Methodology, L.A.-M. and M.D.-A.; Writing—Original Draft Preparation, L.A.-M.; Writing—Review and Editing, L.A.-M. and M.D.-A.; Funding Acquisition, M.D.-A. All authors have read and agreed to the published version of the manuscript.

Funding: This work was funded by the European Research Council (ERC) under the European Union's Horizon 2020 Research and Innovation Programme, through the Artsoundscapes Advanced ERC project (Grant Agreement No. 787842, Principal Investigator: ICREA Research Professor Margarita Díaz-Andreu).

Acknowledgments: The editors would like to extend their gratitude to all the authors who contributed their exceptional research to make this special edition a success. We also wish to express our sincere appreciation to the diligent reviewers whose meticulous efforts and insightful feedback strengthened the quality of the articles presented in this issue. Furthermore, we are indebted to the dedicated editorial and production teams whose support has been pivotal in bringing this special edition to fruition.

Conflicts of Interest: The authors declare no conflicts of interest.

List of Contributions

1. Alvarez-Morales, L.; da Rosa, N.S.; Benítez-Aragón, D.; Macías, L.F.; Lazarich, M.; Díaz-Andreu, M. The Bacinete Main Shelter: A Prehistoric Theatre? *Acoustics* **2023**, *5*, 299–319. https://doi.org/10.3390/acoustics5010018.
2. Foteinou, A.; Murphy, D.; Cooper, J.P.D. An Acoustic Reconstruction of the House of Commons, c. 1820–1834. *Acoustics* **2023**, *5*, 193–215. https://doi.org/10.3390/acoustics5010012.
3. Kearney, G.; Daffern, H.; Cairns, P.; Hunt, A.; Lee, B.; Cooper, J.; Tsagkarakis, P.; Rudzki T.; Johnston, D. Measuring the Acoustical Properties of the BBC Maida Vale Recording Studios for Virtual Reality. *Acoustics* **2023**, *4*, 783–799. https://doi.org/10.3390/acoustics4030047.

4. Weber, A.; Katz, B.F.G. Sound Scattering by Gothic Piers and Columns of the Cathédrale Notre-Dame de Paris. *Acoustics* **2023**, *4*, 679–703. https://doi.org/10.3390/acoustics4030041.
5. Cabrera, D.; Lu, S.; Holmes, J.; Yadav, M. Sound Reflections in Indian Stepwells: Modelling Acoustically Retroreflective Architecture. *Acoustics* **2023**, *4*, 227–247. https://doi.org/10.3390/acoustics4010014.
6. Đorđević, Z.; Novković, D.; Dragišić, M. Intangible Mosaic of Sacred Soundscapes in Medieval Serbia. *Acoustics* **2023**, *5*, 28–45. https://doi.org/10.3390/acoustics5010002.
7. Almagro-Pastor, J.A., García-Quesada, R.; Vida-Manzano, J.; Martínez-Irureta, F.J., Ramos-Ridao, A.F. The Acoustics of the Palace of Charles V as a Cultural Heritage Concert Hall. *Acoustics* **2023**, *4*, 800–820. https://doi.org/10.3390/acoustics4030048.

References

1. UNESCO. *Convention for the Safeguarding of the Intangible Cultural Heritage*; Technical Report for UNESCO: Paris, France, 2018. Available online: https://ich.unesco.org/en/convention (accessed on 23 April 2024).
2. UNESCO. *The importance of sound in today's world: Promoting best practices.*; Technical Report for UNESCO: Paris, France, 2017. Available online: https://unesdoc.unesco.org/ark:/48223/pf0000259172 (accessed on 23 April 2024).
3. Gibson, J.J.; Carmichael, L. *The Senses Considered as Perceptual Systems*; Houghton Mifflin: Boston, MA, USA, 1966; Volume 2.
4. Katz, B.F.G.; Murphy, D.; Farina, A. The Past Has Ears (PHE): XR explorations of acoustic spaces as cultural heritage. In *International Conference on Augmented Reality, Virtual Reality and Computer Graphics*; Springer International Publishing: Cham, Switzerland, 2020; pp. 91–98.
5. Zhu, X.; Oberman, T.; Aletta, F. Defining acoustical heritage: A qualitative approach based on expert interviews. *Appl. Acoust.* **2024**, *216*, 109754. [CrossRef]
6. Cook, I.A.; Pajot, S.K.; Leuchter, A.F. Ancient Architectural Acoustic Resonance Patterns and Regional Brain Activity. *Time Mind* **2008**, *1*, 95–104. [CrossRef]
7. Fazenda, B.; Scarre, C.; Till, R.; Jiménez Pasalodos, R.; Rojo Guerra, M.; Tejedor, C.; Ontañon Peredo, R.; Watson, A.; Wyatt, S.; García Benito, C.; et al. Cave acoustics in prehistory: Exploring the association of Palaeolithic visual motifs and acoustic response. *J. Acoust. Soc. Am.* **2017**, *142*, 1332–1349. [CrossRef] [PubMed]
8. Boren, B. Acoustic simulation of J.S. Bach's Thomaskirche in 1723 and 1539. *Acta Acust.* **2021**, *5*, 14. [CrossRef]
9. De Muynke, J.; Baltazar, M.; Monferran, M.; Voisenat, C.; Katz, B.F.G. Ears of the past, an inquiry into the sonic memory of the acoustics of Notre-Dame before the fire of 2019. *J. Cult. Herit.* **2024**, *65*, 169–176. [CrossRef]
10. Rindel, J.H. Roman Theatres and Revival of Their Acoustics in the ERATO Project. *Acta Acust. United Acust.* **2013**, *99*, 21–29. [CrossRef]
11. Navas-Reascos, G.; Alonso-Valerdi, L.M.; Ibarra-Zarate, D.I. Archaeoacoustics around the World: A Literature Review (2016–2022). *Appl. Sci.* **2023**, *13*, 2361. [CrossRef]
12. Aletta, F.; Kang, J. *Historical Acoustics: Relationships between People and Sound over Time*, 1st ed.; Acoustics, MDPI: Basel, Switzerland, 2020; p. 234. [CrossRef]
13. *ISO 3382-1*; Acoustics-Measurement of Room Acoustic Parameters—Part 1: Performance Spaces. International Organization for Standardization: Geneva, Switzerland, 2009.
14. Postma, B.N.; Katz, B.F. Creation and calibration method of acoustical models for historic virtual reality auralizations. *Virtual Real.* **2015**, *19*, 161–180. [CrossRef]
15. Martellotta, F.; Liuzzi, S.; Rubino, C. Reviving the Low-Frequency Response of a Rupestrian Church by Means of FDTD Simulation. *Acoustics* **2023**, *5*, 396–413. [CrossRef]
16. Adeeb, A.H.; Sü Gül, Z. Investigation of a Tuff Stone Church in Cappadocia via Acoustical Reconstruction. *Acoustics* **2022**, *4*, 419–440. [CrossRef]

Article

Sound Reflections in Indian Stepwells: Modelling Acoustically Retroreflective Architecture

Densil Cabrera [1,*], Shuai Lu [1], Jonothan Holmes [1] and Manuj Yadav [1,2,*]

[1] Sydney School of Architecture, Design and Planning, The University of Sydney, Sydney, NSW 2006, Australia; shuai.lu@sydney.edu.au (S.L.); jonothan.holmes@sydney.edu.au (J.H.); manuj.yadav@akustik.rwth-aachen.de (M.Y.)

[2] Institute of Hearing Technology and Acoustics, RWTH Aachen University, Kopernikusstr. 5, 52074 Aachen, Germany

[*] Correspondence: densil.cabrera@sydney.edu.au

Abstract: Retroreflection is rarely used as a surface treatment in architectural acoustics but is found incidentally with building surfaces that have many simultaneously visible concave right-angle trihedral corners. Such surfaces concentrate reflected sound onto the sound source, mostly at high frequencies. This study investigated the potential for some Indian stepwells (stepped ponds, known as a kund or baori/baoli in Hindi) to provide exceptionally acoustically retroreflective semi-enclosed environments because of the unusually large number of corners formed by the steps. Two cases— Panna Meena ka Kund and Lahan Vav—were investigated using finite-difference time-domain (FDTD) acoustic simulation. The results are consistent with retroreflection, showing reflected energy concentrating on the source position mostly in the high-frequency bands (4 kHz and 2 kHz octave bands). However, the larger stepped pond has substantially less retroreflection, even though it has many more corners, because of the greater diffraction loss over the longer distances. Retroreflection is still evident (but reduced) with non-right-angle trihedral corners ($80°–100°$). The overall results are sufficiently strong to indicate that acoustic retroreflection should be audible to an attuned visitor in benign environmental conditions, at least at moderately sized stepped ponds that are in good geometric condition.

Keywords: architectural acoustics; room acoustics; early reflections; reflection; retroreflection; stepwell; intangible heritage; FDTD

Citation: Cabrera, D.; Lu, S.; Holmes, J.; Yadav, M. Sound Reflections in Indian Stepwells: Modelling Acoustically Retroreflective Architecture. *Acoustics* **2022**, *4*, 227–247. https://doi.org/10.3390/acoustics4010014

Academic Editors: Margarita Díaz-Andreu and Lidia Alvarez Morales

Received: 23 January 2022
Accepted: 28 February 2022
Published: 2 March 2022

Publisher's Note: MDPI stays neutral with regard to jurisdictional claims in published maps and institutional affiliations.

1. Introduction

In architecture, there are some notable acoustic phenomena that can provide fascinating auditory experiences for a listening visitor. Such phenomena include long echoes, extreme reverberation (e.g., in large, hard-surfaced rooms and cisterns), whispering walls and domes (e.g., Gol Gumbaz's dome, Bijapur, India) [1]. Steps can be a source of acoustic intrigue, as exemplified by the chirp-like diffraction effects at the Mayan Chichén Itzá pyramid [1–3], or the efficient sound transmission over arrays of stepped seats in the Epidaurus theatre [4,5]. These and other acoustic phenomena can take the listener's spatio-temporal experience beyond the limits of vision, opening up another mode of experiencing space and time as an expression of the architectural form. As well as a source of intrigue for tourists, such experiences provide concrete exemplars of sound propagation phenomena which may have broader applications in architectural acoustics design.

The present paper investigated another step-related acoustic phenomenon in distinctive and sometimes monumental and ancient architecture: acoustic retroreflection in some Indian stepwells. This phenomenon arises in cases where there are many simultaneously visible concave trihedral right-angle corners and has been previously studied in building façades [6–8]. It is characterised by a multitude of reflections returning to the source position (Section 1.1). Certain stepwells might have the potential to create stronger retroreflective effects than façades partly because of the sheer number of simultaneously visible corners that partially surround accessible vantage points (Section 1.2). With each corner reflecting to the source location, the result is a cluster of reflections that can sum to yield substantial reflected energy, albeit restricted to the high-frequency range. Retroreflection also means that comparatively little acoustic energy is reflected to locations that are not near the source.

The scope of this study is limited to modelling and simulation. This includes image-source-based calculations, and simulation using finite-difference time-domain (FDTD) modelling. This wave-based simulation technique has been demonstrated to be effective in other studies of retroreflective architectural surfaces [7–9], as well as in other stepped monumental architecture [5]. Two sites are taken as exemplars of potentially retroreflective architecture: Panna Meena ka Kund and Lahan Vav.

1.1. Introduction to Acoustic Retroreflection in Architecture

A retroreflector returns the incident sound to the direction from which it came. This is distinct from specular reflections, which only reflect to the incident direction at normal incidence, and it is different to scattered reflections, which reflect to all directions (including the incident direction). Physically, acoustic retroreflectors can be formed by right-angle concave corners (both dihedral and trihedral). In optics, the trihedral retroreflector is often referred to as a 'cube corner reflector', 'corner-cube reflector' or just a 'corner cube'. Arrays of corner cubes are used for some optical treatments (e.g., photo-electric sensor reflectors, bicycle reflectors and lunar ranging reflectors [10]), so that the reflected light around the source is intensified as each trihedron in the array reflects back [11,12]—this provides a model for analogous acoustic treatment. An important consideration in audible-range acoustics is that the wavelengths involved can be large, and thus the trihedron size needs to be correspondingly large to perform as a retroreflector. As a rule of thumb, an indicative frequency (f_R) above which a reflector at normal incidence has little or no diffraction loss can be calculated from the speed of sound (c), distance from the collocated source-receiver to the reflector (r) and the reflector width (d) as per Equation (1) [13,14]. For a square reflector at normal incidence, the denominator is its surface area, S. Below $f_R/2$, the diffraction loss slope is 6 dB/octave, and Rindel [14] suggests $f_R/2$ as a practical frequency above which diffraction loss at normal incidence is of minor importance.

$$f_R \approx \frac{cr}{d^2} \tag{1}$$

Arguably, the simple rectangular room—one of the most-studied forms in theoretical room acoustics—is retroreflective, because every reflection (planar, dihedral and trihedral) is returned to the source location. However, it is only trivially so, since every reflection fills the entire room volume, meaning the source location is not a focus point from retroreflection. For a room to be non-trivially retroreflective, the retroreflectors must be smaller than the room's basic surfaces (but sufficiently large for a useful f_R), so that the reflected wavefronts are returned to the source without filling the rest of the room. With arrays comprising many retroreflectors, each returning sound to the source, there is an energy focus at any sound source, with comparatively little acoustic energy reflected elsewhere in the room. This phenomenon has been demonstrated physically and computationally using a small specially designed retroreflective room [9]. The audible result is that high-frequency phonemes from one's own voice are particularly strong, a phenomenon which is envisaged for practical application in room acoustics to influence relaxed voice projection [15]. Considering the scarcity of room acoustics research literature on retroreflection, the question arises as to whether larger-scale pre-existing cases of acoustically retroreflective rooms exist, and how the phenomenon of retroreflection is manifest within them.

While pre-existing cases of acoustic retroreflection in architecture have been identified in building façades [6–8], finding pre-existing extensive cases in room acoustics is more challenging. Indoors, candidates for incidental retroreflection could include rooms with simple coffered ceilings (where surfaces are planar and perpendicular) [16]. Retroreflection has been used intentionally in some auditoria to support performers on stage [17,18]. However, such retroreflective treatment does not dominate the room's acoustic design—which is mostly motivated to provide high-quality sound to the audience.

Steps can provide dihedral and trihedral concave corners, and thus an environment with many steps could be acoustically retroreflective. Indian stepped ponds are some of the most extensive cases of step-based architecture. Furthermore, unlike stepped pyramids, the overall form of a stepped pond is concave—the stepped pond is a type of room. They provide vantage points from which hundreds of trihedral corners are simultaneously visible. In this way, stepped ponds provide potentially rich cases for retroreflective room acoustics, notwithstanding issues such as diffraction loss, atmospheric dissipation, variable wind and geometric deviations that could reduce retroreflectivity. Most importantly, Equation (1) raises the question of whether the steps are just too small for significant acoustic retroreflection over typical distances from vantage points within the frequency range important for humans.

1.2. Introduction to Indian Stepped Ponds

The term stepwell is used for a wide range of architectural forms that are built into the ground with steps leading to water. In northern and western parts of India, where the structures studied in this paper are located, they can be referred to using several terms, which may depend on their history, purpose, colloquial usage, use in the literature, etc. Some common terms include baori/baoli (बावड़ी/बावली in Hindi), vav (વાવ in Gujarati), vapi (वापी in Sanskrit and Hindi), barav (बारव in Marathi) and kund (कुंड in Hindi). It must be noted, however, that these terms are oftentimes interchangeable. Still, a baori/vav/vapi generally refers to an underground building, which, in some instances, is extensive and elaborate, with one or more staircases leading to water [19]. A kund, which is the type of stepwell that this paper is concerned with, is typically a 'stepped pond' or 'stepped tank', often with many staircases leading to a pool of water, sometimes resembling an inverted pyramid [20]. Kunds have been built for religious reasons (*Yagna kund*), for medicinal bathing (*Brahma kund*) and for general bathing (*Snan kund*), and some may be used for drinking water [20–24]. Some are associated with temples, some are in public places and some are in private residences. Many larger stepped ponds have a rectangular plan with arrays of intersecting steps on three sides, a more vertical structure on the fourth side and a pool at the base. In more elaborate cases, this fourth side may have arched loggias and rooms, and a platform from which the architectural spectacle of the three other sides can be fully viewed. The water level can vary greatly with the season, and thus part of the ingenuity of the design is that the steps lead to the water regardless of its level.

More than 30 stepped ponds that approximately or exactly follow the above form are documented by the main sources referred to in this paper [19–25], mostly in Rajasthan and Gujarat. Many more are documented in the crowd-sourced Atlas of Stepwells [26]. Their scale varies from intimate to monumental—with Chand Baori (Abhaneri, Rajasthan) being one of India's largest and best known, dating from around 800 CE. Chand Baori has many hundreds of steps which could contribute to acoustic retroreflection over its three intricately stepped faces, but smaller stepped ponds also have good prospects for acoustic retroreflection—and may be better for retroreflection because of their smaller size. Table 1 provides some examples of stepped ponds for which basic architectural data are documented (the data in the table were extracted from the cited architectural drawings).

Table 1. Ten examples of stepped ponds that follow the form described, with architectural drawing sources cited. Some have alternative names, and in such cases, the name in the table is taken from a recent source, such as [23]. The top plan area is of the basic well opening, excluding surrounds. The number of steps only counts those in the main surfaces, quantified as the number of discrete horizontal walking surfaces.

Name (Location)	Top (m^2)	Pool (m^2)	Depth	Steps
Bala Kund (Bundi) [20] (p. 83)	160	7	30 steps	179
Champa Bagh ka Kund (Bundi) [20] (p. 87)	96	14	28 steps	91
Chand Baori (Abhaneri) [22] (p. 41)	1376	53	24.5 m, 91 steps	1803
Gangvo Kund (Dedadara) [22] (p. 37)	184	32	20 steps	170
Hadi Rani (Todaraisingh) [22] (p. 120)	1194	150	85 steps	1300
Idar Stepped Pond [22] (p. 42)	145	7	12.5 m, 61 steps	206
Jaipura Kund (Bundi) [20] (p. 86)	323	21	49 steps	280
Lahan Vav (Basantgarh) [22] (p. 6)	237	21	6.75 m, 25 steps	300
Nagar Sagar Kund [1] (Bundi) [20] (p. 84)	439	15	103 steps	554
Panna Meena ka Kund (Amer) [22] (p. 110)	608	84	14.25 m, 56 steps	798

[1] Nagar and Sagar are twin stepped ponds in close proximity.

Table 1 indicates that there is typically about one step per square metre of top area (ranging from 0.87 at Jaipura Kund to 1.42 at Idar Stepped Pond). This ratio is mostly governed by the horizontal area of individual steps and platforms and the area of the pool (which is the minimum for the Table 1 data). The average step height for the four cases with depth in metres spans the range 0.21–0.28 m (mean 0.27 m). The number of trihedral concave corners in a stepped pond is greater than the number of step horizontal surfaces and includes large trihedra where the main faces intersect. The number of large trihedra depends mainly on the number of levels in the stepwell. Some stepped ponds have extra corners from niches in their walls (e.g., Panna Meena ka Kund). Many of these examples have an approximately constant slope in the stepped surfaces, but the slope of Idar steepens dramatically towards the pond, whereas the slope of Hadi Rani is steepest around its upper part.

This paper investigated Panna Meena ka Kund (Amer, Rajasthan), which is a well-known stepped pond, dating from the 16th century. Not as deep as Chand Baori, it nevertheless is a large and deep stepped pond with hundreds of crisscrossing steps that lead to the pool—including some on the fourth side (Figure 1a). Mostly, the walls are steps, with only one platform near the base of the stepped pond (submerged in Figure 1 photograph). It has 1.31 steps per square metre of top area. The vertical surfaces are rendered with a form of polished plaster [22]. The stepwell is in good condition and potentially provides an excellent case for acoustic retroreflection.

As an instance of a smaller stepped pond, this paper also focused on Lahan Vav (Basantgarh, Rajasthan, Figure 1b)—which is not well known. Dating from 976–999 CE, this stepped pond has bare stone block surfaces, some of which are damaged or displaced. It was chosen because of its smaller size and the availability of architectural drawings, and in its original state, it would be expected to have similar acoustic characteristics to other similarly sized stepped ponds, some of which are in better condition. It has a step-to-area ratio of 1.27.

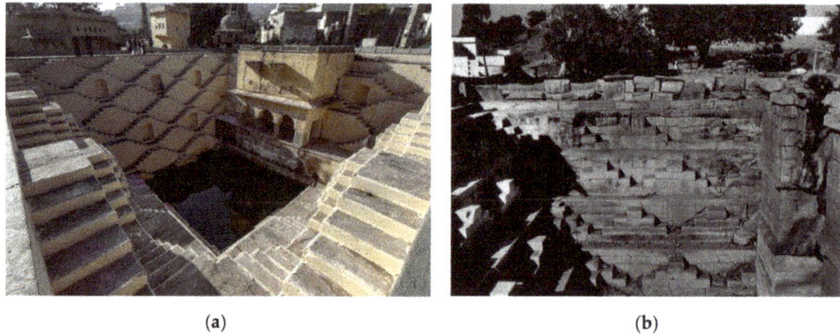

(a)   (b)

Figure 1. Photographs of the two stepped ponds used as exemplars in this paper: (**a**) Panna Meena ka Kund (also known as Panna Mia Kund and Panna Mian Kund); (**b**) Lahan Vav (also known as Lahini Vapi and Vasantgadh Stepped Pond). Source: American Institute of Indian Studies [25], used with permission.

2. Materials and Methods

2.1. Modelling and Simulation

The main predictive method used in this study was finite-difference time-domain (FDTD) simulation. This approach, which is widely used in architectural acoustics research, inherently accounts for the diffraction and higher-order propagation effects from architectural forms. Its biggest limitation is the size of the calculation, which requires large computational resources when large volumes and/or high frequencies are simulated.

Computer models of the exemplar stepwells were created based on published architectural drawings and photographs [22,25]. Due to limitations in computer resources, simulations of Panna Meena ka Kund were conducted using a partial model of the stepwell (Figure 2a). Simulations of Lahan Vav were conducted using the entire stepwell (Figure 2b). Simulations were also conducted with an anechoic environment the same size as each model for each source position, allowing the direct sound wave to be removed by subtraction. Values for reflected sound energy levels are expressed relative to the free field energy level 1 m from the source.

Omnidirectional point source positions were chosen for a commanding view of the stepwells, at a height of 1.5 m above the tower platforms (Figure 2). Sources were spaced linearly at 0.5 m intervals, with eight positions at Panna Meena ka Kund and six positions at Lahan Vav. The source simulation used a soft source [27], with a Gaussian pulse waveform.

The simulation volume for Panna Meena ka Kund was 18.78 m × 12.64 m × 8.82 m = 2093 m^3, excluding the perfectly matched layer. It was voxelised with a cubic voxel width (dx) of 9 mm, yielding 2.15×10^9 voxels. The basic receiver grid was a horizontal plane at source height (extending to the edges of the simulation), spaced at $4 \times dx$, yielding 151,206 receivers. A sampling rate of 60 kHz was used. Simulations were run for 13,200 time samples (0.22 s).

The simulation volume for Lahan Vav was 12.58 m × 18.12 m × 10.05 m = 2291 m^3 (2.35×10^9 voxels), excluding the perfectly matched layer. The same dx and sampling rate were used, and the basic receiver plane was designed in the same way as Panna Meena ka Kund (145,326 receivers over a horizontal plane at source height). Basic simulations were run for 12,000 time samples (0.2 s), and longer simulations (30,000 time samples, 0.5 s) were also conducted to include more of the stepwell's reverberation.

The open parts of the models were surrounded with a 10-voxel perfectly matched layer, as described by Chern [28], providing anechoic boundaries. The simulated stepwell surfaces were hard, with no sound absorption introduced.

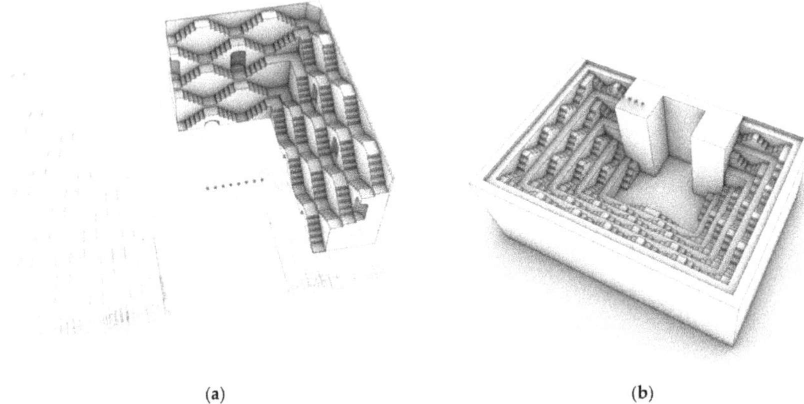

(a)   (b)

Figure 2. Perspective views of the stepwell models: (**a**) the partial model of Panna Meena ka Kund, extended with the whole stepwell lightly traced; (**b**) full model of Lahan Vav. Source positions are indicated by red dots, which are 1.5 m above the platform. Sources are numbered from the one closest to the central plane of symmetry.

2.2. Prediction of Retroreflected Energy from Trihedral Corners

A simple method to predict the retroreflected sound energy level returned to the source was presented by Cabrera et al. [8]. This method treats each visible concave trihedral corner as an equivalent mirror facing the source and uses first-order image-source calculation accounting for diffraction loss in the frequency domain. The complex transfer function for each reflector depends on geometric dispersion and delay (distance travelled is twice the distance, r, from the source to the reflector; k is the wave number), and the diffraction coefficient, K. The diffraction loss depends on the size and distance of the reflector (as indicated indirectly by Equation (1)) and can be calculated using the Kirchhoff–Fresnel approximation (used here), or alternatively Rindel's more efficient approximation [14]. The total retroreflected energy level, $L_{\text{retro (re 1 m)}}$, is obtained by summing the transfer functions of all n reflectors (Equation (2)). Octave band transfer functions can then be derived from the constituent spectrum components.

$$L_{\text{retro (re 1 m)}} = 20 \log \left| \sum_{i=1}^{n} \frac{\sqrt{K_i}}{2r_i} e^{-jk2r_i} \right| \tag{2}$$

This is a simplification of the theory in [8], neglecting the absorption coefficient of the surface and atmospheric loss, to match the FDTD simulation.

For a given source position, the first stage is to identify the visible concave trihedral corners—since occluded corners will not act as retroreflectors. For a given corner with its particular geometry, the size of the equivalent mirror is hard to precisely determine. The simplified approach taken in this study is to start with the boresight shadow of a square trihedron of the same edge length. For diffraction coefficient calculation, an equivalent area square reflector is used, multiplied by the cosine of the incidence angle. For the stepwells, there are many small trihedral corners of individual steps, as well as a smaller number of large corners. For the modelling in this paper, a conservative approach was taken, i.e., the equivalent square trihedron is the largest square trihedron that fits the corner, meaning that the edge length is the minimum edge of the three surface edges. This means that for steps, the step height is taken as the trihedron edge length (step heights of $l = 0.254$ m for Panna Meena ka Kund and 0.27 m for Lahan Vav). For the large corners, again, the minimum edge length is taken, which is generally the width of the walking surface ($l = 0.917$ m for Panna Meena ka Kund and 0.546 m for Lahan Vav). The boresight shadow area then is

$S = \sqrt{3} \times l^2$, which can then be adjusted by the cosine of the incidence angle. It should be borne in mind that trihedral reflections are considerably more complicated than this simplification [29,30], but the point of this model is to provide a simple estimate without the large computational demands of wave-based simulation or higher-order image-source modelling. The model provides a rough theory incorporating diffraction loss that can help interpret simulation results.

This modelling is expected to underestimate the reflected energy for several reasons. It only considers the retroreflected energy from individual reflectors, with no other reflected energy included (e.g., scattered reflections, specular reflections). It does not include di-hedral corner reflections. Furthermore, it only includes first-order reflections and has no diffuse reverberation contribution. The effective size of the reflectors remains constant with the frequency, even though small geometric deviations become insignificant at long wave-lengths, potentially increasing the reflectors' effective size. The potential for larger faces to contribute to trihedral retroreflection at oblique angles of incidence [31] is neglected.

Despite these limitations, the modelling allows some analysis of the contribution of various retroreflectors to the sound returned to the source. It is expected to be most useful at high frequencies, where retroreflection is expected to dominate the sound returned to the source.

3. Results

3.1. Spatial Distribution of Reflected Energy

Figure 3 shows the spatial distribution of reflected energy for the eight source positions at Panna Meena ka Kund from the FDTD simulation. Retroreflection is clearly evident in the high-frequency octave bands from the concentration of reflected energy on the respective source position. Retroreflection is clear in the upper three bands evaluated but also more weakly suggested in the 500 Hz and 250 Hz bands. The top row of Figure 3 shows the spatial distribution of reflected energy for a flat-surfaced simplification of the Panna Meena ka Kund partial model, which has energy distributed much more evenly across the receiver plane, and no concentration at the source position.

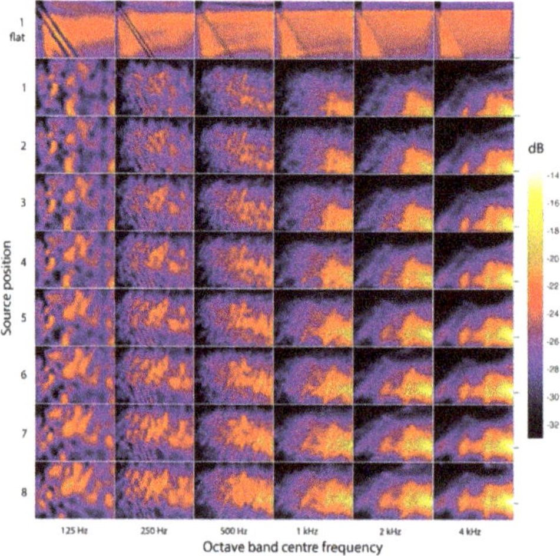

Figure 3. Spatial distribution of the octave-band reflected energy level for the Panna Meena ka Kund simulation over a horizontal plane of 18.1 m $\times$ 12.6 m for the eight source positions. The source position is on the right edge of each subplot, indicated by the marker on the right of the 4 kHz band subplots. Values are expressed relative to the emitted sound energy level at 1 m.

Figure 4 shows the spatial distribution of reflected energy for the six source positions at Lahan Vav. An important difference between this and the Panna Meena ka Kund visualisation (Figure 3) is that the ground reflection from the tower platform is included—which is clearly evident in many of the subplots. The top row of Figure 4 shows the spatial distribution of reflected energy for a flat-surfaced simplification of Lahan Vav, which helps to disambiguate the ground-reflected energy (seen in all subplots) from the retroreflected energy (seen in the non-flat simulation subplots, especially in the higher-frequency bands). Retroreflection is evident in the 1–4 kHz octave bands, most strongly in the 4 kHz band.

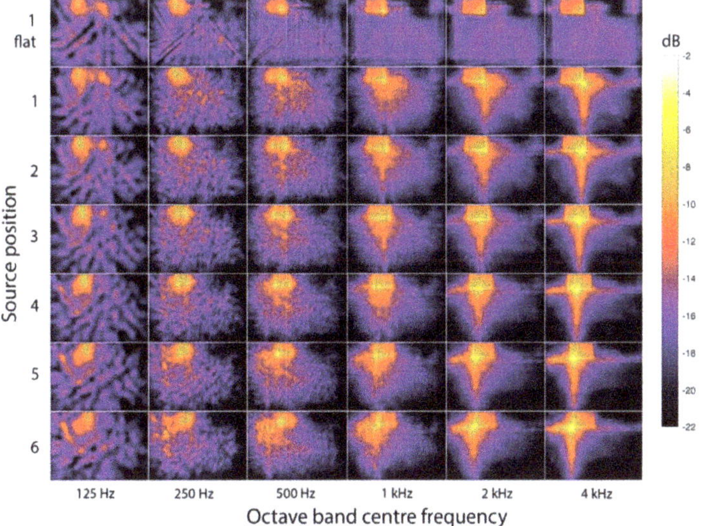

Figure 4. Spatial distribution of the octave-band reflected energy level for the Lahan Vav simulation over a horizontal plane of 18.8 m x 12.7 m for the six source positions. The source positions are evident from the maxima, especially in the 4 kHz band (the two towers are at the top of each subplot). Values are expressed relative to the emitted sound energy level at 1 m.

3.2. Reflected Energy at Collocated Source–Receivers
3.2.1. FDTD Simulation Results

Reflected energy levels at the source–receiver positions are shown in Figure 5. Values are greater in the higher-frequency bands for both stepwell models, especially the 2 kHz and 4 kHz bands. Values in these high-frequency bands are about 11 dB greater at Lahan Vav than at Panna Meena ka Kund. At both stepwells, there is a tendency for greater reflected energy levels at lateral positions (higher numbered positions). At both stepwells, this may be influenced by the increased proximity to the side face steps. Another feature of Figure 5 is that it shows the reflected energy levels at Lahan Vav with and without the ground reflection (which was absent at Panna Meena ka Kund). The contribution of the ground reflection is small in the high-frequency bands, for which retroreflection occurs, and larger at lower frequencies. A 3 dB difference would indicate that the ground reflection had equal energy to the subsequent reflections, and some of the positions have differences greater than or equal to 3 dB in the 125 Hz and 250 Hz bands.

The analytically calculated reflection energy level from an extensive ground plane for a source–receiver height of 1.5 m is $20\log(1/3) = -9.5$ dB. This does not take the edges of the plane into account, which would be relevant for the simulated stepwell vantage points. Nevertheless, many of the Lahan Vav reflected energy levels (including the ground reflection) are similar to this value in the 125 Hz and 250 Hz bands. Values are considerably greater in the upper octave bands. The reflected energy levels in every octave band at Panna Meena ka Kund are all lower than the analytically calculated ground reflection.

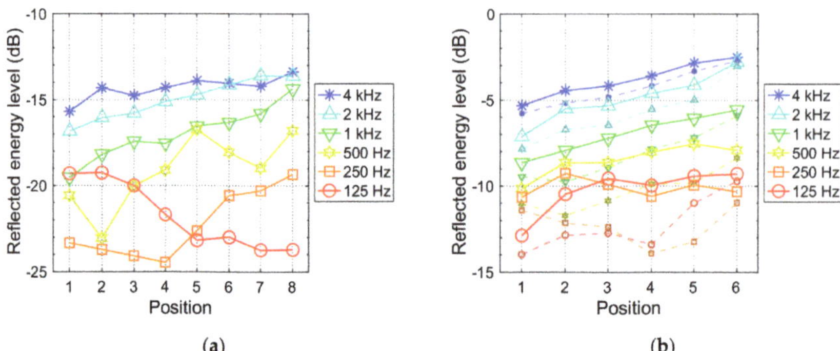

(a)   (b)

Figure 5. Reflected octave-band energy levels returned to the source position from FDTD simulation: (**a**) Panna Meena ka Kund (partial model, without ground reflection); (**b**) Lahan Vav (full model)—for which large markers include the ground reflection and small markers (dashed lines) exclude it. Values are expressed relative to the emitted free field energy at 1 m.

3.2.2. Equivalent Reflector Model Prediction

The results from Equation (2), evaluated as described in Section 2.2 (at a 10 Hz resolution, combined into octave bands), are shown in Figure 6. For Panna Meena ka Kund, the predicted reflected energy levels in the 4 kHz band are greater than the simulated results (Figure 5), which is not surprising considering that a partial model was used for the simulation, while a full model was used for evaluating Equation (2). At Position 1, the reflected energy levels increase due to in-phase summation from symmetry. Reflected energy levels in the low-frequency bands are much lower than the simulation results because the calculation only models first-order retroreflection from the trihedra and neglects all other reflections and reverberation. Bands tend to be spaced at about 6 dB intervals, indicating that diffraction loss is important, and that band frequencies are mostly below $f_R/2$ of most reflectors.

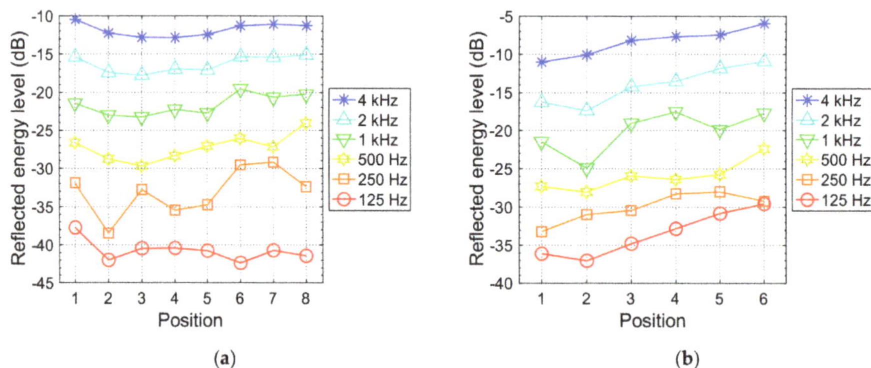

(a)   (b)

Figure 6. Retroreflected octave-band energy levels returned to the source position predicted by Equation (2): (**a**) Panna Meena ka Kund (full model); (**b**) Lahan Vav. Values are expressed relative to the emitted free field energy at 1 m.

The calculated values for Lahan Vav are 3–6 dB less than the simulation results in the 4 kHz band but diverge more at lower frequencies. Again, the excessive loss at lower frequencies is expected because only first-order retroreflection is included in the calculation. The calculation does not include the ground reflection because it would obscure the much lower values in the lower-frequency bands. The band results are separated by about 6 dB, reflecting the fact that most bands are below $f_R/2$.

This is illustrated directly in Figure 7, which plots the calculated values of $f_R/2$ for individual reflectors as a function of distance from the source. The large reflectors have low values and hence are expected to have little diffraction loss in the upper octave bands. The small reflectors have some values within or above the 4 kHz octave band at Lahan Vav Position 6, or entirely above the 4 kHz octave band at Lahan Vav Position 1 and at both illustrated positions at Panna Meena ka Kund. For a given distance and reflector size, values vary because of variation in incidence angles.

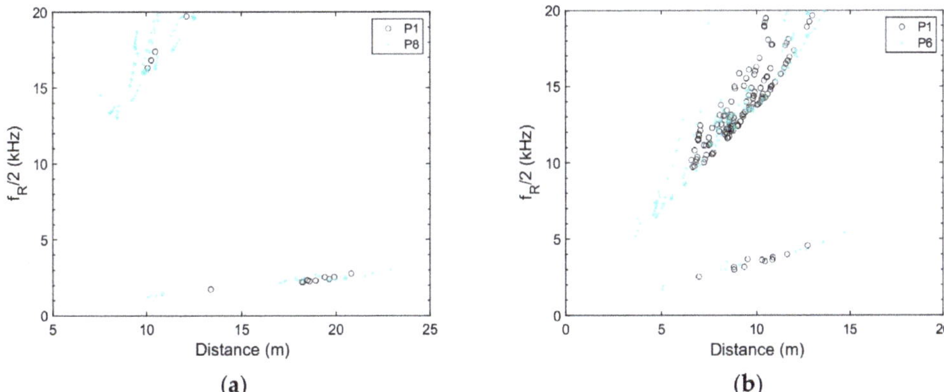

Figure 7. Calculated values of $f_R/2$ based on Equation (1) for individual equivalent reflectors: (**a**) Panna Meena ka Kund (full model, Positions 1 and 8); (**b**) Lahan Vav (Positions 1 and 6).

The 4 kHz band was chosen for further analysis because it is the band with the greatest retroreflection (both modelled and simulated). The energy contribution of individual reflectors is shown in Figure 8 as a function of distance for the two extreme positions in each stepwell, for the 4 kHz octave band only. This highlights the importance of proximity to retroreflectors. The combined effects of diffraction loss and geometric dispersion should yield an individual reflector spatial decay rate of -12 dB per distance doubling for $f < f_R/2$, which is seen for the small reflectors. However, more distant reflectors of a given size can be more numerous, and thus their combined spatial decay rate should be less severe. When $f > f_R/2$, the spatial decay rate is mainly affected by geometric dispersion (-6 dB per distance doubling)—hence the shallower slopes for the large reflectors. While there is a much smaller number of big reflectors than small reflectors, the big reflectors can make an out-sized contribution, as seen at Panna Meena ka Kund (evident in the cumulative distribution curves). On the other hand, at Lahan Vav, the small reflectors make a greater contribution than the big ones.

The half-energy (-3 dB) distances of the cumulative sums in the 4 kHz band are as follows: Panna Meena ka Kund, 18.2 m for P1 (312 reflections, out of 380) to 10.8 m for P8 (71 reflections, out of 399); Lahan Vav, 8.7 m for P1 (61 reflections, out of 142) to 5.0 m for P6 (20 reflections, out of 155). While the reported values of the half-energy distance are for the 4 kHz band, the half-energy distances are about the same in the other octave bands. For both stepwells, there is a monoclinal transition of the half-energy distance through the intermediate positions (not illustrated). For comparison, the mean distances to open trihedra are as follows: at Panna Meena ka Kund, 16.0 m for P1 and 15.8 m for P8; and at Lahan Vav, 9.3 m for both P1 and P6.

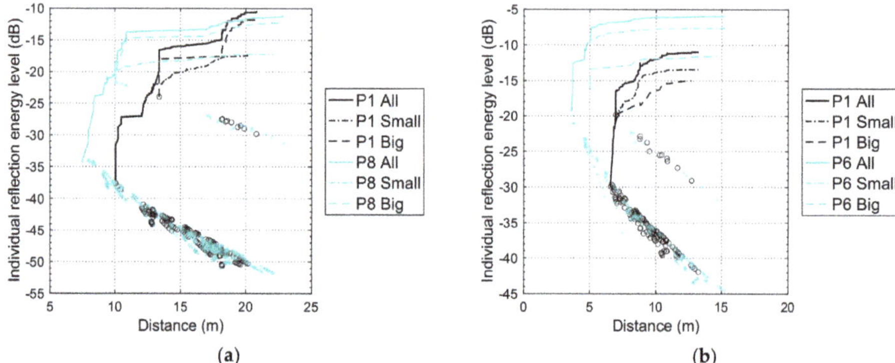

Figure 8. Retroreflected octave-band energy levels (4 kHz octave band) returned to the source position of individual equivalent reflectors: (**a**) Panna Meena ka Kund (full model); (**b**) Lahan Vav. Markers show the individual reflection values; lines show the cumulative complex sum of reflections as a function of distance. Values are expressed relative to the emitted free field energy at 1 m. Note that all markers for Panna Meena ka Kund P1 are superimposed pairs from symmetry.

3.2.3. Temporal Characteristics of Impulse Responses at Lahan Vav

Being based on a partial model, the Panna Meena ka Kund simulation lacks many reflections and reverberant decay that would be seen in the full stepwell, and thus this section just focuses on Lahan Vav. Previous studies of retroreflective façades showed a prominent cluster of reflections due to retroreflection in measured and simulated impulse responses for collocated source–receivers. This is also seen at Lahan Vav (Figure 9). With the source at Position 1, the retroreflection cluster starts at 40 ms and is increasingly prominent with the frequency. With the source at Position 6, the retroreflection cluster starts at 22 ms. Based on geometry, the first-order reflections should be finished by 90 ms, and thus subsequent energy is from higher-order reflections (or reverberation). At Position 6, a persistent 19 ms flutter echo (unrelated to retroreflection) is evident, most obviously in the 1 kHz octave band. For comparison, the figure also shows impulse responses for receiver positions distant from the source—these distant positions lack the retroreflection cluster.

3.3. Sensitivity to Geometric Error

The models used for the main simulations were created from perfectly flat planes with exact right-angle corners. Real cases are not that simple, and hence the question arises as to whether retroreflection focusing remains when deviations from the ideal geometry are introduced. The trihedral step corners of the Panna Meena ka Kund and Lahan Vav models were manipulated to angles other than 90°. Simulations were run for the source at Position 1. The results (Figure 10) show the largest effect of angle deviation is at high frequencies, for which deviations from 90° reduce the reflected energy level at the source. In the 4 kHz octave band, the reduction at Panna Meena ka Kund is 2.7 dB for a ±10° deviation; at Lahan Vav, it is 1.6 dB. Reflected energy level reductions from angle deviations are not evident in the bands 1 kHz and below. Even with a ±10° deviation, the values of the reflected energy returned to the source are still much greater than for a flat-surfaced model (shown in the top rows of Figures 3 and 4)—which for 4 kHz is −33.3 dB at Panna Meena ka Kund and −13.2 dB at Lahan Vav.

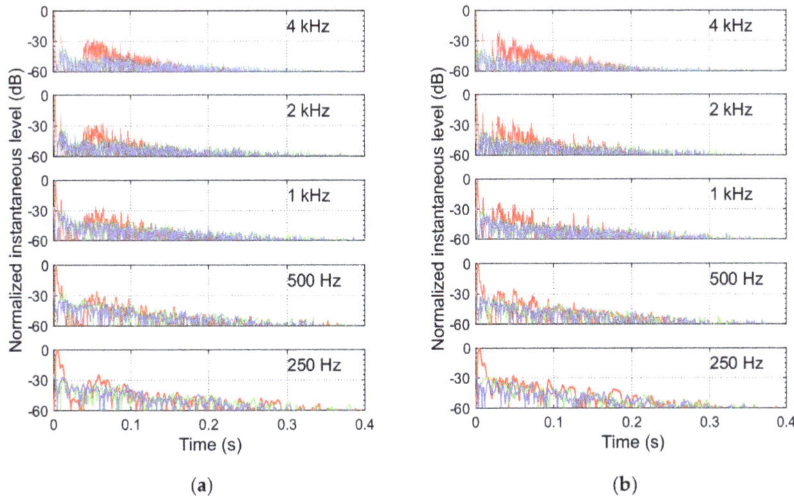

Figure 9. Lahan Vav octave-band impulse responses: (**a**) source at Position 1; (**b**) source at Position 6. The red line shows the impulse response for a receiver 0.5 m from the source; the blue and green lines show impulse responses for receivers about 10 m from the source. Arrivals are time-aligned, and levels are with respect to the red line's direct sound peak. The visual display is simplified by taking the absolute value of the Hilbert transform.

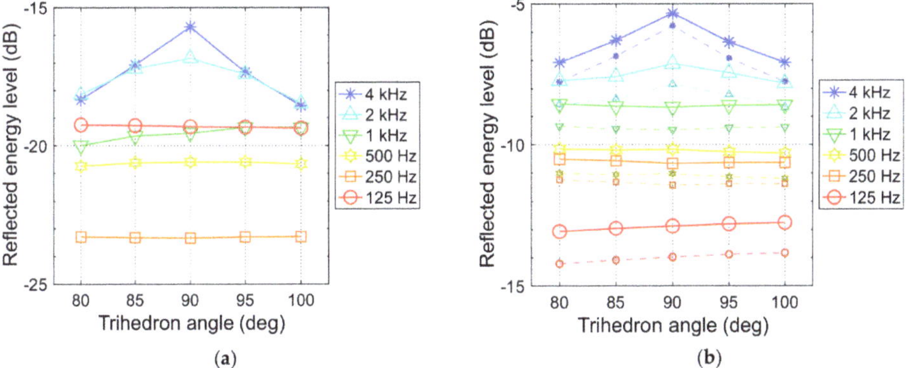

Figure 10. Reflected energy level at the source position (P1 only) from FDTD simulations, for concave trihedral corner angles varied from 80° to 100°: (**a**) Panna Meena ka Kund (partial model, without ground reflection); (**b**) Lahan Vav (full model)—for which large markers include the ground reflection and small markers (dashed lines) exclude it. Values are expressed relative to the emitted free field energy at 1 m.

A more complete picture of the effect of angle deviation comes from examining the sound field across the whole receiver plane (Figure 11). As well as a reduction in the reflected energy around the source position, non-right-angled trihedra result in greater reflected energy at locations that receive relatively little in the corresponding right-angle trihedra model simulations. The effect of angle deviation is considerably greater for the Panna Meena ka Kund model than the Lahan Vav model. Lahan Vav has a maximum positive deviation of 2.5 dB (4 kHz band, 80°), whereas the maximum positive deviation for the Panna Meena ka Kund model is 6.6 dB (4 kHz band, 80°). The smaller changes at

Lahan Vav are likely partly attributable to its semi-enclosed form (the Panna Meena ka Kund model being partial).

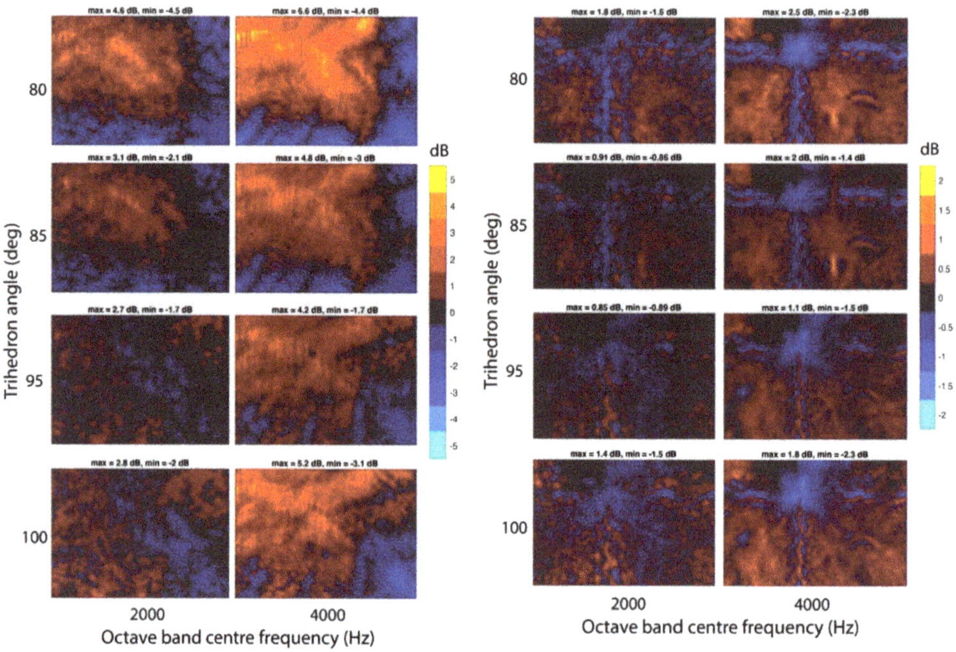

Figure 11. Change in reflected energy level over the horizontal receiver plane from trihedron angles other than 90°, with the source at Position 1: (**a**) Panna Meena ka Kund (partial model, without ground reflection); (**b**) Lahan Vav (full model). Subplots match the orientation of Figures 3 and 4 (for a and b, respectively).

3.4. Effect of Temperature Gradient

It is frequently observed that the temperature at the base of a stepped pond may be less than the general temperature on a hot day. Reports of a 5 °C reduction are common, even if anecdotal. The authors are not aware of detailed data on this, but similar phenomena have been studied in basins and sinkholes, in which cool air pools at the base [32]. This introduces the possibility of acoustic refraction, as the speed of sound reduces deeper into the well. This should not necessarily degrade retroreflection, because reciprocity still applies (the path from the source to the reflector is the same as the path from the reflector to the source). Conceivably, this may increase the well's acoustic effect by reducing the amount of sound lost to the sky.

A simulation was conducted using the Panna Meena ka Kund model, with a temperature gradient of 0.5 °C/m for a source at Position 1. The resulting sound returned to the source is essentially the same as the simulation without a temperature gradient. Deviations of the reflected energy level across the receiver plane were < 0.1 dB in all octave bands.

4. Discussion

4.1. The Retroreflective Potential of Stepped Ponds

Stepwells are designed for water access, not for acoustics; they are certainly not designed for optimum acoustic retroreflection. Yet, this study shows that stepped ponds can be good candidate cases of incidentally acoustically retroreflective rooms—at least when the water level is low. In this respect, stepped ponds are very distinctive acoustic environments, considering that predominantly retroreflective rooms are rarely seen in architecture. Steps are designed for people to ascend and descend, but fortuitously, the steps used in these and similar stepped ponds are larger than modern standard steps. Nevertheless, from an acoustics perspective, the steps are undersized for retroreflection, especially in larger stepped ponds such as Panna Meena ka Kund, which, at any vantage point, will have large distances to numerous step trihedral corners. This is indicated by the values of f_R, which generally predict large diffraction loss over the important frequency range for humans. Conceivably, a stepped pond smaller than Lahan Vav (with shorter distances) should perform even better as a retroreflective environment for autophonic sound such as speech and clapping. As a guide, Figure 12 shows the calculated values of $f_R/2$ for steps of various sizes and distances from the source: for example, a somewhat impractical stepped pond with 0.5 m steps and most distances within 10 m would have diffraction loss mostly limited to frequencies below 4 kHz.

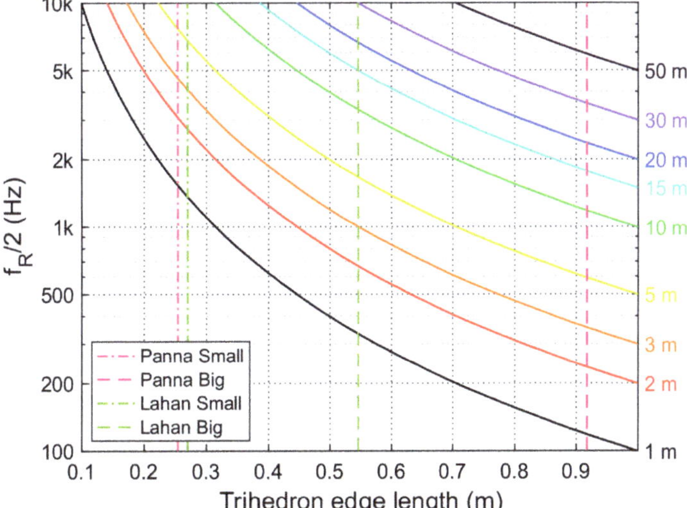

Figure 12. Evaluation of $f_R/2$ values using Equation (1) for the boresight shadow of square trihedra of various sizes (*x*-axis) at various distances (lines). Edge lengths of small and large trihedra used for modelling Panna Meena ka Kund and Lahan Vav are shown.

Given the above, it is remarkable that retroreflection is strongly evident in the spatial distribution of reflected energy in both stepwell models studied in the 4 kHz and 2 kHz bands and is even seen to a smaller extent at relatively low frequencies (e.g., 500 Hz). Based on Figure 12, the steps at Panna Meena ka Kund are too small for the distances involved. However, considering that the FDTD simulation of Panna Meena ka Kund only included two big trihedra, the simulation results show that the steps nonetheless function as an ensemble of weak retroreflectors.

In Appendix A, simplified models of ten stepped ponds are evaluated (based on the features of those listed in Table 1), suggesting that a smaller top area may be a good simple indicator of a stepped pond's retroreflective potential. The size disadvantage of a large stepped pond may be overcome at a source position lower into the well, such as at a lower platform or loggia. At lower positions, the number of visible concave trihedra reduces, but this should be more than compensated for by the reduced distance to the remaining reflectors. This concept is developed in Appendix B, which shows an increase in retroreflected energy at lower source positions in models of large stepped ponds.

4.2. Considerations with Real Stepwells

Real stepwells are more complicated than the models used in this paper. Geometric deviations from the ideal modelled surfaces are inevitable and, in some cases, are likely too great for retroreflection. This might be an issue affecting the real Lahan Vav (chosen because of the availability of architectural drawings), which now has somewhat rough and irregular stones. Hence, the modelling and simulation of this case are a representation of the stepped pond's original acoustics, but not necessarily of its present-day acoustics. The simulation study of sensitivity to geometric error indicates that perfect trihedra are not required for a stepped pond to be significantly retroreflective—with an angular error of $10°$ causing a focus loss less than 3 dB in the 4 kHz band, and less than 2 dB in the 2 kHz band (Panna Meena ka Kund), and smaller losses for the Lahan Vav simulations.

The simulations and modelling used in this study assume no sound absorption by the surfaces—all sound absorption is by the open space (which is anechoic). The smooth plaster and stone surfaces used in some stepped ponds would absorb little sound and thus would not be expected to be substantially different to the modelling. However, some stepped ponds have surfaces that are expected to absorb and scatter more (e.g., exposed rubble masonry).

Acoustic conditions at real stepwells differ from the models in other ways too. Wind and temperature may affect sound propagation. Based on the simulated case, a simple temperature gradient does not damage the retroreflective quality of a stepped pond. However, turbulent windy conditions would be expected to variably refract, and hence scatter, high frequencies, severely degrading retroreflection. Background noise affects the audibility of a room's impulse response, including retroreflection. It is obvious that quiet and still conditions would be needed to experience retroreflection on site.

The findings in this study apply to stepped ponds that have about one step per square metre of top area. Generally, this requires that the pool area not be very extensive. Cases with a larger pool (e.g., Roda [22] p. 33) would be expected to have less retroreflection evident.

4.3. Audibility of Retroreflection

The most prominent features observed in this paper are in the 4 kHz octave band. While retroreflection focusing may be theoretically stronger at higher frequencies (e.g., 8–16 kHz bands), higher frequencies are of less interest because atmospheric propagation losses become significant, they are more sensitive to surface roughness and they are towards the upper limit of human sound production and hearing. Even the 4 kHz band is an unusually high frequency range to concentrate on. For people with noise-induced hearing loss, acuity in the 4 kHz band is typically reduced [33], which would limit their potential experience of retroreflective focusing.

The audibility of acoustically retroreflective architecture depends on the acoustic excitation, the retroreflected energy quantity and distributions over time and frequency, the background noise conditions and the listener's interest in and attentiveness to sound. Experience with other retroreflective environments is varied. In some, handclaps are effective, even though a typical clap has an approximately pink power spectrum [34,35] (whereas a white or high-frequency-dominated spectrum would be preferable). Clapping is most likely to be effective when there is an audibly significant time delay for the retroreflection cluster arrival. Another autophonic excitation method that can make retroreflection obvious is the use of high-frequency phonemes or tongue clicks. The directionality of such excitation may help in suppressing the ground reflection and other unwanted reflections from surfaces near the person. When retroreflection is strong, it may be audible directly from one's own speech, with the reflections giving an unusual 'crisp' quality to the sound [9]. Considering that the partial model of Panna Meena ka Kund yielded quite low reflected energy levels, the noticeability of retroreflection from the vantage points at that site needs further investigation. A positive indicator is that the first-order image-source calculation from equivalent reflectors for Panna Meena ka Kund yielded higher reflection levels because of the full stepped pond being included, with the high frequency values at Position 1 quite close to those for Lahan Vav's Position 1. Furthermore, the relatively long time delays from the large distances at Panna Meena ka Kund favor audibility. The retroreflection at a stepped pond such as Lahan Vav, but with surfaces in good geometric condition, is expected to be audible considering that the reflected energy levels and distances are similar to those at the previously studied Ainsworth Building, where retroreflection is directly audible [8]. The presence of room reverberation is a difference between the stepped ponds and the previously studied retroreflective building façades.

4.4. Future Study

On-site acoustic measurements for this project have been delayed due to COVID-19 disruption. Future study incorporating in situ measurements (similar to [8]) will be conducted when possible.

5. Conclusions

From an acoustics perspective, the broad research problem addressed by this paper was to find and characterise pre-existing instances of predominantly retroreflective rooms—and it has shown that some Indian stepwells are good candidates for this. While many types of buildings have concave trihedral corners, the stepped pond type of stepwell sometimes takes this to an extreme. As demonstrated in this paper, these sometimes ancient buildings potentially provide intense instances of acoustically retroreflective room acoustics, albeit with an open-sky ceiling in the exemplar cases. With the ravages of time, some sites will have lost their original acoustic characteristics, but wave-based computational acoustics allows their acoustic restoration for analysis. Other sites appear to be sufficiently well preserved to retain their retroreflective characteristics, which should be observable in good environmental conditions to an attuned visitor.

The main findings are as follows:

- Acoustic retroreflection in stepped ponds can be substantial in the high-frequency range, resulting in reflected sound focusing onto the source position—which is seen as a dense cluster of high-frequency reflections in the early part of the impulse response;
- Both small trihedral corners (from steps) and large trihedral corners (from wall intersections at each level) contribute to retroreflection, with the balance of them depending on the scale of the stepped pond;
- Retroreflection is not reduced greatly with angular distortion of trihedra of up to $10°$;
- Smaller stepped ponds tend to be more retroreflective, because the effect of shorter distances is stronger than the effect of the smaller number of reflectors; however, lower positions in a large stepped pond see increased retroreflection.

Overall, the acoustic result of the stepped pond form is distinctively bright and strong reflections to the source position, while positions away from the source receive much less sound from early reflections, especially at high frequencies.

Author Contributions: Conceptualisation, D.C. and M.Y.; methodology, D.C., S.L. and J.H.; software, J.H. and S.L.; formal analysis, D.C., S.L. and J.H.; writing—original draft preparation, D.C.; writing—review and editing, D.C., S.L., J.H. and M.Y.; visualisation, D.C., S.L. and J.H.; funding acquisition, D.C., S.L., J.H. and M.Y. All authors have read and agreed to the published version of the manuscript.

Funding: Fieldwork for this research was funded by the Australian Acoustical Society via its Education Grant scheme. However, the fieldwork is yet to be conducted due to COVID-19 disruption.

Institutional Review Board Statement: Not applicable.

Informed Consent Statement: Not applicable.

Data Availability Statement: The data presented in this study are openly available in the Sydney eScholarship Repository, available online at: https://doi.org/10.25910/npyj-cq27 (accessed on 1 March 2022), item name: 'Acoustic modelling and simulation of Indian stepwells: Panna Meena ka Kund and Lahan Vav'.

Acknowledgments: Photographs and permission to use them were provided by the American Institute of Indian Studies.

Conflicts of Interest: The authors declare no conflict of interest.

Appendix A

Appendix A.1 Simplified Model of Stepped Pond Retroreflection

A simplified generic model of stepped ponds was constructed to examine the relationship between acoustic retroreflection and architectural parameters. The model assumes that a stepped pond takes the form of an inverted frustrum, with one of its faces steeper or vertical. The other three faces are the stepped surfaces. The ten stepped ponds listed in Table 1 were represented using this simplified modelling approach. Some advantages of the approach are as follows: (i) stepped ponds with incomplete architectural data could be quickly modelled, considering that only plans are available for most cases; (ii) the modelling focuses on the general form without quirks of particular cases; and (iii) the reflector size can be held constant, allowing the effects of larger-scale architectural features to be seen more clearly.

While the exact locations of the steps might be hard to determine, they are numerous and fairly uniformly distributed over the surfaces, and thus, here, the steps are statistically modelled by evenly distributing small retroreflectors across the three 'stepped' frustum surfaces, with alternating boresight directions. The number of distributed retroreflectors equals the total number of steps (also derived from the plan, as per Table 1). Then, the visibility of the retroreflectors with respect to a particular source position is determined, and the distance and the incidence angle of each visible retroreflector are calculated. Using these data, the retroreflected energy contributed by each small retroreflector is calculated as per Section 2.2. Similarly, large retroreflectors are also included in the modelling, determined as four per stepwell level, at the four intersections of the stepped surfaces. Their visibility, distance and incidence angle are determined using the same method as the small retroreflectors to calculate their retroreflected energy.

The concept is illustrated in Figure A1. In the modelling reported here, the source–receiver position is placed at the centre of the front edge of the platform, 1.5 m above the ground, as with Position 1 in the Panna Meena ka Kund case (shown in Figure 2a). A trihedron edge length of 0.25 m was used for all small retroreflectors, and 0.7 m was used for big ones.

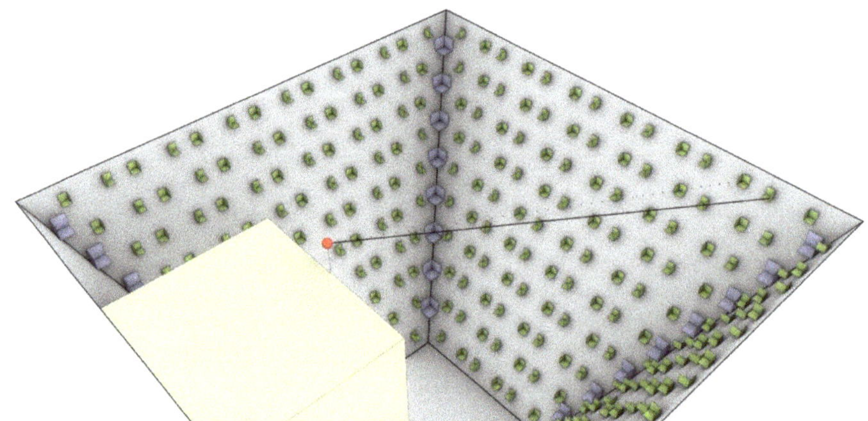

Figure A1. An example of the generic model. Green: small reflectors; blue: large reflectors; red dot: source position; grey: inverted frustum; yellow: platform. Two small reflectors connected with solid and dotted lines are examples of visible and obscured reflectors, respectively.

The key parameters used in the estimations, along with the estimated retroreflected energy, are provided in Table A1. Because the model is a general approximation rather than an exact replica of a given stepwell, reflections are summed energetically rather than using complex values. The energetic average of the 1–4 kHz octave bands is used as an indicator of the strength of the retroreflected sound returned to the source. On average, the 1–4 kHz value is 3.3 dB less than the 4 kHz octave band value in these cases (ranging from -2.9 dB for Bala Kund to -3.6 dB for Hadi Rani).

Table A1. Parameters used for simplified retroreflected energy level estimates based on the ten example stepped ponds from Table 1. Symbols: r is the distance of visible retroreflectors, θ is the incidence angle from the trihedron boresight and n is the number of visible reflectors. Subscripts: s denotes small reflectors, and b denotes big reflectors. 'Levels' is the number of levels in the well. The table also shows the estimated retroreflected energy level results for the cases (L) averaged over the 1–4 kHz octave bands. Note that the average values of the distance and incidence angle are provided here, but the underlying individual values are used in the estimations of retroreflected energy.

Name	$\overline{r}_s$ (m)	$\overline{\theta}_s$	n_s	Levels	$\overline{r}_b$ (m)	$\overline{\theta}_b$	n_b	L (dB)
Bala Kund	7.2	26.4°	86	4	7.9	21.5°	10	-10.7
Champa Bagh ka Kund	7.2	24.5°	47	2	8.7	19.5°	4	-14.9
Chand Baori	20.5	26.9°	905	13	23.9	20.9°	30	-21.2
Gangvo Kund	8.5	24.2°	84	3	9.6	14.5°	6	-14.7
Hadi Rani	23.8	26.0°	657	7	26.4	21.7°	16	-25.3
Idar Stepped Pond	9.6	27.2°	109	6	10.9	25.3°	16	-13.1
Jaipura Kund	11.9	26.2°	143	6	13.2	24.4°	14	-15.8
Lahan Vav	9.2	25.9°	155	5	10.1	17.4°	12	-12.9
Nagar Sagar Kund	17.1	28.8°	290	7	19.9	26.2°	16	-20.4
Panna Meena ka Kund	15.4	25.8°	401	8	18.2	20.0°	18	-18.8

Appendix A.2 Simplified Model Predictions

The modelling indicates that the simplified model of Bala Kund has the greatest prospects for retroreflection, followed by Lahan Vav and Idar Stepped Pond (Table A1). The large stepwell models have the weakest prospects, even though they have the greatest number of steps. The results span a wide range of retroreflective conditions and suggest that the Lahan Vav case chosen for detailed analysis represents relatively strong retroreflective conditions.

The physical parameters are mutually correlated—larger-area wells tend to have more steps, a greater depth, more levels, a larger pool, etc. In the following, the architectural parameters are used logarithmically, i.e., the natural logarithm of each parameter is used. This provides more evenly distributed values that have a better linear fit with the reflected sound energy level. Table A2 shows the correlation coefficients between parameters.

Table A2. Correlation matrix for the reflected energy level (L, in decibels) and the logarithm of the architectural parameters (top area, depth in steps, total number of steps, number of levels, pool area). The top right of the table shows pairwise linear correlation coefficients. The bottom left of the table shows p-values.

Parameter	L	Top	Depth	Steps	Levels	Pool
L		-0.86	-0.76	-0.83	-0.58	-0.80
Top	<0.01		0.73	0.99	0.85	0.80
Depth	0.01	0.02		0.76	0.80	0.33
Steps	<0.01	<0.01	<0.01		0.89	0.74
Levels	0.07	<0.01	<0.01	<0.01		0.44
Pool	<0.01	<0.01	0.34	0.01	0.20	

The best single architectural predictor of the estimated retroreflective energy level is the stepwell top area. A linear fit of the logarithm of the top area is shown in Figure A2 (R^2 0.754, $p = 0.001$, RMSE 2.4 dB). However, if we were to only consider the six smaller cases (top area ≤ 323 m^2), there would be no apparent relationship between the top area and estimated retroreflection. Alternatively, the smallest case (Champa Bagh ka Kund) might be considered an outlier, hinting that there might be a small area limit on the negatively sloped relationship between stepped pond size and retroreflection strength.

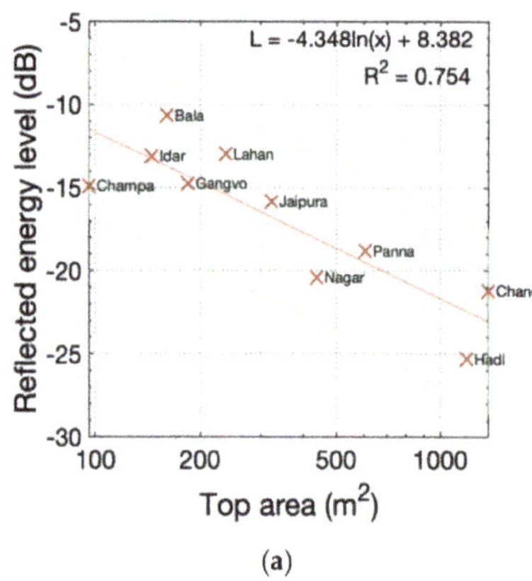

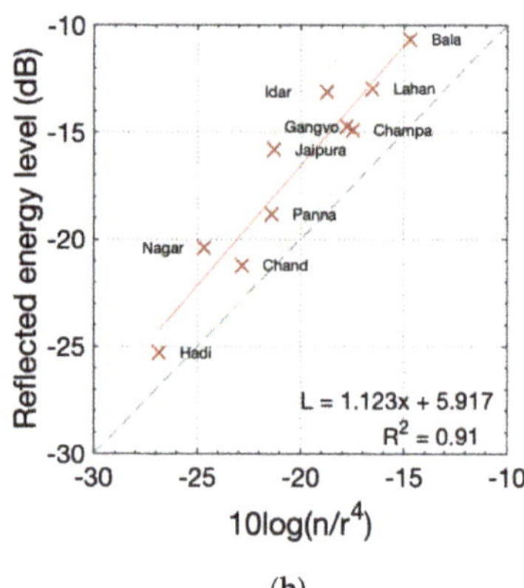

Figure A2. Estimated reflected energy level (1–4 kHz octave band mean, calculated as described in this appendix) from retroreflection in relation to the top area of the 10 stepped ponds from Table 1: (**a**) calculated values in relation to the top area of the stepped pond; (**b**) calculated values in relation to $10\log\left(n/(\bar{r})^4\right)$ for visible reflectors. Regression lines and 95% confidence intervals are shown.

A further simplified way of understanding the results is that the expected reflected energy level should increase by 3 dB for doubling of the number of reflectors but decrease by up to 12 dB for doubling of the distance (accounting for geometric spreading and maximum diffraction loss). If the average distance is used as the single-number distance, then this is represented by $10\log\left(n/(\bar{r})^4\right)$, shown in Figure A2b. Although this very simple formulation is only designed to approximate the slope, it returns values not far from the 1–4 kHz reflected energy level values.

Appendix B

Appendix B.1 Retroreflection for Lower Positions in Large Stepped Ponds

While the results in the main paper, together with those in Appendix A, suggest that large stepped ponds have weak retroreflection, this finding applies to source positions at the top of the well. Another way of thinking about a large stepped pond is that it is a small stepped pond extended upwards (except that a larger well may have a larger pool). A source position lower into the well loses visibility of trihedral corners at upper levels, but the distances to the smaller number of visible trihedra should decrease. This trade-off is examined briefly in this appendix by varying the source–receiver height in two large stepped pond examples.

The first example is Panna Meena ka Kund, which was modelled as described in Section 2.2. The eight original source positions were lowered to be 1.5 m above the upper and lower loggias (height reduced by 4 m and 8.2 m, respectively). Note that the upper loggia's arches are filled in in the present day, but here, we imagine that they are open. The mean numbers of visible concave trihedra on the three levels for the eight positions are: 385 (top level), 331 (upper loggia) and 180 (lower loggia). The mean distances of the visible trihedra are: 15.9 m (top level), 13.9 m (upper loggia) and 11.5 m (lower loggia).

The second example is a generic stepwell model (using the Appendix A method) based on the main features of Chand Baori. The real Chand Baori has a complicated building on its non-stepped face, but for this example, source–receiver positions are simply in a vertical line, with one per stepwell platform level (14 positions at 1.88 m intervals).

Appendix B.2 Results

The results for Panna Meena ka Kund (Figure A3a) are consistent with the hypothesis that lower positions in a large stepped pond have greater retroreflective potential. The mean reflected energy levels are -14.8 dB, -13.0 dB and -12.0 dB for the top level, upper loggia and lower loggia, respectively. Values at the two loggias are similar to those at Lahan Vav in Table A1, and only Bala Kund has a retroreflected energy level greater than that of the lower loggia.

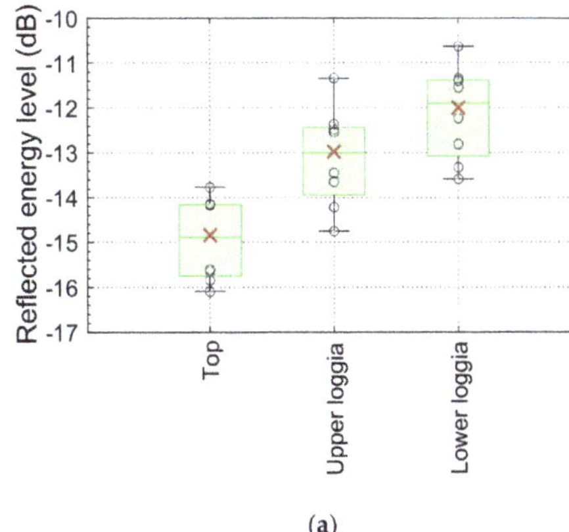
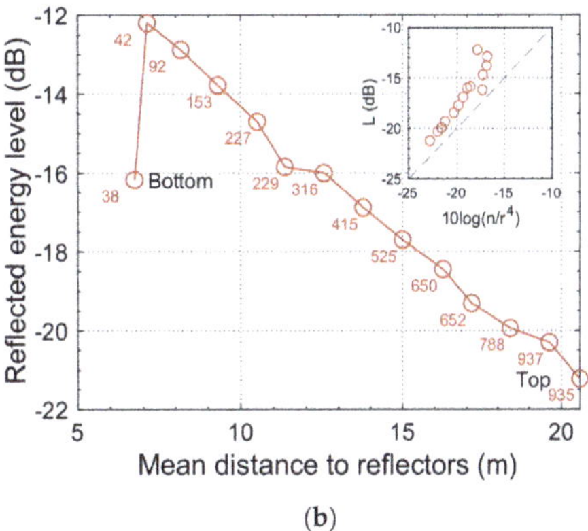

(a) (b)

Figure A3. Estimated retroreflected energy level (1–4 kHz octave band energy average) at various heights in the two examples: (**a**) Panna Meena ka Kund for various heights of a horizontal line of eight source–receiver positions ('X' markers show energy average); (**b**) generic stepwell model based on features of Chand Baori for one position per stepwell platform level, with results shown in relation to the mean distance to visible reflectors (each marker is labelled with the number of visible reflectors). The relationship between the reflected energy level and $10\log\left(n/(\bar{r})^4\right)$ for visible reflectors is inset.

The results for the simplified Chand Baori calculation (Figure A3b) are also consistent with the hypothesis. Excluding the bottom position, there is a highly linear relationship between the height of the source position and the reflected energy level ($r = -0.998$), and between the mean distance to visible reflectors and the reflected energy level ($r = -0.996$). Furthermore, the number of visible reflectors is similarly negatively correlated with the reflected energy level ($r = -0.980$)—which, of course, is the opposite to what would be expected if all other factors were held constant. Overall, this demonstrates the importance of the distance to reflectors in determining the retroreflected energy level, consistent with the relationship $10\log\left(n/(\bar{r})^4\right)$, introduced in Appendix A (R^2 of 0.88 for all datapoints, or R^2 of 0.93 for datapoints excluding the bottom position).

The drop in reflected energy at the bottom position of the Chand Baori approximate model appears to be from the loss of access to big reflectors (six on the level above, two on the bottom level). The mean incidence angle also increases from 26.9° to 39.3° between the level above and the bottom level. The value at the second lowest position is again similar to that at the much smaller Lahan Vav in Table A1.

References

1. Cox, T. *The Sound Book: The Science of the Sonic Wonders of the World*, 1st ed.; W.H. Norton & Company: New York, NY, USA, 2014.
2. Lubman, D. Archaeological acoustic study of chirped echo from the Mayan pyramid at Chichén Itzá. *J. Acoust. Soc. Am.* **1998**, *104*, 1763. [CrossRef]
3. Declercq, N.F.; Degrieck, J.; Briers, R.; Leroy, O. A theoretical study of special acoustic effects caused by the staircase of the El Castillo pyramid at the Maya ruins of Chichen-Itza in Mexico. *J. Acoust. Soc. Am.* **2004**, *116*, 3328–3335. [CrossRef] [PubMed]
4. Declercq, N.F.; Dekeyser, C.S. Acoustic diffraction effects at the Hellenistic amphitheater of Epidaurus: Seat rows responsible for the marvelous acoustics. *J. Acoust. Soc. Am.* **2007**, *121*, 2011–2022. [CrossRef] [PubMed]
5. Lokki, T.; Southern, A.; Siltanen, S.; Savioja, L. Acoustics of Epidaurus–studies with room acoustics modelling methods. *Acta Acust. United Acust.* **2013**, *99*, 40–47. [CrossRef]
6. Crawford, F.S. Cube corner retroreflectors for sound waves. *Am. J. Phys.* **1991**, *59*, 176–177. [CrossRef]
7. Cabrera, D.; Holmes, J.; Caldwell, H.; Yadav, M.; Gao, K. An unusual instance of acoustic retroreflection in architecture–Ports 1961 Shanghai flagship store façade. *Appl. Acoust.* **2018**, *138*, 133–146. [CrossRef]
8. Cabrera, D.; Yadav, M.; Holmes, J.; Fong, O.; Caldwell, H. Incidental acoustic retroreflection from building façades: Three instances in Berkeley, Sydney and Hong Kong. *Build. Environ.* **2020**, *172*, 106733. [CrossRef]
9. Cabrera, D.; Holmes, J.; Lu, S.; Rapp, M.; Yadav, M.; Hutchison, O. Voice support from acoustically retroreflective surfaces. In Proceedings of the Euronoise 2021, Madeira, Portugal, 25–27 October 2021.
10. Currie, D.; Dell'Agnello, S.; Delle Monache, G. A lunar laser ranging retroreflector array for the 21st century. *Acta Astronaut.* **2011**, *68*, 667–680. [CrossRef]
11. Barrett, H.H.; Jacobs, S.F. Retroreflective arrays as approximate phase conjugators. *Opt. Lett.* **1979**, *4*, 190–192. [CrossRef]
12. Wang, T.; Wang, W.; Du, P.; Geng, D.; Kong, X.; Gong, M. Calculation of the light intensity distribution reflected by a planar corner-cube retroreflector array with the size of centimeter and above. *Optik* **2013**, *124*, 5307–5312. [CrossRef]
13. Jacobs, S.F. Experiments with retrodirective arrays. *Opt. Eng.* **1982**, *21*, 212281. [CrossRef]
14. Rindel, J.H. Attenuation of sound reflections due to diffraction. In Proceedings of the Nordic Acoustical Meeting, Aalborg, Denmark, 20–22 August 1986.
15. Rapp, M.; Cabrera, D.; Yadav, M. Effect of voice support level and spectrum on conversational speech. *J. Acoust. Soc. Am.* **2021**, *150*, 2635–2646. [CrossRef] [PubMed]
16. Caldwell, H.P. An Investigation into Ceiling Geometries for Acoustic Control: Spatial Configurations for Absorption and Retroreflection. Master's Thesis, The University of Sydney, Sydney, Australia, 2019. Available online: http://hdl.handle.net/2123/20765 (accessed on 1 March 2022).
17. Toyota, Y.; Komoda, M.; Beckmann, D.; Quiquerez, M.; Bergal, E. *Concert Halls by Nagata Acoustics: Thirty Years of Acoustical Design for Music Venues and Vineyard-Style Auditoria*, 1st ed.; Springer Nature: Cham, Switzerland, 2021.
18. Tuominen, H.T.; Rämö, J.; Välimäki, V. Acoustic retroreflectors for music performance monitoring. In Proceedings of the Sound and Music Computing Conference, Stockholm, Sweden, 30 July–3 August 2013; KTH Royal Institute of Technology: Stockholm, Sweden, 2013; pp. 443–447.
19. Jain-Neubauer, J. *The Stepwells of Gujarat: In Art-Historical Perspective*, 1st ed.; Abhinav: New Delhi, India, 1981.
20. Gupta, D. *Baolis of Bundi: The Ancient Stepwells*, 1st ed.; INTACH: New Delhi, India, 2015.
21. Jain-Neubauer, J. (Ed.) *Water Design: Environment and Histories*, 1st ed.; Marg Foundation: Mumbai, India, 2016.
22. Livingston, M. *Steps to Water: The Ancient Stepwells of India*, 1st ed.; Princeton Architectural Press: New York, NY, USA, 2002.
23. Lautman, V.; Gupta, D. *The Vanishing Stepwells of India*, 1st ed.; Merrell Publishers Ltd: London, UK, 2017.
24. Bahadur, V.N. *Stepwells of Rajasthan*, 1st ed.; Shubhi Publications: Gurgaon, India, 2016.
25. Center for Art and Archaeology. Virtual Museum of Images and Sounds. Available online: https://vmis.in (accessed on 10 January 2022).
26. Earis, P. Atlas of Stepwells. Available online: https://stepwells.org/ (accessed on 10 January 2022).
27. Botts, J.; Savioja, L. Effects of sources on time-domain finite difference models. *J. Acoust. Soc. Am.* **2014**, *136*, 242–247. [CrossRef] [PubMed]
28. Chern, A. A reflectionless discrete perfectly matched layer. *J. Comp. Phys.* **2019**, *381*, 91–109. [CrossRef]
29. Eckhardt, H.D. Simple model of corner reflector phenomena. *Appl. Opt.* **1971**, *10*, 1559–1566. [CrossRef]
30. Li, S.; Tang, B.; Zhou, H. Calculation on diffraction aperture of cube corner retroreflector. *Chin. Opt. Lett.* **2008**, *6*, 833–836. [CrossRef]
31. Doerry, A.W.; Brock, B.C. A better trihedral corner reflector for low grazing angles. *Proc. SPIE* **2012**, *8361*, 83611B. [CrossRef]
32. Whiteman, C.D.; Eisenbach, S.; Pospichal, B.; Steinacker, R. Comparison of vertical soundings and sidewall air temperature measurements in a small Alpine basin. *J. Appl. Meteorol. Climatol.* **2004**, *43*, 1635–1647. [CrossRef]
33. Le, T.N.; Straatman, L.V.; Lea, J.; Westerberg, B. Current insights in noise-induced hearing loss: A literature review of the underlying mechanism, pathophysiology, asymmetry, and management options. *J. Otolaryngol. Head Neck Surg.* **2017**, *46*, 41. [CrossRef]
34. Fletcher, N.H. Shock waves and the sound of a hand-clap—A simple model. *Acoust. Aust.* **2013**, *41*, 165–168.
35. Papadakis, N.M.; Stavroulakis, G.E. Handclap for Acoustic Measurements: Optimal Application and Limitations. *Acoustics* **2020**, *2*, 224–245. [CrossRef]

Article

Sound Scattering by Gothic Piers and Columns of the Cathédrale Notre-Dame de Paris

Antoine Weber * and Brian F. G. Katz *

Institut Jean Le Rond d'Alembert UMR 7190, Sorbonne Université, CNRS, 75005 Paris, France
* Correspondence: antoine.weber@dalembert.upmc.fr (A.W.); brian.katz@sorbonne-universite.fr (B.F.G.K.)

Abstract: Although the acoustics of Gothic cathedrals are of interest to researchers, the acoustic impact of their many columns is often neglected. The construction of the Cathédrale Notre-Dame de Paris spanned several centuries, including a wide variety of architectonic elements. This study investigates the sound scattering of a selection of seven designs that are relevant to this building as well as to the architectural style itself. These were measured on scale models (1:8.5 to 1:12), using a subtraction method, for receivers at about 3 m at full scale and a far-field source. They were also numerically simulated using a finite-difference time-domain method in two-dimensional space with an incident plane wave. The method integrates a finite volume framework to employ an unstructured mesh conforming to the complex geometries of interest. The two methods are in strong agreement for the considered configurations. Relative levels to the direct sound of backscattered reflections between −10 dB and 2 dB and between −15 dB and −6 dB in the transverse directions were estimated for the dimensions considered, relative to reported reflection audibility thresholds. Cross-sections with smaller scale geometrical elements on their perimeter can produce diffuse reflections similar to those of surface diffusers.

Keywords: sound scattering; sound diffusers; room acoustics; FDTD; archaeoacoustics; scale model

Citation: Weber, A.; Katz, B.F.G. Sound Scattering by Gothic Piers and Columns of the Cathédrale Notre-Dame de Paris. *Acoustics* **2022**, *4*, 679–703. https://doi.org/10.3390/acoustics4030041

Academic Editor: Margarita Díaz-Andreu and Lidia Alvarez Morales

Received: 19 July 2022
Revised: 13 August 2022
Accepted: 17 August 2022
Published: 26 August 2022

Publisher's Note: MDPI stays neutral with regard to jurisdictional claims in published maps and institutional affiliations.

1. Introduction

In large ancient buildings such as Gothic churches and cathedrals, columns and piers (In art history, a column refers to a support based on a circular section, while a pier is a generic term.) with different shapes can be present in a large number. These architectonic elements are obstacles that can scatter a wave in all directions when it reaches it. In this case, the term volumetric diffuser has been proposed [1]. Examples of modern applications are the canopy of reflectors suspended above some concert hall stages [2] or hanging panels in reverberation chambers [3]. As they are finite-sized objects, they usually have less effect on long wavelengths [4], and the effect of the curvature of reflector panels on this frequency limit has been considered [5,6]. For wavelengths of comparable size, the waves can propagate around the obstacle and are strongly diffracted, notably in the shadow zone [7]. The rows of columns surrounding the nave in a church, delimiting the subspaces, can be seen as lateral reflectors. In the Cathédrale Notre-Dame de Paris, cylindrical obstacles are found with cross-sectional shapes that are representative of different Gothic styles as its construction spanned several centuries. Among them, some are concave, star-shaped or not, involving intersecting circles, and outer and inner corners that in the end, form grooves and cavities; others are formed with multiple cylinders. The present study aims to examine the scattered reflections of different selected geometries that are relevant regarding the origin and the evolution of Gothic architecture, through numerical simulations and physical scale model measurements.

In room acoustics, studies concerning scattered reflections have often been interested in that of wall surfaces whose properties can be characterized in the far field using standardized procedures to determine the scattering [8] and diffusion [9] coefficients. Their effects on the sound field of concert halls have been evaluated perceptually and objectively by examining the room acoustic parameters with scale [10–12] and computer [11,13,14] models. Similarly, the effect of circular columns placed close to the walls of a concert hall has been studied in [15]. In contrast, we propose here to study the reflections of obstacles in isolation, similarly to [6], with a far-field source and a receiver at distances representative of a listening situation in the cathedral.

Wave scattering by cylindrical objects has been of interest for a long time, and different methods have been applied to address different geometries, boundary conditions, and frequency ranges. Lord Rayleigh was the first to derive wave scattering by small obstacles compared to the wavelength [16]. Subsequently, analytical solutions using partial wave series expansion methods have been derived for simple shapes [17,18]. The study of more arbitrary geometries involving complex shapes and/or multiples bodies can be achieved using measurements on full-size [5] or scale models [19] and simulations with appropriate computational methods. Frequency methods solving the Helmholtz equation, such as the boundary element method [20,21] or the finite element method [22] have been employed. Time domain methods solving the wave equation allow for the study of scattering on a wide frequency band through the use of pulse excitation and to evaluate the temporal spreading occurring. They are widely used in the field of electromagnetism for the study of antennas [23]. In the field of acoustics, they have been applied to the characterization of sound diffusers [24] and sonic crystals [25,26]. Nevertheless, they suffer from numerical dispersion causing the anisotropy of phase velocities in the discrete space [23]. Moreover, if the boundaries of the scattering object are meshed from a regular grid, which in most cases does not conform to it, staircasing artifacts appear [27]. In this work, a finite difference scheme solving the two-dimensional wave equation by operating on a hexagonal grid is used in conjunction with a finite volume method operating on an unstructured mesh in proximity to the boundaries [28]. These artifacts are thus eliminated, the isotropy is improved, and the dispersion is reduced compared to other compact Cartesian schemes [29].

The audibility of early reflections is usually expressed in terms of detection threshold or masked threshold, defined as the highest level of a reflection just before it becomes inaudible, relative to the direct sound [30]. This depends on many factors, including the delay of the reflection, its direction, the signal type, its spectrum and its sound level, or the environment. Their influence has been addressed in several studies [30–37]. A practical "rule of thumb" has been proposed [35], such that early reflections will be inaudible if their levels are less than -22 dB relative to the direct sound for a 3 ms delay, lowered to -31 dB for 15 ms to 30 ms, and that a modest amount of reverberation added in the stimuli increases the thresholds by up to 11 dB. This criterion has been adapted to discuss the audibility of reflections from panels with various curved edges [6]. These thresholds are also related to the human ability to echolocate dicrete objects [38,39]. The effect of diffusion has been studied by Robinson et al. [40]. White Gaussian noise was multiplied with gamma distributions of different parameters to emulate the temporal spreading and envelope of diffuse reflections. In comparison to specular reflections with the same peak amplitude, diffuse reflections were more detectable, indicating that integrated power of the reflection is probably a better indicator to predict its audibility. Wendt and Höldrich [41] considered reflections from a finite wall modeled with Lambertian surface. They found similar results that they attributed to the temporal position of the energy centroid for a diffuse reflection in relation to the forward masking pattern of the direct sound. In addition, they provide relationships linking the masking thresholds to the logarithm of the time differences of arrival, with excellent correlations to the experiments. Our ability to perceptually discriminate between different reflections from surfaces with respect to their topology inducing spectral coloration has been studied in [42], and with finite-difference time-domain method simulations in [43,44].

A brief introduction reviewing the long construction of Notre-Dame de Paris and its acoustics is given first in Sections 1.1 and 1.2, and the cross-sections of the columns and piers studied are then described in Section 2.1. The numerical methods and set-ups used are presented in Section 2.2. Sound scattering is additionally measured on 1:8.5 to 1:12 scale models of the selected piers and columns. The protocol and post-processing methods used to isolate the scattered pressure are described in Section 2.3. Measurements and simulations are compared for cross-validation in Sections 3.1 and 3.2. Simulation results are analyzed in the time-frequency domain and presented in a perceptually relevant way in Section 3.3 Finally, we discuss their efficiency in terms of volumetric diffusers and their audibility with respect to the thresholds reported in the literature in Section 4.

1.1. Introduction to the History of the Construction of Notre-Dame de Paris

The history of the construction of Notre-Dame de Paris has been treated in detail by several authors [45–48]. We recall here the main stages of it, as well as some background information.

The construction of the Cathédrale Notre-Dame de Paris began in 1163 under the responsibility of Maurice de Sully, bishop of Paris. Figure 1a shows the principal phases of construction. The first part to be built was the choir, completed in 1182, as shown at the top of Figure 1b. At this date, the western wall of the transept was already erected. The choir had a double ambulatory and tribunes as with nowadays, but did not have its radiating chapels yet. The elevation of the nave started in 1180 while the vaults were missing in the choir. Several construction campaigns will have been necessary for the western part of the cathedral to be completed. The first bay of the nave, connecting it to the facade composed of the bases of the towers, was completed around 1220. While the construction of the towers continued, changes were made to the building from 1225, following a fire between the roof and the vaults according to Viollet-le-Duc [47]. At that time, the work had already been going on for more than 60 years, a period during which the techniques of masonry knew many innovations. Thus, the hypothesis that, by rivalry with other dioceses building cathedrals at the same time, the bishop would have decided to bring it up-to-date has also been advanced [47]. The towers were completed around 1240, at the same time that the lateral chapels were added to the nave. Around 1250, the rose window of the north arm of the transept, extended by one bay, was built by the master mason Jean de Chelles and that of the south, added 10 years latter, is attributed to his successor, Pierre de Montreuil. In the meantime, a wooden spire was added above the crossing. The radiating chapels were built between 1290 and 1330 under the supervision of Pierre de Chelles and his successor, Jean Ravy.

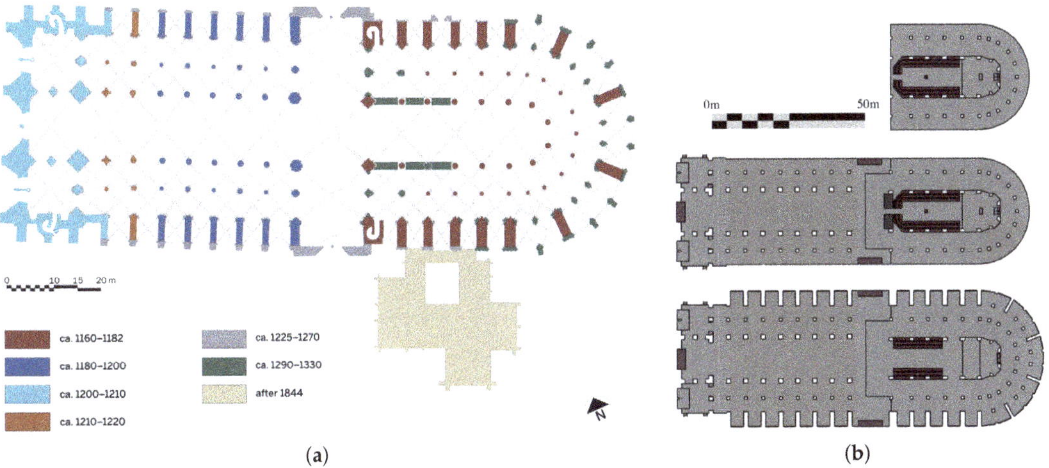

Figure 1. Floor plans of Notre-Dame de Paris: (**a**) with principal phases of construction, from [47]; (**b**) after the main campaigns, from [49]. *From top to bottom: ca. 1182, ca. 1230, ca. 2015.*

It took almost two centuries for the cathedral to reach a shape close to the one we know today, and it is only from the 18th century that major modifications were made to the interior. After the Vow of Louis XIII in 1638, the cathedral underwent a period sometimes qualified as a Baroque transformation, with the addition of the still present polychromatic marble pavement, Pietà, and oak stalls. After the French Revolution, the building was requisitioned and rededicated as the Temple of Reason on 10 November 1793. It returned to the clergy less than 10 years later, on 10 April 1802, in a more deteriorated state than it already was. In 1843, Lassus and Viollet-le-Duc won the tender for a renovation project [50], being the only ones to submit on time. Their wish was to return to a state closer to the cathedral's origins. They removed some of the additions from the Baroque period but renovated some of them, such as the pavement. They rebuilt a spire above the crossing in imitation of the 13th century one that had to be dismantled at the end of the 18th century because it showed signs of fragility and was in danger of collapsing. Oculi were introduced at the level of the triforium (Triforium: Narrow level below the clerestory.) on the walls of the transept, on the last bay of the nave and the first of the choir. These renovations, lasting for almost 20 years, were conscientiously consigned in a daily work journal [51].

1.2. General Acoustics of Notre-Dame de Paris

The Cathedral suffered a major fire in April 2019. However, measurements were taken almost 4 years earlier [52]. The reverberation time T_{20} measured at that time was about 7 s at 500 Hz. Additional measurements were subsequently made by the authors after the fire in 2020 [53]. Among the damages, several vaults collapsed, including the one at the crossing, created large openings. The reduction in reverberation time was estimated to be 8% on average. More information on the acoustics in past states, obtained by simulations, is available in [49].

2. Materials and Methods

2.1. Columns and Piers of Interest

The many successive construction and renovation campaigns can be seen in part through the geometry of the many piers and columns in the cathedral (Figure 1a). The study of the plinths, bases, and capitals contributed notably to the sequencing of the building site [46]. We restrict ourselves here to the study of the scattering properties of the shafts, being the major part of a column. In total, seven geometries were retained according to architectural criteria, such as their location or frequency, and historical criteria such as their place among the different Gothic styles or their links of influence with later or earlier architectural styles. The groups of columns they define are shown in Figure 2 with a label attributed to each. There are five compound piers, consisting of a core flanked by engaged columns and/or pilasters. These elements extend the arches and ribs to take some of their loads and articulate the structure vertically. These principles were already used in Romanesque architecture [54,55]. Their section is formed of a single closed shape. This distinguishes them from the piers with *colonnettes*, where long thin *en-délit* circular columns flank without contact with a central part, in this case, they have a decorative function; two were selected. They are all located in the nave, except one. The columns with circular sections present in this part of the cathedral are also indicated in Figure 2, with their diameters. The piers that are not included in the current study, i.e., not colored, are generally formed with shafts of similar geometries, some of which are visible in [48].

Figure 3 shows the cross-sections of the studied shafts, with their dimensions given in cm. They were drawn based on orthoimages extracted from the interactive 3D visualization environment developed by the Modèles et simulations pour l'architecture et le patrimoine (MAP) laboratory in the framework of the "digital data" working group of the scientific project for the restoration of the cathedral supported by the Centre National de Recherche Scientifique (CNRS) and the French Ministry of Culture [56]. This numerical tool integrates the 3D point clouds obtained by several laser survey campaigns conducted notably by Andrew Tallon in 2010 [47], and also by the company Art Graphique et Patrimoine (AGP) just after the fire of 15 April 2019. They are described in more details in the following.

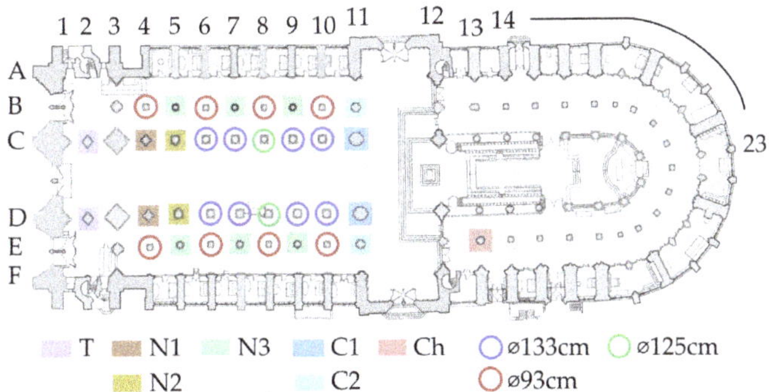

Figure 2. Floor plan with piers, columns, and responds locations. The shafts of identical cross-sections from the selected groups under study are highlighted with a common color. The columns of the nave with circular shafts are indicated with a circle.

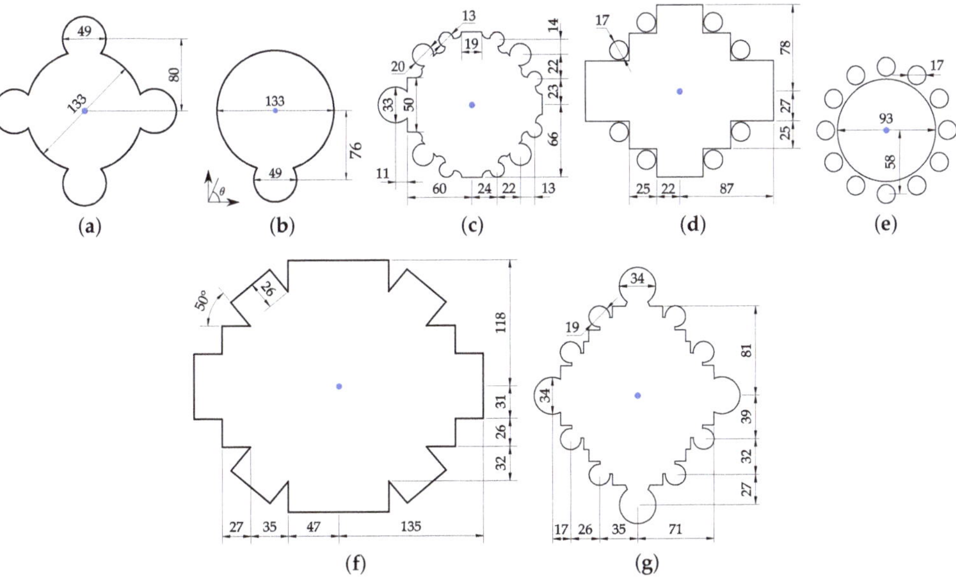

Figure 3. Cross-sections of the selected shafts. Piers of the western bay of the nave, (**a**) **N1**, (**b**) **N2**. Pier (**c**) **Ch** in the southern ambulatory. Piers with detached colonnettes (**d**) **C1**, western wall of the transept, and (**e**) **N3**, nave aisles. Western crossing piers (**f**) **C1** at the arcade level. Piers (**g**) **T**, supporting the tribune between the towers. Dimensions are given in cm. A cylindrical coordinate system is assigned to each, centered on the blue point, and the directions of propagation of the incident waves are indicated with respect to the abscissa, as shown in (**b**) in the following.

2.1.1. Compound Piers

The first bay of the nave, connecting it with the frontispiece, was built last (see Figure 1a). It resulted in specificity on its columns, because more stiffness was needed in this part, according to Bruzelius [46]. The shafts **N1**, located at C/D4 in Figure 2, are engaged with four circular colonnettes on a circular core, as shown in Figure 3a. Just besides, the piers **N2**, located at C/D5, are engaged, with a single one facing the nave (Figure 3b).

The central part is the same diameter as the other columns of the nave arcades, except the columns at C/D8 that follow the principle of strong and weak piers, as their diameter is 125 cm. The colonnettes are engaged by less than a quarter of their diameter, penetrating 11 cm and 15 cm, respectively. This type of circular lobe shaft is also found engaged in every responds (Respond: Half-pier or half-column embedded in a wall.). They are *piliers cantonnés*, a type of pier already used in Romanesque architecture where massive rectangular piers are flanked by semicircular columns, as with Saint-Étienne de Caen Church, Santiago de Compostela or Mainz Cathedrals. Geometries similar to these two couples may be found in other High Gothic cathedrals, as with the naves of Notre-Dame de Chartres and Notre-Dame de Reims, or in the choir of Notre-Dame de Noyon. They are more robust variations of the columns known as *"soissonnaises"*, which first appeared in Soissons Cathedral, where a single colonnette is engaged in a circular core on the part facing the nave and rises to the vault [57].

Another selected geometry are the transept crossing piers on the nave side **C1**, located at C/D11. They are actually the union of several pilasters, each one receiving a transverse or diagonal rib of the nave or crossing vaults. At the arcade level, this results in an asterisk-shaped section, as shown in Figure 3f. They are the largest piers of the selection, and, with the two previous ones, they directly surround the nave where listeners are located.

Typical of Gothic architecture, it is also found in the cathedral compound piers formed by a cluster of coursed shafts shaped with relatively thin engaged circular parts that extend vertically the arcs and the ribs of the vaults. The supports of the towers, located at C/D3, are built with this method. Their sides facing the central vessel extend up to high capitals at the base of the sexpartite vaults, forming a diffusing surface close to the Grand Organ. This is also the case for its intermediate piers. The piers **T**, located at C/D2, are selected to study the influence of such shafts. Their cross-section is shown in Figure 3g. At each corner of the diamond shape is engaged a wider column of diameter 34 cm, and on the sides, there are alternately right corners and engaged colonnettes of diameter 19 cm. This pattern is repeated on the wall and outer aisle responds, between each chapels, of the choir [58].

The eastern transept crossing piers, located at C12 and D12, on either side of the current altar, are the oldest of the building with such shafts. When the chapels of the choir were nearly completed, Pierre de Chelles renewed the eastern wall of the transept around 1315 [59], introducing foliate gables to the arcades of the ambulatory entrance and the piers located at B12 and E12 were modified. The North and South parts are different; the second, probably built in first, is less massive and less prismatic. Their shapes are similar to those of the intermediate piers of the chapels, located from A18 to A23, as well as the responds located in the outer direction to them, at the back of the chapels, and their counterpart in the southern half. The shafts have, as in Early Gothic, engaged columns and colonnettes, but we find some of several diameters revealing a more advanced Gothic style, closer to the *flamboyant* Gothic, and the corners are no longer all straight, as he introduced curved faces where the colonnettes are flanked. The piers, located at B13 and E13 in the ambulatory, and the second one of the south side arcade, located at D14, were also built around this moment, and are similar. They replaced circular piers following structural problems that led to ruptures, and it is not clear who between Pierre de Chelles and Jean Ravy was leading these repairs [59]. The pier in the southern part of the ambulatory is selected to study a shaft more representative of a latter Gothic style. It is labeled **Ch** and its cross-section is shown in Figure 3c.

2.1.2. Piers with Detached *colonnettes en délit*

The use of *colonnettes en délit* is widespread in the cathedral. They divide the tribune openings, in two in the choir and in three in the nave, except in the first bay. They are in the responds of the nave, in the central vessel from the clerestory (Clerestory: Upper row of bays of a nave located above the triforium and the tribunes.) to the tribunes, extending in a uniform way the ribs of the sexpartite vaults and in the outer aisles. They are 17 cm in diameter. This principle is found at the arcade level on the columns **C2**, located at B/E11, at the entrance of the double aisles at the western wall of the transept. It is represented in Figure 3d. The prismatic piers formed by the union of the arcades and vaults dosserets (Dosseret: Pilaster used as a straight jamb for an arch.) are supplemented by detached colonnettes at each re-entrant corner, separated by a distance of 1 cm.

Separating the double aisles, every other circular pier **N3**, located at B/E5/7/9, is surrounded by 12 detached colonnettes, as shown in Figure 3e. Viollet-le-Duc [60] explained this difference with the single circular cylinder neighbors by considerations of structural strength and stability. In particular, since these piers are in line with the most heavily loaded columns of the nave, they had to take the load of the buttresses, which had existed before 1220, and were allowed to counter the thrust from the sexpartite vaults. However, this has been challenged since [58], and Viollet-le-Duc himself acknowledged that they have a decorative function when installed at the responds after settlement of the building. A couple of them are also included among the coursed shafts that form the piers supporting the towers, located at B/E3. Many examples of such elements can be found in other cathedrals at the time, including, not exhaustively, the cathedrals Notre-Dame de Noyon, Saint-Étienne de Bourges, Notre-Dame de Dijon, and Notre-Dame de Laon [61]. Their use facilitated the way in which the walls are vertically articulated, with the vaults compared to coursed shafts with circular shapes, as in the choir responds. They could be manufactured in mass by standard processes while the walls were built with stones cut in regular rectangular shapes [62]. The piers surrounded by colonnettes are also at the origin of a whole architectural style in England [63]. We can give the examples of the Canterbury Cathedral, the Lincoln Cathedral, or the Salisbury Cathedral, where the colonnettes are in marble of a different color from the central part [64].

2.2. Numerical Methods

Scattering problems can be treated with numerical simulations such as finite-difference methods, which are particularly convenient for solving the wave equation and thus working directly in the time domain. However, in the context of complex geometries, as studied here, particular care must be taken with the boundary conditions to avoid the staircase approximation usually employed when the numerical scheme is derived from a regular spatial grid. In this study, this issue is addressed by implementing a hybrid time-domain method using finite-difference and finite-volume methods detailed in the following.

The studied objects are relatively long straight cylinders; it is then relevant to restrict the problems to a two-dimensional space. Thus, the second-order wave equation in two space dimensions is solved using a finite difference scheme operating on a hexagonal grid. The Laplacian is approximated by the centered difference operator using seven spatial points: the central point and its six nearest neighbors [65]. Similarly, the time derivative is approximated by a centered finite difference resulting in a fully explicit second-order accurate scheme. The update equation for $p_{i,j}^n$, $(i,j) \in \mathbb{Z}^2$, $n \in \mathbb{N}$, approximating the pressure $p(ih\mathbf{e_1}, jh\mathbf{e_2}, n\Delta t)$ with Δt the time step, h the grid spacing, $\mathbf{e_1} = \begin{bmatrix} 1 & 0 \end{bmatrix}^T$, $\mathbf{e_2} = \begin{bmatrix} -1/2 & \sqrt{3}/2 \end{bmatrix}^T$, is

$$p_{i,j}^{n+1} = 2p_{i,j}^n - p_{i,j}^{n-1} + \lambda^2 \left[\frac{2}{3}\left(p_{i+1,j}^n + p_{i-1,j}^n + p_{i,j+1}^n + p_{i,j-1}^n + p_{i+1,j+1}^n + p_{i-1,j-1}^n \right) - 4p_{i,j}^n \right], \tag{1}$$

where $\lambda = c\Delta t/h$ is the Courant number. This is a simple way to reduce the numerical dispersion [23] and to improve the isotropy of the propagation, compared to other compact schemes operating on a rectilinear grid [29].

When a wave is scattered by an obstacle, part of it circumvents the latter and propagates as a circumferential wave, so it is crucial to take into account the specific contour geometry of the objects. Depending on the scheme used, a boundary mesh based on a regular grid does not always allow for consistency [66]. This is because using a staircase approximation for a closed curved boundary converges in area but not in perimeter at the limit of small spatial steps [28]. Several locally conformal FD schemes have been proposed [27,67] to handle curved boundaries. Here, the structured hexagonal grid is modified over a thin layer around the object boundaries into an unstructured mesh conforming to it. The approximation of the wave equation on this part of the space is then treated by a finite volume method [68]. The boundary condition of the objects studied are considered to be rigid, which is consistent with the strong impedance contrast existing between air and rocky materials that constitute historic buildings, and the 40 µm average roughness of Lutetian limestone measured in [69]. The update equations in matrix form for the pressures $\mathbf{p} = \begin{bmatrix} p_1 & \cdots & p_l & \cdots & p_N \end{bmatrix}^T, l = 1, \ldots, N$, associated with the N finite volume cells are thus

$$\mathbf{p}^{n+1} = 2\mathbf{p}^n - \mathbf{p}^{n-1} - c^2 \Delta t^2 \mathbf{D}^T \mathbf{D} \mathbf{p}^n, \tag{2}$$

where

$$\mathbf{D} = \mathbf{S}^{1/2} \mathbf{H}^{-1/2} \mathbf{Q}^T \mathbf{V}^{-1/2}, \tag{3}$$

with $\mathbf{Q}$ is the matrix form of an oriented adjacency tensor $Q_{l,e}$, where $e = 1, \ldots, N_e$, N_e is the number of internal edges forming the cells. For a given edge e adjacent to two cells of index l_e^+ and l_e^-, such that $l_e^+ > l_e^-$, the tensor is defined as

$$Q_{le} = \begin{cases} 1 & \text{if } l = l_e^+, \\ -1 & \text{if } l = l_e^-, \\ 0 & \text{otherwise}. \end{cases} \tag{4}$$

$\mathbf{V}$ is a diagonal matrix with the cell volumes V_l, $\mathbf{S}$ and $\mathbf{H}$ are diagonal matrices with the edge lengths S_e and inter-cell distances H_e, respectively. $\mathbf{D}$ is the FV approximation of the gradient operation, and $\mathbf{D}^T \mathbf{D}$ approximates the Laplacian operation.

The FV mesh is obtained from the Voronoi diagram of the set of grid points enclosing the object completely. The achieved polygons are, in a second step, clipped by the boundaries of the scatterer. An example is shown in Figure 4a. The FV approximation is equivalent to the FD one when applied to a regular mesh [70], as represented with black lines and dots. The red dots are on the FD hexagonal grid. They are used to bound the Voronoi diagram in the outer direction to the scattering object and thus obtain their adjacency relation with the Voronoi cells. It is then possible to establish the discrete Laplacian to link the two meshes when updating the equations.

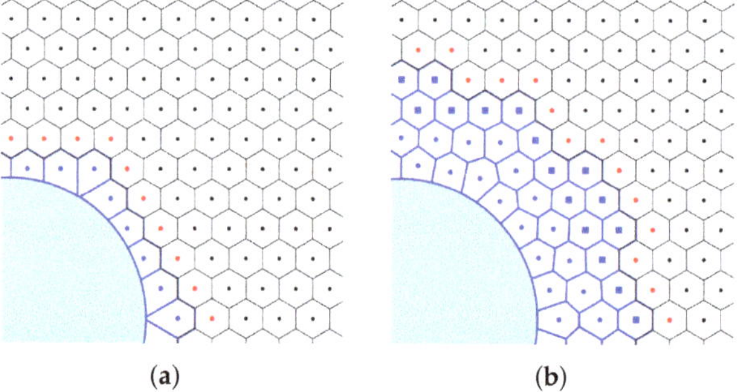

(a)   (b)

Figure 4. Examples of hybrid meshes in proximity to a curved boundary. (**a**) Voronoi diagram of grid points enclosing the boundary. (**b**) Centroidal Voronoi diagram after 10 iterations of Lloyd's method. Voronoi cells are shown with blue edges, and their corresponding generating sites with blue dots and square. They are clipped by the boundary of the object shown in cyan, and bounded in the outer direction by the regular grid points enclosing them, shown with red dots. Regular hexagonal grid is shown in black.

Depending on its geometry, i.e., whether it is smooth with respect to the spatial step or sharp with re-entrant corners; such a mesh can result in very strict stability conditions in the second case [71], which is not computationally efficient. Therefore, in a third step, a new mesh is generated from the Voronoi diagram of the centroids of the Voronoi polygons obtained previously and clipped as before. This process is repeated to approximate a centroidal Voronoi diagram with Lldoyd's method [72]. Figure 4b shows the finite volume mesh obtained after 10 iterations. Here, Lldoyd's method is only applied on the points less than two spatial steps away from the boundary at the first iteration and not on the sites marked with a blue square. In other words, these points remain at their original position on the regular grid, ensuring complete regularity beyond.

Setup

All simulations are performed with a grid spacing about $h = 5$ mm in the regular part, corresponding to 10 points per wavelength at 7 kHz. The hybrid mesh is obtained with 5 iterations of Lloyd's method. Each contour is discretized by a set of linear elements using GMSH (version 4.7.1) [73], so that the error on the perimeter is less than 0.1%. The time step Δt is limited by the stability condition from the FV formulation derived in [68], which does not reduce to the proper stability condition for the hexagonal scheme. It is therefore set at

$$\Delta t = \min\left\{\frac{2}{\sqrt{\beta_{\max}}c}, \sqrt{\frac{2}{3}\frac{h}{c}}\right\}, \tag{5}$$

where $\beta_{\max}$ is the largest eigenvalue of $\mathbf{D}^T\mathbf{D}$. The speed of sound c for the simulations carried out to be compared with the measurements is estimated using [74], according to the temperature and humidity measured in the experimental room. The resulting Courant number λ of each simulation is given in Table 1. All simulations realized for the time-frequency analysis are performed with $c = 344$ m s^{-1}.

Table 1. Experimental set-ups and parameters for the measurements on scale models with the Courant number λ used in each corresponding simulation.

Label (Fig.)	Scale Factor	Incidence Angle θ_0	Distance from Center [1]		c [m s^{-1}]	λ
			Source [cm]	Receiver [cm]		
N1 (Figure 3b)	1:12	90°	31	307	346.2	0.755
N2 (Figure 3a)	1:12	90°	32	307	345.7	0.751
		45°	32	307	345.8	0.731
C1 (Figure 3f)	1:12	0°	32	306	345.7	0.753
		90°	32	306	346.2	0.742
T (Figure 3g)	1:10	90°	31	319	346.1	0.744
Ch (Figure 3c)	1:10	0°	31	319	346.5	0.743
C2 (Figure 3d)	1:8.5	90°	33	307	346.1	0.753
		0°	33	307	345.9	0.739
N3 (Figure 3e)	1:8.5	0°	33	338	346.6	0.745
		15°	37	317	346.6	0.762

[1] Represented by a blue dot in Figure 3a–e.

The source signal is a Ricker wavelet with a central frequency of 2 kHz injected as a soft source [75] over a line of grid points to have a plane wave incidence. A uniaxial perfectly matched layer is introduced at the end of the computational domain in the direction of propagation of the plane wave. Moreover, periodic boundary conditions are used to prevent edge effects. The scattered field p_s is defined as

$$p_s = p - p_i, \tag{6}$$

where p is the total field and p_i is the incident field. Therefore, a free-field simulation, without the cylinder, is performed in parallel to the total field simulation, on the regular grid used to generate the hybrid mesh with, consequently, strictly identical parameters, allowing to isolate the scattered field similarly to [24]. Examples of pressure field simulations are included as videos in the Supplementary Materials.

2.3. Experimental Methods

In spite of the progress of numerical methods, physical scale models are still a widespread tool in architectural acoustics, and in particular for investigating the acoustics of complex rooms such as concert halls or for the characterization of scattering from surfaces. They offer the advantage of accounting for all wave phenomena occurring in full-scale problems, provided that they are representative of them. In the case of airborne sound scattering by a rigid object, if the thermo-viscous and molecular relaxation effects are neglected, the full-scale results are obtained, depending on the domain considered, by frequency transposition or time dilation of the scale model results. Moreover, they can be used as validation tools by providing reference results from measurements.

The scattering of the different geometries was measured experimentally by a subtraction method with scale models. They are, for the most part, made of an assembly of long rigid PVC tubes and/or dense wooden boards and cleats, as shown in Figure 5a–d. The detached colonnettes are positioned at the right distance from their core with the help of wedges. Their total length is about 2 m. Both models of compound piers with clustered engaged shafts are made of staff, a plaster-based material. The fresh material is spread in successive layers with a comb whose shape is the negative of one symmetric part of the section. These parts are eventually attached together to form the cylinder, as shown in Figure 5e. Those are 1 m long. Their scaling factor, given in Table 1, is determined by the manufacturing constraints, i.e., according to the standard dimensions of the PVC tubes and wooden cleats, and for those in plaster, it is chosen as a compromise between the size and weight of the model and the minimum size required by the technique to achieve the details.

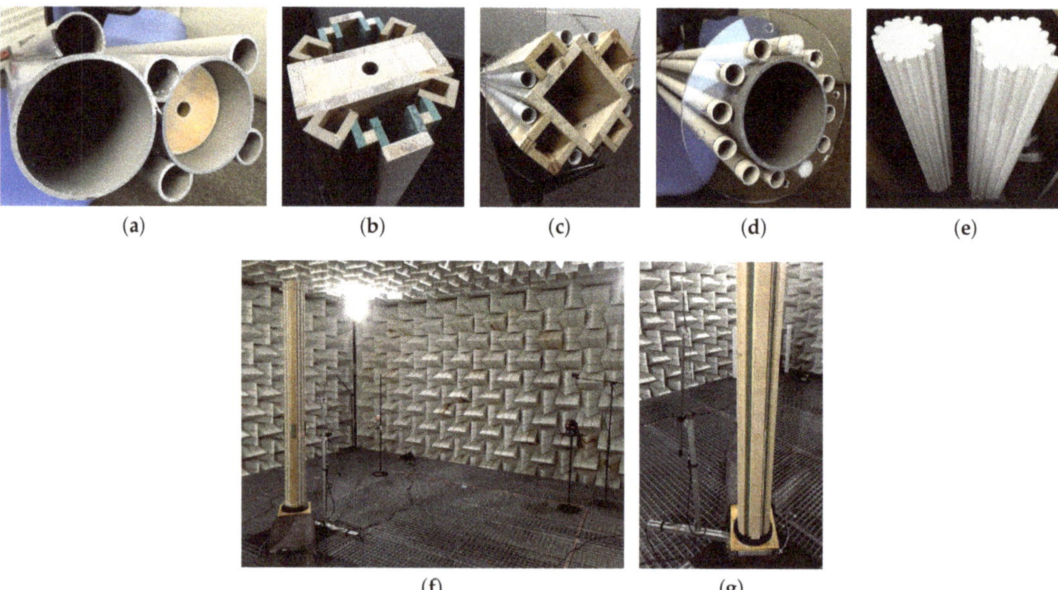

Figure 5. Photographs of scale models and experimental set-up. Piers and columns: (**a**) **N2** and **N1**, (**b**) **C1**, (**c**) **C2**, (**d**) **N3**, and (**e**) **Ch** and **T**. Experimental set-up: (**f**) overview with the sound source on the right, (**g**) view of the platform, turntable, and microphone mounted on an articulated arm.

The measurements were carried out in the anechoic chamber of Sorbonne Université (Figure 5f). The sound field was measured on a circular arc around the cylinder. For that, a microphone was attached to an articulated arm allowing for positioning in space, which was mounted on a turntable (Brüel & Kjær Turntable System Type 9640). The cylinders were positioned on a platform above the turntable that was not in contact with it, to allow the arm with the microphone to rotate around. In practice, the legs of the support platform prevented measurement for about a quarter of a circle (Figure 5g). Since all diffusers have at least one plane of symmetry, recording the signals on an arc greater than a semicircle that includes the forward and backscattering positions allows for full measurement of the scattering in the case where the source is included in this plane of symmetry and orientated towards the center of the scatterer.

The source was a 20 mm diameter dome tweeter (Audax TM020G3) driven by an amplifier (Samson Servo 120a) positioned at a corner of the chamber, as may be seen on the right of Figure 5f. The signals were recorded using a miniature microphone (Feichter Audio M1). Its axis was parallel to the cylinder axis to minimize variations due to its directivity. All were connected to an audio interface (RME Babyface) configured at a sample rate of 192 kHz. The exponential swept sine method [76] was used with signal spanning frequencies from 2 kHz to 95 kHz over 3 s. The exploitable frequency band was eventually identified from 2 kHz to 30 kHz, limited by the source, with a drop in the signal-to-noise ratio (SNR) at 20 kHz. The measurements were carried out with an angular step of $5°$, and the emission and acquisition of signals, as well as the control of the turntable, were performed with MATLAB 2020a through an automatic procedure. The source, the microphone, and the cylinder were positioned with the help of laser levels visible in Figure 5f. The different set-ups are summarized in Table 1. The shortest distance from the source to the columns is nearly 3 m, well beyond the Rayleigh distance of the source, $\pi a_{source}^2 f / c$, where a_{source} is the source radius at any frequency f.

For each tested configuration involving a geometry and an angle of incidence, a series of measurements with, then without the cylinder have been made, taking care not to change the positions of the source and the receiver during the removal. It is then possible to isolate the scattered pressure by subtracting the incident pressure to the total pressure recorded respectively without and with the cylinder present. This method is very sensitive to variations of environmental conditions. Changes in temperature and humidity in the chamber, microphone positioning, or response of the equipment due to electrical deviations and Joule heating cause disparities in time and amplitude between measurements. Several methods have been proposed to compensate for them in post-processing [77]. At low-frequency, amplitude variation is the dominant error factor, while at high-frequency, it is the time shift. Here, we compensate only for the difference of time of arrival. It is obtained by a cross-correlation between the free-field signal and the total pressure signal. As it is generally a fraction of a sample, the estimation is refined by interpolation on a Gaussian curve, as proposed in [78]. Since this method is not applicable to signals measured in the shadow zone, these are corrected with the average of the estimated time shifts for the neighboring visible positions of the same series.

3. Results

3.1. Validation of the Experimental Methods with a Rigid Circular Cylinder

In order to validate the proposed measuring system and post-processing to obtain the scattered pressure, measurement series have been made with a PVC tube of outer diameter 110 mm to model an infinite rigid circular cylinder. The thickness of the pipe is 3.2 mm, which is sufficient to assume such a boundary condition in the case of airborne propagation [79]. The scale factor is set to 1:9.1 in order that it represents a cylinder of 1 m in diameter at full scale. Three repetitions of the same measurement have been made. The source and microphone positions are not changed between them. They are at 310 cm and 42 cm from the cylinder axis, respectively. The humidity and room temperature have been measured for each repetition. The average estimated speed of sound for the three measurements was 345.5 ± 0.1 m s^{-1} as the maximum absolute difference. The variations of c between the repeated series are small, so they are compared to the analytic solutions for the plane and spherical incidences calculated with this average value.

Figure 6 compares the measured and analytic results for the scattered sound in the frequency domain. Figure 6a shows the polar diagrams as functions of the scattering angle $\theta_s = \theta - \theta_0$ of scattered pressure levels relative to the incident field at the discrete scaled frequencies 250 Hz, 500 Hz, 1000 Hz, and 2000 Hz, compared to analytic solutions with plane wave and spherical incidence. The corresponding Helmholtz number ka is also indicated for generalization, where k is the wavenumber and a the cylinder radius. For the source distance chosen, the scattered relative level for the plane wave incidence is about 1.8 dB higher than the spherical incidence in the backscattering direction for all the frequencies considered here. For other angular positions, the differences are lower. In the forward scattering direction, the plane wave incidence leads to a relative scattered level about 0.6 dB lower than the monopole source. Overall, the measurements are in a better agreement with spherical incidence than the plane wave. A very good agreement is observed at 500 Hz and 1000 Hz. At 250 Hz, the back and transverse relative scattered pressure levels are lower than the analytic solutions and the variations between series are high. This is also the case at 2000 Hz in the transverse direction. These frequencies are close to the limit of the sound source where the signal-to-noise ratio is lower, degrading the accuracy. Additionally, the proposed correction for positions in the shadow zone leads to a good estimate, given the excellent agreement found for the forward scattering peak and the observed repeatability between the series. Finally, the rigidity hypothesis for this scale model seems acceptable.

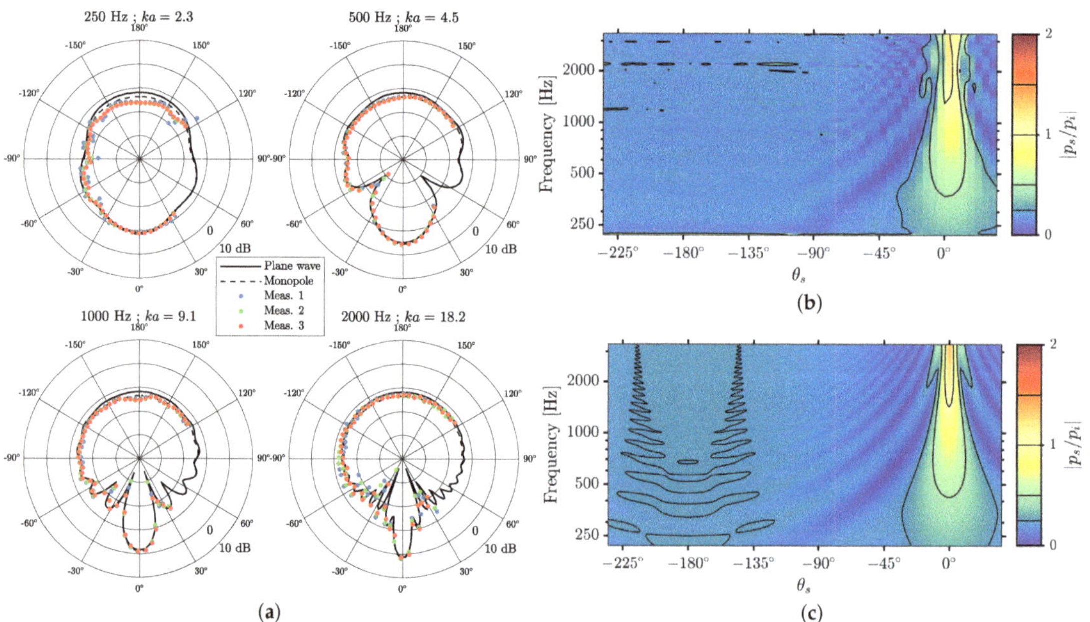

Figure 6. Measured scattered fields (3 repetitions) for a rigid circular cylinder compared to analytic solutions in frequency domain. (**a**) Polar diagrams of relative scattered pressure level, $20\log_{10}|p_s/p_i|$, as functions of the scattering angle θ_s at different central frequencies of octave bands with the corresponding Helmholtz numbers ka indicated, where k is the wavenumber and a the radius. Magnitude ratio, $|p_s/p_i|$, from (**b**) measurements (Meas. 3) with scaled frequency, and (**c**) analytic solutions for plane wave incidence.

Figure 6b shows the magnitude of the measured scattered pressure relative to the incident pressure for Meas. 3 over all the available frequency ranges, scaled according to the factor, for each angular position. Comparing it to the plane wave solution represented in Figure 6c confirms the underestimation in the backscattering region, and the slight overestimation in the forward direction. The interference pattern in the transverse direction is less visible at high frequency in Figure 6b compared due to the low angular sampling. The drop of signal-to-noise ratio around 20 kHz is visible at the scaled frequency of 2200 Hz, manifested by an horizontal line of high values, mainly over the backscattering positions of the arc. The proposed methods for correcting time shifts appear to be suitable for the visible and shaded positions. In the following, the measurements on the columns of interest are compared to simulation results where the source is a plane wave. The present results show that this assumption is likely to affect mainly the backscattering region.

3.2. Measurements Compared to Simulations

Figure 7 shows the magnitude ratios between the experimentally measured scattered and incident pressures and their numerically simulated counterparts for three of the configurations; results of other configurations are provided in Appendix A. The proposed subtraction method was applied to the signals and the spectra were obtained by Fast Fourier Transform; no correction for excess air absorption is made. The results are compared in the frequency domain and the vertical axis of the surface plots of the experimental results has been scaled according to the factor of each given Table 1. Overall, a good agreement between measurements and simulations can be observed, with constructive and destructive interference appearing within the scattered pressure in steady state match in frequency and space. In particular, the post-processing method for obtaining the scattered pressure in the shadow area based on neighboring positions is suitable, as shown by the comparison of the figures around the direction $\theta_s = 0°$. In the backscattering region, the magnitude ratios are systematically slightly lower for the measurements on scale models compared to simulations, in agreement with the difference expected between a plane and a spherical incidence. The drop of SNR around 20 kHz is visible in some of the measurement results, manifested by an horizontal line of high values on all positions of the arc; for example, in Figure 7a at around 2400 Hz, or in Figure A2c at around 1700 Hz.

In addition to the error due to the nature of the incident field, other sources are that, for some configurations, the centering of the cylinder and the perpendicularity of the measurement plane to the cylinder axis are not perfectly achieved. For this latter, a part of the wave is thus scattered out of the plane corresponding to oblique incidence. This can be seen when the measured backscattered and forward scattered pressures do not exactly match the expected diametrically opposite directions $\theta_s = -180°$ and $\theta_s = 0°$, respectively, that are supposed to be symmetry lines in the figures for such configurations as shown in the simulated results. This is particularly striking in Figure A1e compared to Figure A1f, where the cylinder is probably leaning in the transverse direction. If the cylinder is slightly tilted forward or backward from the source, then it is not visible in this way, but the deflection is still present.

Another one is the geometrical differences that can exist between a hand-made physical scale model and a perfect numerical model. For geometries with outward or inward corners, the scale models will have rounded corners compared to their digital counterparts, where no rounding was introduced afterwards. The effect of rounding corners on the scattered field by concave cylinders with one, two, or four corners has been studied numerically in [80], and they found that maximum differences occurred in the backscattering region with expected dependencies on the radius of curvature, the wavelength, and the angle of incidence, with respect to the position of the corners. Here, the scale models made of staff (Figure 5e) have more rounded edges compared to the wooden models, because of the viscosity of the material having a surface tension, and also affecting both the outer and re-entrant corners. For manufacturing reasons, the angles existing at the intersections of the circles for **N2** and **N1** (Figure 5a) are also rounded. In addition, for geometries with several elements, a bad straightness, and therefore, the positioning of the small cylinders leads to a different scattering, such as in Figure A2e in comparison to Figure A2f.

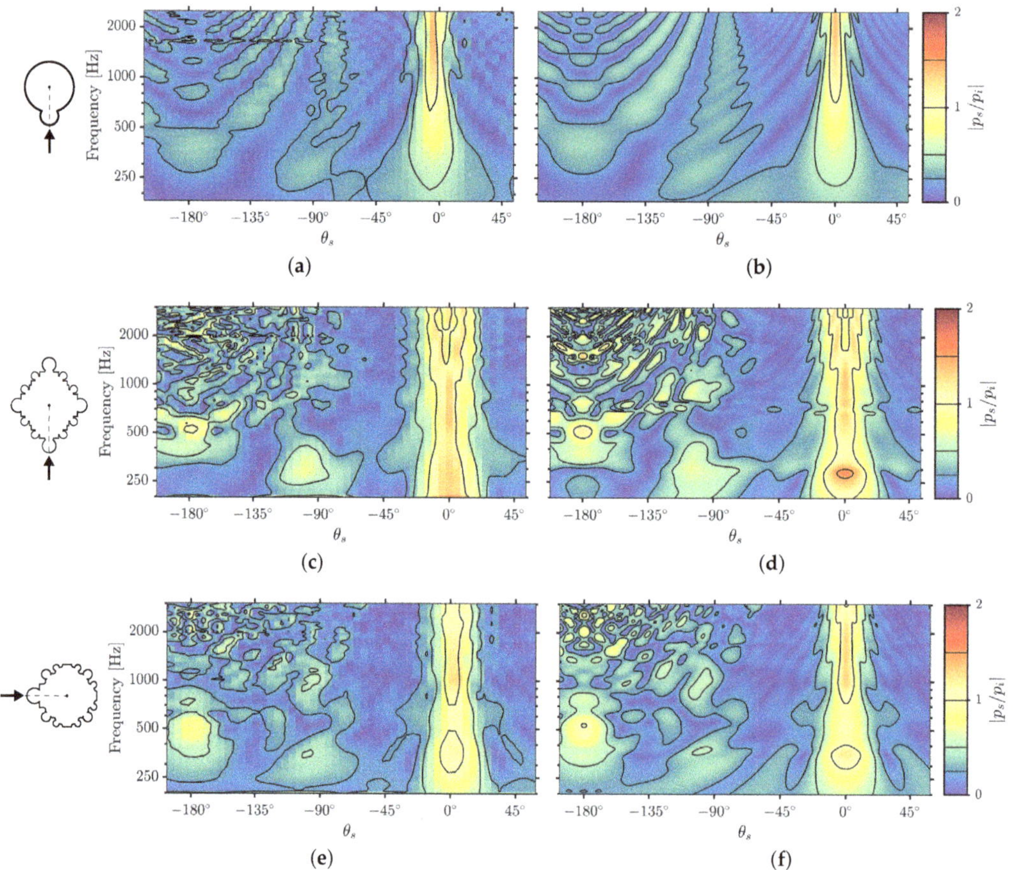

Figure 7. $|p_s/p_i|$ for **N1** with $\theta_0 = 90°$ (**a**,**b**), **T** with $\theta_0 = 90°$ (**c**,**d**), and **Ch** with $\theta_0 = 0°$ (**e**,**f**). Scale model measurements (**left**) compared to simulations (**right**). The frequency axis of the measurements is scaled according to the factors given in Table 1.

3.3. Simulation Results

Comparisons between measurements and simulations have shown that the latter faithfully represent the situation of a far-field source for the receiver positions considered. They allow to overcome the experimental difficulties and limitations, and thus to enlarge the accessible frequency range. They are nevertheless subject to their own limitations; however, with the parameters used, the scattering can be studied on the octave bands from 125 Hz to 4000 Hz without the numerical dispersion affecting the accuracy significantly. In addition, they are free of background noise and, consequently, allow to reveal the phenomena of low sound levels.

3.3.1. Time-Frequency Analysis

Figure 8 shows the simulated pressure signals at 4 m from the centroid of six different piers in the backscattering direction, and their wavelet scalograms in dB, where each scale has been normalized by the maximum value obtained for the incident pulse. The continuous wavelet transform has been performed using the function cwt implemented in the Wavelet Toolbox version 5.5 of MATLAB 2020a. It has been computed using Morse wavelets, which can be expressed in the frequency domain

$$\Psi_{\beta,\gamma}(\omega) = H(\omega)a_{\beta,\gamma}\omega^{\beta}e^{-\omega^{\gamma}}, \tag{7}$$

where H is the Heaviside step function, ω is the angular frequency, $a_{\beta,\gamma}$ is a normalizing constant, β is a compactness parameter, and γ characterizes the symmetry [81]. The signals are transformed using 10 voices per octave, with parameters set to $\beta\gamma = 25$ and $\gamma = 3$.

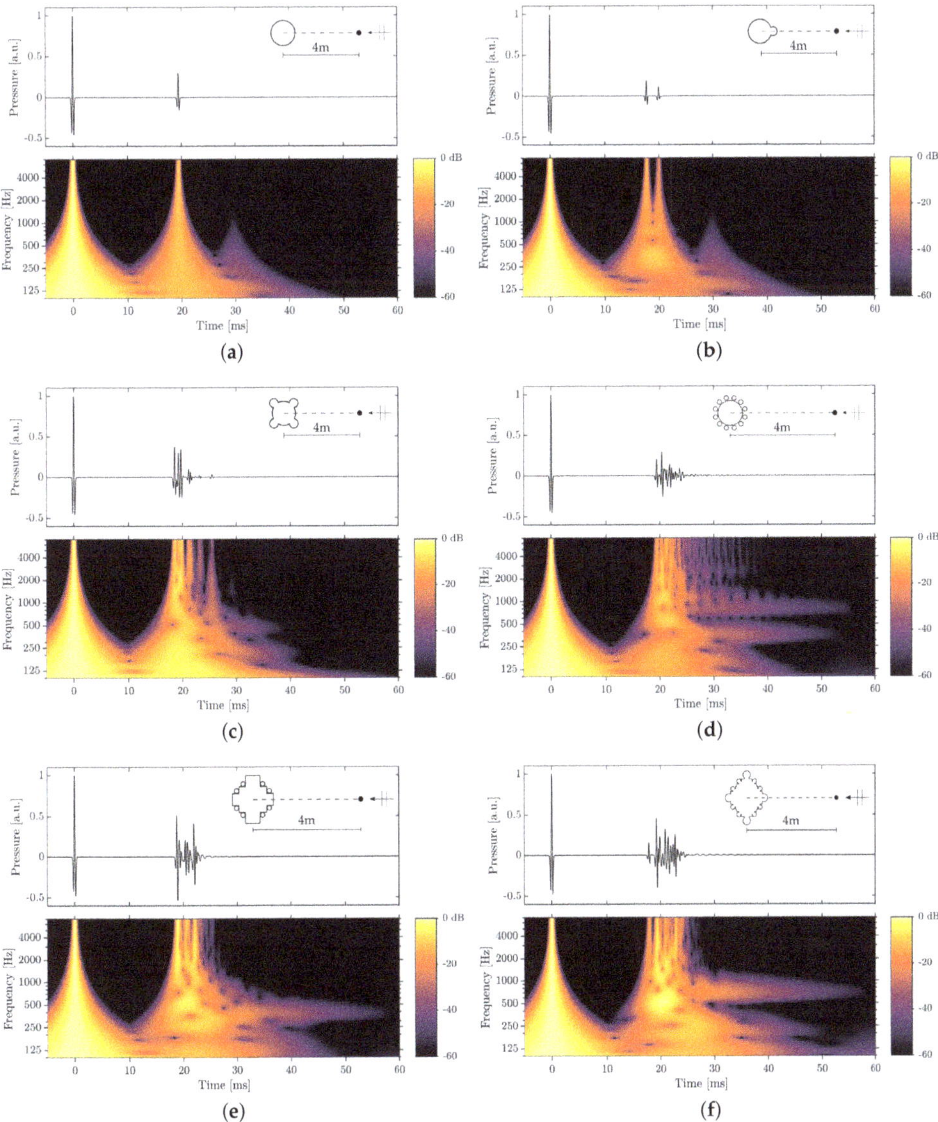

Figure 8. Pressure signal (**top**) and corresponding wavelet scalogram (**bottom**) normalized by the free-field maximum for (**a**) a circular cylinder of diameter 133 cm, (**b**) **N2** with $\theta_0 = 90°$, (**c**) **N1** with $\theta_0 = 45°$, (**d**) **N3** with $\theta_0 = 0°$, (**e**) **C2** with $\theta_0 = 90°$, and (**f**) **T** with $\theta_0 = 0°$ at 4 m from their center in the backscattering direction.

As expected, the temporal structure of the backscattered pulses depends strongly on the geometry of the diffuser. Figure 8a represents the results obtained for a circular cylinder of diameter 133 cm, corresponding to the columns of the nave arcade columns, and to the central part of **N1** and **N2**. In the time domain, the first arrival after the direct sound has a relative peak level of -10.5 dB, which is approximately the value found in the scalogram. A second arrival is visible in the scalogram at low frequency, 30 ms after the direct sound, corresponding to the creeping waves circumventing the cylinder that are strongly attenuated at high frequency. It is not visible in the signal as its relative peak level is -62 dB. For **N2** with $\theta_0 = 90°$ (Figure 8b), there are two visible arrivals, corresponding to the reflections on the two circular parts constituting the cross-section, with relative peak levels of -14.5 dB and -18.9 dB. The arrival due to the creeping waves is still visible in the scalogram, and has a level and frequency range similar to the circular cylinder. For **N1** with $\theta_0 = 45°$ (Figure 8c), the pressure signal is composed of a three localized pulses between 18 ms and 20 ms after the direct sound with -9.5 ± 1.1 dB relative peak levels. The latter arrivals are due to higher-order reflections between the different part of the cross-section that account for about 4% of the cumulative energy of the backscattered pulse. Its normalized wavelet scalogram also shows a spreading of low frequency, similarly to the previous geometries. For **C2** with $\theta_0 = 90°$ (Figure 8e), the wave packet has visible pulses at its onset and offset. They are attributed to the reflections on the plane faces of the cylinder whose normal is colinear with the direction of propagation. In comparison, those of **N3** (Figure 8d) and **T** (Figure 8f) look more like diffuse reflections [82].

Resonance tails are visible in the scalograms of the sections with smaller scale geometrical features, i.e., for **N3**, **C2**, and **T**, thus favoring multiple interactions during scattering. The column **N3** (Figure 8d) has two resonances over the considered frequency range. The first one occurs at around the same frequency as **C2** (Figure 8e), around 400 Hz. The second one is around 850 Hz and seems to decay slightly slower. The resonance of **T** (Figure 8f) is around 700 Hz. They all have low amplitudes, so they hardly appear on the linear scale of the pressure signals and account for a very small part of the cumulative energy of the backscattered pulses, less than 0.2% for **T**, for instance.

3.3.2. Reflected-to-Direct Level Differences

To analyze the reflected signals in a way that is relevant to our perception of sound, the Reflected-to-Direct Level Differences (RDLDs) [38] are calculated for each one-third octave band, in order to better observed the spectral and strength differences depending on the geometry of the cylindrical obstacle, the incidence angle, and the direction of scattering. They are calculated in the frequency domain with the discrete equivalent of

$$\mathrm{RDLD}_f = 20 \log_{10} \sqrt{\frac{1}{f_2 - f_1} \int_{f_1}^{f_2} \left| \frac{p_s(\omega)}{p_i(\omega)} \right|^2 d\omega}, \tag{8}$$

where f_1 and f_2 are the edge frequencies of the one-third octave band f. Moreover, to be compared with the audibility thresholds reported in the literature, a single-number RDLD [38] is also calculated, taking into account the spectral sensitivity of the ear, with

$$\mathrm{RDLD} = 10 \log_{10} \left(\frac{\sum_{f=1}^{N} 10^{\frac{\mathrm{RDLD}_f + T_{W,f}}{10}}}{\sum_{f=1}^{N} 10^{\frac{T_{W,f}}{10}}} \right), \tag{9}$$

where RDLD_f are the one-third octave band RDLDs, and $f = 1, \ldots, N$, and $T_{W,f}$ are the weights obtained by inverting the equal loudness curve at 40 phons according to ISO 226:2003. Note that no additional weighting is applied, contrary to [38], which is equivalent to considering a flat source magnitude spectrum.

Figure 9 represents the results obtained for **N2**, **C1**, **C2**, and **N3** for two incidence angles for each one. We recall that they are derived from simulations representing the experimental set-ups reported in Table 1, where the distances of the receivers to the center of the section are indicated for each one. The one-third octave band RDLDs are shown on a semicircle only, since the configurations are symmetrical. Moreover, the RDLDs for the positions located in shadow zone are also represented; however, they can not be interpreted as such, because of the interference between the incident and scattered pressures occurring in this region.

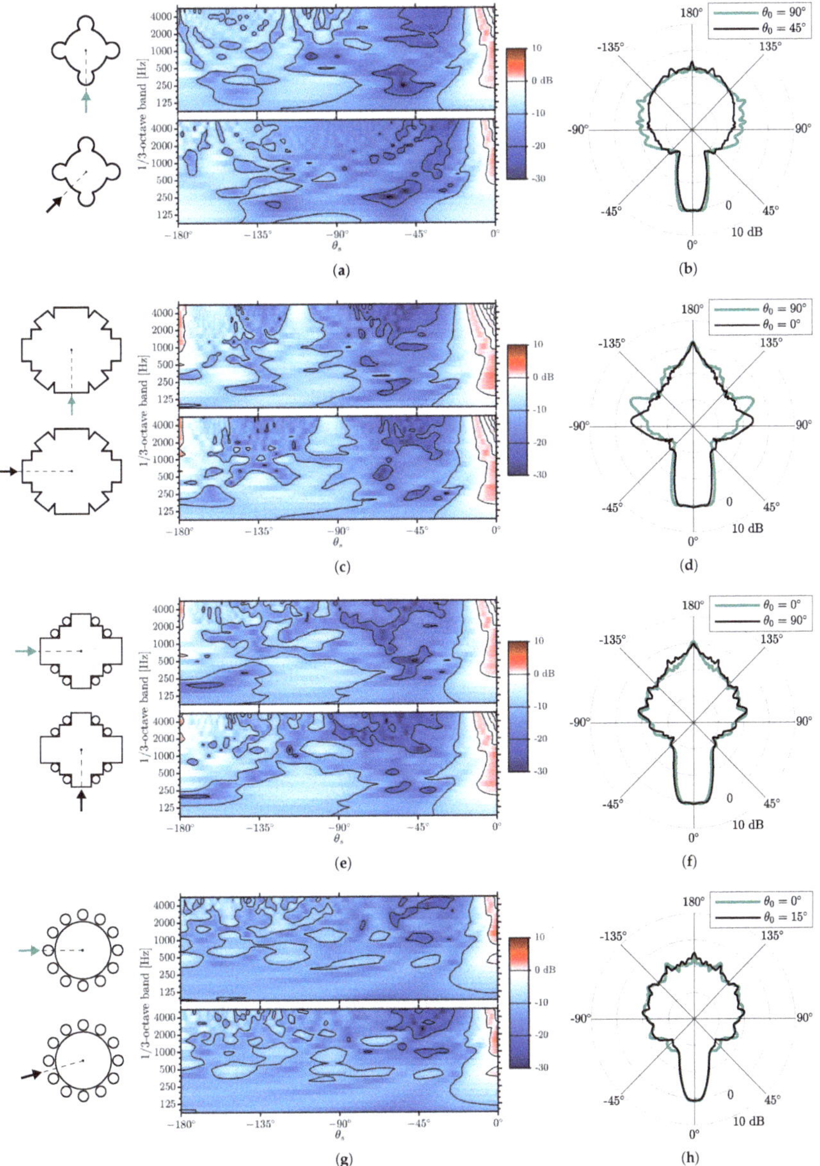

Figure 9. RDLDs for **N2** (**a,b**), **C1** (**c,d**), **C2** (**e,f**), and **N3** (**g,h**) for different plane wave incidence angles θ_0 as functions of the scattering angle θ_s: One-third octave bands (**a,c,e,g**) and overall (**b,d,f,h**) results. The receiver positions are reported in Table 1.

For the piers **N2**, the two incidence angles considered result in strong spectral and strength differences across the scattering directions, as shown in Figure 9a. For $\theta_0 = 90°$, the overall RDLDs represented in Figure 9b are around 3 dB higher in the transverse directions compared to $\theta_0 = 45°$. For the piers **C1**, the RDLDs are very similar, up to 1 kHz, as shown in Figure 9c. Above, the large planar parts of the section favor some directions, according to ray acoustics. These particular directions are therefore highlighted in the overall RDLDs, represented in Figure 9d, where they found their second maximum at around −4 dB and −6 dB for $\theta_0 = 90°$ and 0°, respectively. Their maximum of about 2 dB is found in the backscattering direction, as these incidences are normal to the large plane faces of the cylinders. This is also the case for **C2**, as shown in Figure 9e, where a positive value of nearly 1 dB is found in the backscattering direction for $\theta_0 = 0°$. For the two incidences considered, the overall RDLDs, represented in Figure 9f, differ mainly in this region, for $\theta_s \geq 150°$. Compared to the other section, the RDLDs for **N3** seems to depend less on the incidence angle, as shown in Figure 9g.

4. Discussion

4.1. Gothic Piers as Volumetric Diffusers

The function of a surface diffuser is to redirect a sound wave in directions other than the specular one, and to spread it out in time. From these perspectives, a simple circular cylinder alone achieves the spatial spreading, but is not really a good diffuser as it produces a high-pass strongly correlated reflection in the backscattering region [1]. Moreover, its finite size allows it to interact only in a limited way with the wave that impinges on it. Even in its resonant regime, we have seen that the circumferential waves have a very low level and only exist at low frequency. But as soon as discontinuities are included in the geometry, they are potential sources of scattering that produce additional wavefronts.

The compound piers studied here are all concave and some of them are star-shaped. This allows potentially several interactions of a wave scattered by a part of a shape with another of these parts. This is also true for the piers with *colonnettes* **C2** and **N3**, especially as the small cylinders are close to the central part and to each other. This effect is particularly visible through the existence of resonance frequencies revealed for the latter, as well as for the compound piers with geometric elements of small size, such as **T**. They are probably the result of coupling between the small cavities formed on the surface of the cylinders creating surface waves, as described in [83,84]. They are, by definition, localized in frequency, and in the cases studied here, count very little for the total energy of the reflections. However, around these frequencies, where the wavelengths are of the order of magnitude of the geometrical elements, i.e., up to about 1 kHz for the geometries considered here, the scattered power is increased without strongly favoring any particular direction. Contrary to beyond, in the limit of small wavelengths, the scattering directions can be determined according to the acoustic ray model, and the overall scattered power is related to the size of the shadow. See [85] for more detailed simulation results of different column geometries analyzed in terms of classical scattering quantities.

We have considered here the obstacles alone, but one may wonder if volumetric diffusion could be possible by multiple scattering between columns. In the cathedral, the piers of the nave are approximately 5.5 m between centers. Therefore, based on the RDLDs obtained, following the decreasing of intensity of a spherical wavefront, the level of a re-scattered wave would be too low compared to the other wall reflections. However, this is valid for a far field source, and it would be interesting to study the effect of an obstacle near a source for a distant listener. Moreover, their influence on the late reverberation, especially on the modes at low frequency, could also be a topic future investigations, considering that, in this case, an image-source of a high order of reflections has interacted with a lot of small obstacles, similar to the propagation within a sonic crystal.

4.2. Audibility of Scattering by Cylindrical Obstacles

The simulations assumed a source positioned at infinity; however, they were shown to be equivalent to the experiments where the source was in the far field, at about 30 m at full scale, and as the receiver is near the obstacle, about 3 m from their center here. This is even more true for the transverse directions, which better correspond to a listening situation with a listener facing the source and with an obstacle near him on its side. In this case, the results presented in Figure 9 indicate that the reflections have sufficient levels to be audible. For the configurations shown here, RDLDs in the direction $\theta_s = 90°$ are greater than -20 dB, which is approximately the thresholds of audibility reported in [38] or the criterion used in [6]. They are between -15 dB and -10 dB, except for **C1**, where values exist up to -6 dB and -4 dB, for $\theta_0 = 0°$ and $90°$, and $\theta_s = 95°$ and $113°$, respectively.

For such levels, the reflections are audible through changes in coloration rather than loudness. Humans are more sensitive to spectral overlap below 1 kHz [31], which is the range that is best scattered in all directions. Furthermore, in a binaural context, it has been shown that a reflection coming from a lateral direction was more audible than if it came from the same direction or from behind [30,35,36]. However, as already mentioned, for positions more distant in the transverse directions, the decreasing of intensity becomes important enough so that the reflected wave becomes undetectable. Comparing the one-third octave band RDLDs between piers, the spectral differences seem significant enough to discriminate between these reflections. Sound examples of impulses are included in the Supplementary Materials for back and transverse scattering. With preliminary listening, columns inducing diffuse reflections are discernible from a simple circular cylinder, in agreement with the results of [44], but these are monoaural responses and further perceptual studies, and measurements on scale models or three-dimensional simulations of the binaural responses, for example, would be necessary to be able to conclude as to the other shapes between them, in scenarios approaching real conditions for isolated columns.

A listener in the cathedral receives several early reflections, from the columns, but also from the walls. The question arises of a possible masking happening systematically for the positions considered here. For a source located in the choir between the stalls, the reflection off the side wall arrives between approximately 25 ms to 60 ms for the receivers located respectively from the back to the front of the nave. This leaves an interval before which these early column reflections can be significant. Furthermore, if we consider a realistic source, the relative positions will be decisive. Based on scattering theory, a spherical or cylindrical source will be scattered more in the forward and backward directions, and less transversely, compared to a plane wave incidence, and the closer it is, the greater the effect [86]. However, such sources also imply a decay of intensity due to the spatial spread of their wavefront, which could result in a lower relative level of the reflections [6]. Further investigations on room impulse responses and simulations could evaluate these effects. Moreover, several studies have investigated the effect of surface diffusers [10–14] or columns [15] on room acoustic parameters in concert halls, and it would be interesting to conduct similar studies, especially in relation to the previous discussions on the relative positions of the source, obstacles, and listener, and on their collective effects.

5. Conclusions

The purpose of this work was to investigate the sound scattering by obstacles of complex geometries that are the piers and columns of the Cathédrale Notre-Dame de Paris. This heritage monument has evolved across the centuries, and is marked by several Gothic styles, which allowed us to select typical geometries that are relatively different, reflecting the evolution of this medieval architecture. Their scattering has been simulated up to 6 kHz using a low dispersion and anisotropy finite difference scheme with a pulse excitation. It is modified according to the formalism of the finite volumes around the boundaries to conform to it and to avoid the staircase approximation. The method has been validated by comparison with measurements on scale models and analytic models.

A plane wave incidence has been considered in the simulations, and has been shown to be close to the experimental measurements in the case of a far field source and a receiver close to the cylinder. The simulated scattered fields were consequently analyzed in terms of perceptually relevant quantities. Similarly to reflectors, the reflections from columns and piers are limited by their finite size. However, due to their early arrival before most wall reflections, the scattered field at the evaluated positions revealed that these obstacles could produce audible reflections over all scattering directions, based on thresholds reported in the literature. The temporal spreading strongly depends on the scatterer, i.e., the piers' form. Those with small geometrical features have the ability to produce diffuse reflections similarly to surface diffusers. Low level resonances due to complex forms have also been revealed; however, they represent a very small part of the total reflected energy. Strong spectral differences were observed between piers, such that it is likely possible to be able to discriminate between their reflections. Further studies could evaluate more realistic scenarios with a spherical source and different relative distances between the latter, the obstacle and the listener, numerically and with perceptual testings.

Supplementary Materials: The following supporting information can be downloaded at: https://www.mdpi.com/article/10.3390/acoustics4030041/s1.

Author Contributions: Conceptualization, B.F.G.K. and A.W.; methodology, A.W. and B.F.G.K.; software, A.W.; validation, A.W.; formal analysis, A.W.; investigation, A.W.; resources, A.W.; data curation, A.W.; writing—original draft preparation, A.W.; writing—review and editing, A.W. and B.F.G.K.; visualization, A.W.; supervision, B.F.G.K.; project administration, B.F.G.K.; funding acquisition, B.F.G.K. and A.W. All authors have read and agreed to the published version of the manuscript.

Funding: Funding has been provided in part by the French project PHEND (The Past Has Ears at Notre-Dame, Grant No. ANR-20-CE38-0014, phend.pasthasears.eu (http://phend.pasthasears.eu accessed on 19 July 2022)) and the Chantier Scientifique CNRS/MC Notre-Dame (https://notre-dame-de-paris.culture.gouv.fr/fr/science-et-dame accessed on 19 July 2022).

Institutional Review Board Statement: Not applicable.

Informed Consent Statement: Not applicable.

Data Availability Statement: The data that support the findings of this study are available from the corresponding authors upon reasonable request.

Acknowledgments: The authors are grateful to Cristina Dagalita, Elsa Ricaud, and Dany Sandron for reviewing the parts of the manuscript related to art history and architecture.

Conflicts of Interest: The authors declare no conflict of interest.

Abbreviations

The following abbreviations are used in this manuscript:

FD	Finite difference
FV	Finite volume
RDLD	Reflected-to-Direct Level Difference
SNR	Signal-to-noise ratio

Appendix A. Measurements vs. Simulations: Additional Examples

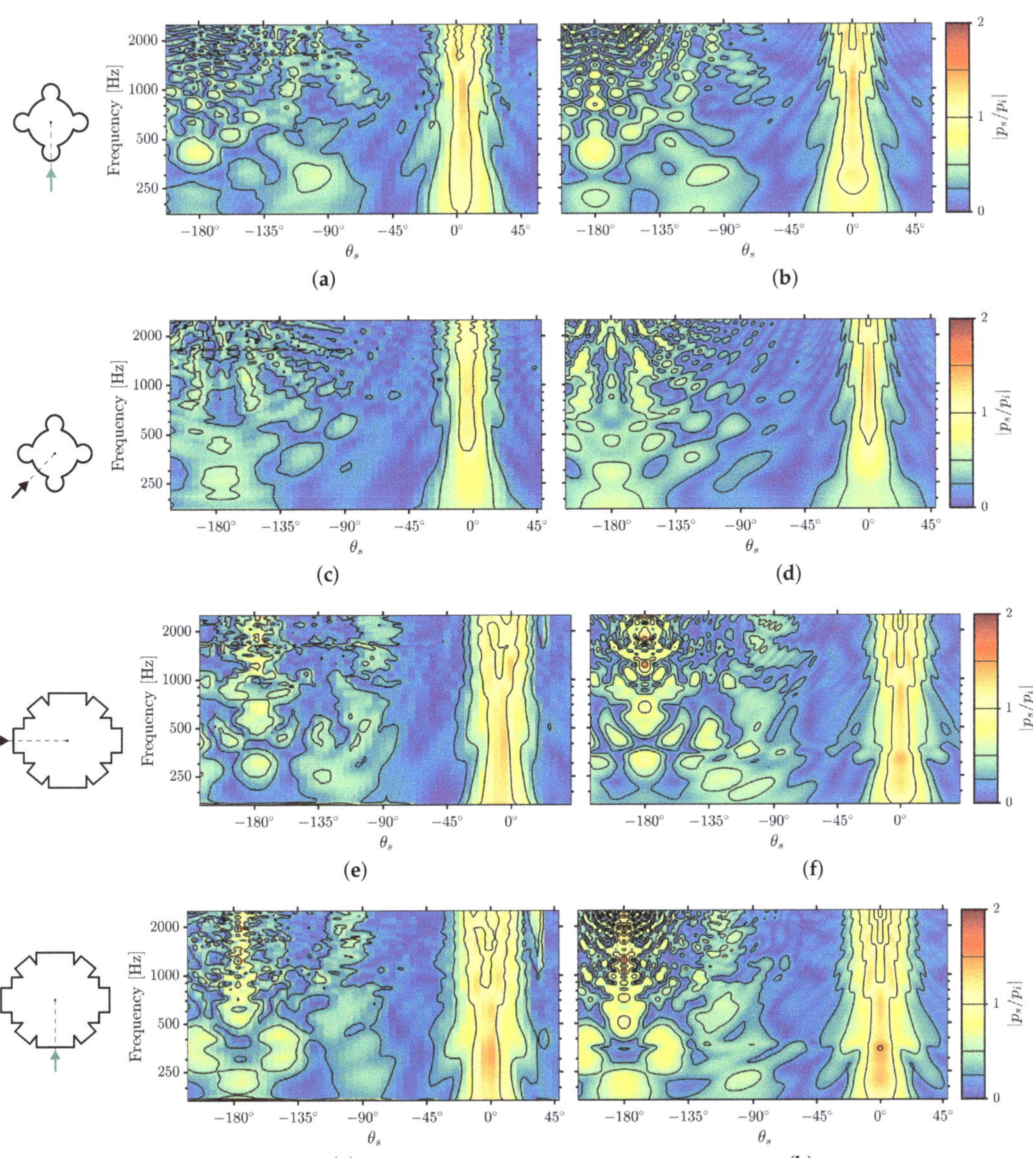

Figure A1. $|p_s/p_i|$ for the compound piers **N1** with (**a**,**b**) $\theta_0 = 90°$ and (**c**,**d**) $\theta_0 = 45°$, and **C1** with (**e**,**f**) $\theta_0 = 0°$ and (**g**,**h**) $\theta_0 = 90°$. Scale model measurements (**left**) compared to simulations (**right**). The frequency axis of the measurements is scaled according to the factors given in Table 1.

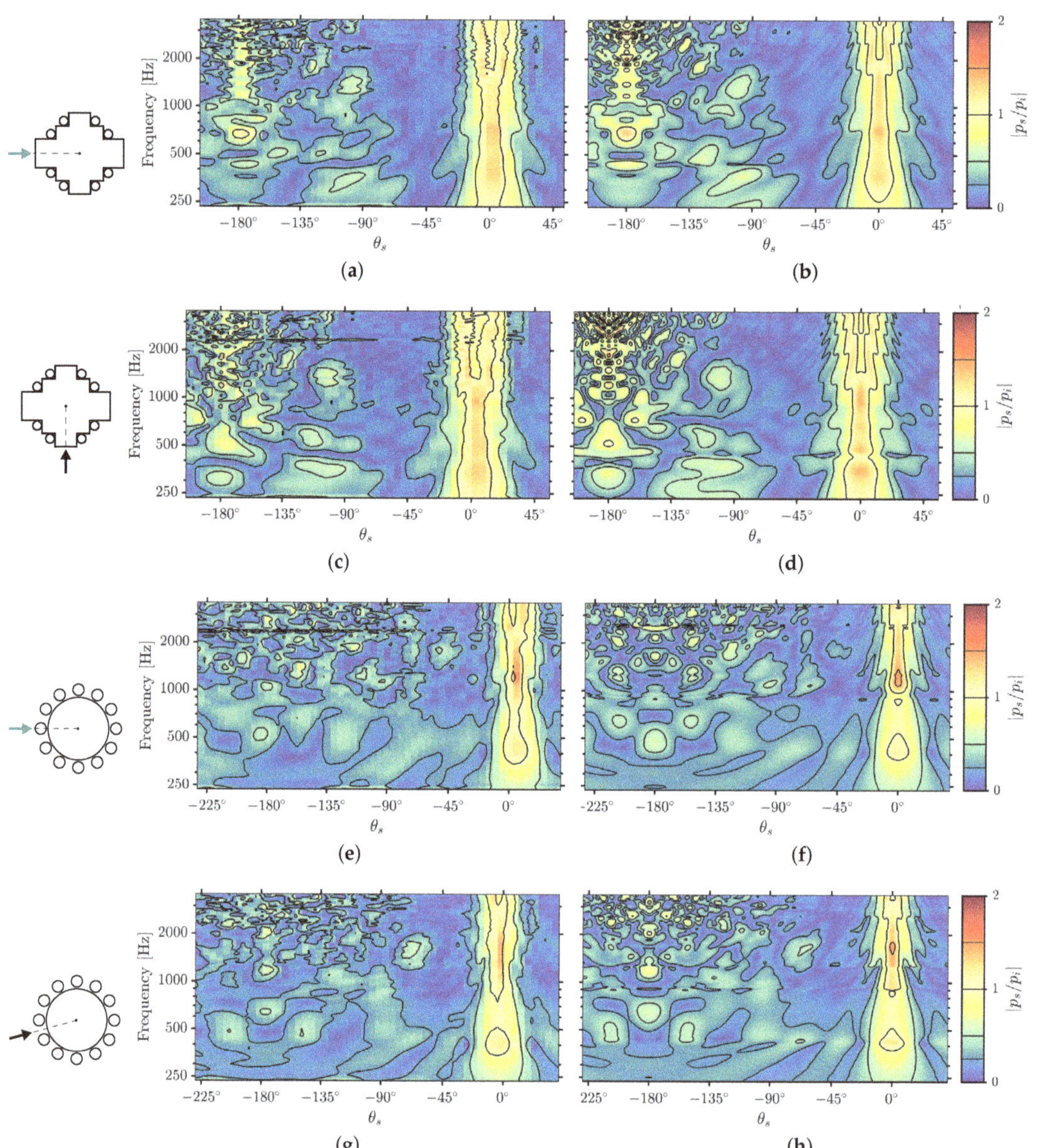

Figure A2. $|p_s/p_i|$ for the piers with *colonnettes* **C2**, with (**a**,**b**) $\theta_0 = 0°$ and (**c**,**d**) $\theta_0 = 90°$, and **N3** with (**e**,**f**) $\theta_0 = 0°$ and (**g**,**h**) $\theta_0 = 15°$. Scale model measurements (**left**) compared to simulations (**right**). The frequency axis of the measurements is scaled according to the factors given in Table 1.

References

1. Cox, T.; D'Antonio, P. *Acoustic Absorbers and Diffusers: Theory, Design and Application*; CRC Press: Boca Raton, FL, USA, 2016. [CrossRef]
2. Rindel, J.H. Design of new ceiling reflectors for improved ensemble in a concert hall. *Appl. Acoust.* **1991**, *34*, 7–17. [CrossRef]
3. Bradley, D.T.; Müller-Trapet, M.; Adelgren, J.; Vorländer, M. Comparison of Hanging Panels and Boundary Diffusers in a Reverberation Chamber. *Build. Acoust.* **2014**, *21*, 145–152. [CrossRef]

4. Rindel, J.H. Attenuation of Sound Reflections due to Diffraction. In Proceedings of the Nordic Acoustical Meeting, Aalborg, Denmark, 20–22 August 1986.
5. Szelag, A.; Kamisiński, T.; Lewińska, M.; Rubacha, J.; Pilch, A. The Characteristic of Sound Reflections from Curved Reflective Panels. *Arch. Acoust.* **2014**, *39*, 549–558. [CrossRef]
6. Rathsam, J.; Wang, L.M. Planar Reflector Panels with Convex Edges. *Acta Acust. United Acust.* **2010**, *96*, 905–913. [CrossRef]
7. Medwin, H.; Clay, C.S. *Fundamentals of Acoustical Oceanography*; Applications of Modern Acoustics, Academic Press: San Diego, CA, USA, 1998. [CrossRef]
8. *Standard ISO 17497-1:2004*; Acoustics—Sound-Scattering Properties of Surfaces–Part 1: Measurement of the Random-Incidence Scattering Coefficient in a Reverberation Room. International Organization for Standardization: Geneva, Switzerland, 2004.
9. *Standard ISO 17497-2:2012*; Acoustics—Sound-Scattering Properties of Surfaces—Part 2: Measurement of the Directional Diffusion Coefficient in a Free Field. International Organization for Standardization: Geneva, Switzerland, 2012.
10. Ryu, J.K.; Jeon, J.Y. Subjective and objective evaluations of a scattered sound field in a scale model opera house. *J. Acoust. Soc. Am.* **2008**, *124*, 1538–1549. [CrossRef]
11. Jeon, J.Y.; Jo, H.I.; Seo, R.; Kwak, K.H. Objective and subjective assessment of sound diffuseness in musical venues via computer simulations and a scale model. *Build. Environ.* **2020**, *173*, 106740. [CrossRef]
12. Kim, Y.H.; Kim, J.H.; Jeon, J.Y. Scale Model Investigations of Diffuser Application Strategies for Acoustical Design of Performance Venues. *Acta Acust. United Acust.* **2011**, *97*, 791–799. [CrossRef]
13. Shtrepi, L.; Astolfi, A.; Pelzer, S.; Vitale, R.; Rychtáriková, M. Objective and perceptual assessment of the scattered sound field in a simulated concert hall. *J. Acoust. Soc. Am.* **2015**, *138*, 1485–1497. [CrossRef]
14. Shtrepi, L.; Astolfi, A.; Puglisi, G.E.; Masoero, M.C. Effects of the Distance from a Diffusive Surface on the Objective and Perceptual Evaluation of the Sound Field in a Small Simulated Variable-Acoustics Hall. *Appl. Sci.* **2017**, *7*, 224. [CrossRef]
15. Suzumura, Y.; Sakurai, M.; Ando, Y.; Yamamoto, I.; Iizuka, T.; Oowaki, M. An evaluation of the effects of scattered reflections in a sound field. *J. Sound Vib.* **2000**, *232*, 303–308. [CrossRef]
16. Strutt, H.J.X. On the light from the sky, its polarization and colour. *Lond. Edinb. Dublin Philos. Mag. J. Sci.* **1871**, *41*, 107–120. [CrossRef]
17. Morse, P.; Ingard, K. *Theoretical Acoustics*; International Series in Pure and Applied Physics; Princeton University Press: Princeton, NJ, USA, 1986.
18. Bowman, J.; Senior, T.; Uslenghi, P.; Asvestas, J. *Electromagnetic and Acoustic Scattering by Simple Shapes*; Taylor & Francis: Abingdon, UK, 1988.
19. Hughes, R.J.; Angus, J.A.S.; Cox, T.J.; Umnova, O.; Gehring, G.A.; Pogson, M.; Whittaker, D.M. Volumetric diffusers: Pseudorandom cylinder arrays on a periodic lattice. *J. Acoust. Soc. Am.* **2010**, *128*, 2847–2856. [CrossRef]
20. Yashiro, K.; Ohkawa, S. Boundary element method for electromagnetic scattering from cylinders. *IEEE Trans. Antennas Propag.* **1985**, *33*, 383–389. [CrossRef]
21. Liu, Y. On the BEM for acoustic wave problems. *Eng. Anal. Bound. Elem.* **2019**, *107*, 53–62. doi: 10.1016/j.enganabound.2019.07.002. [CrossRef]
22. Ihlenburg, F. (Ed.) *Finite Element Analysis of Acoustic Scattering*; Springer: Berlin/Heidelberg, Germany, 1998. doi: 10.1007/0-387-22700-8_2. [CrossRef]
23. Taflove, A.; Hagness, S. *Computational Electrodynamics: The Finite-Difference Time-Domain Method*; Artech House Antennas and Propagation Library, Artech House: Norwood, MA, USA, 2005.
24. Redondo, J.; Picó, R.; Roig, B.; Avis, M.R. Time Domain Simulation of Sound Diffusers Using Finite-Difference Schemes. *Acta Acust. United Acust.* **2007**, *93*, 611–622.
25. Hoare, S.; Murphy, D. Prediction of scattering effects by sonic crystal noise barriers in 2d and 3d finite difference simulations. In Proceedings of the Acoustics 2012, Nantes, France, 23–27 April 2012; Société Française d'Acoustique, Paris, France , 2012.
26. Redondo, J.; Picó, R.; Sánchez-Morcillo, V.J.; Woszczyk, W. Sound diffusers based on sonic crystals. *J. Acoust. Soc. Am.* **2013**, *134*, 4412–4417. [CrossRef]
27. Tornberg, A.K.; Engquist, B. Consistent boundary conditions for the Yee scheme. *J. Comput. Phys.* **2008**, *227*, 6922–6943. [CrossRef]
28. Bilbao, S.; Hamilton, B.; Botts, J.; Savioja, L. Finite Volume Time Domain Room Acoustics Simulation under General Impedance Boundary Conditions. *IEEE/ACM Trans. Audio Speech Lang. Process.* **2016**, *24*, 161–173. [CrossRef]
29. Hamilton, B.; Bilbao, S. Hexagonal vs. rectilinear grids for explicit finite difference schemes for the two-dimensional wave equation. *J. Acoust. Soc. Am.* **2013**, *133*, 3532–3532. [CrossRef]
30. Olive, S.E.; Toole, F.E. The Detection of Reflections in Typical Rooms. *J. Audio Eng. Soc.* **1988**, *37*, 539–553.
31. Buchholz, J.M. A quantitative analysis of spectral mechanisms involved in auditory detection of coloration by a single wall reflection. *Hear. Res.* **2011**, *277*, 192–203. [CrossRef] [PubMed]
32. Burgtorf, W.; Oehlschlägel, H.K. Untersuchungen über die richtungsabhängige Wahrnehmbarkeit verzögerter Schallsignale. *Acta Acust. United Acust.* **1964**, *14*, 254–266.
33. Seraphim, H. Über die Wahrnehmbarkeit mehrerer Rückwürfe von Sprachschall. *Acta Acust. United Acust.* **1961**, *11*, 80–91.
34. Brunner, S.; Maempel, H.J.; Weinzierl, S. On the Audibility of Comb Filter Distortions. In Proceedings of the 122nd AES Convention, Vienna, Austria, 5–8 May 2007; Audio Engineering Society: New York, NY, USA, 2007.

35. Begault, D.R.; McClain, B.U.; Anderson, M.R. Early Reflection Thresholds for Anechoic and Reverberant Stimuli within a 3-D Sound Display. In Proceedings of the 18th International Congress on Acoustics (ICA04), Kyoto, Japan, 4–9 April 2004.
36. Zhong, X.; Guo, W.; Wang, J. Audible Threshold of Early Reflections with Different Orientations and Delays. *Sound Vib.* **2018**, *52*, 18–22. [CrossRef]
37. Walther, A.; Robinson, P.W.; Santala, O. Effect of spectral overlap on the echo suppression threshold for single reflection conditions. *J. Acoust. Soc. Am.* **2013**, *134*, EL158–EL164. [CrossRef]
38. Pelegrín-García, D.; Rychtáriková, M. Audibility Thresholds of a Sound Reflection in a Classical Human Echolocation Experiment. *Acta Acust. United Acust.* **2016**, *102*, 530–539. [CrossRef]
39. Pelegrín-García, D.; Rychtáriková, M.; Glorieux, C. Single Simulated Reflection Audibility Thresholds for Oral Sounds in Untrained Sighted People. *Acta Acust. United Acust.* **2017**, *103*, 492–505. [CrossRef]
40. Robinson, P.W.; Walther, A.; Faller, C.; Braasch, J. Echo thresholds for reflections from acoustically diffusive architectural surfaces. *J. Acoust. Soc. Am.* **2013**, *134*, 2755–2764. [CrossRef]
41. Wendt, F.; Höldrich, R. Precedence effect for specular and diffuse reflections. *Acta Acust.* **2021**, *5*, 1. [CrossRef]
42. Kleiner, M.; Svensson, P.; Dalenbäck, B.I. Auralization of qrd and other diffusing surfaces using scale modelling. In Proceedings of the 93rd AES Convention, San Francisco, CA, USA, 1–4 October 1992; Audio Engineering Society: New York, NY, USA.
43. Meyer, J.; Savioja, L.; Lokki, T. A case study on the perceptual differences in finite-difference time-domain-simulated diffuser designs. In Proceedings of the 146th AES Convention, Dublin, Ireland, 20–23 March 2019; Audio Engineering Society: New York, NY, USA.
44. Kritly, L.; Sluyts, Y.; Pelegrín-García, D.; Glorieux, C.; Rychtáriková, M. Discrimination of 2D wall textures by passive echolocation for different reflected-to-direct level difference configurations. *PLoS ONE* **2021**, *16*, e0251397. [CrossRef]
45. Aubert, M.; Vitry, P. *La Cathédrale Notre-Dame de Paris: Notice Historique et Archéologique*; Notices Historiques et Archéologiques sur les Grands Monuments; D.-A. Longuet: Paris, France, 1919.
46. Bruzelius, C. The Construction of Notre-Dame in Paris. *Art Bull.* **1987**, *69*, 540–569. [CrossRef]
47. Sandron, D.; Tallon, A.; Cook, L. *Notre Dame Cathedral: Nine Centuries of History*; Penn State University Press: University Park, PA, USA, 2020. [CrossRef]
48. Celtibère, M.; Lemercier, M.; Hibon, A.; Ribaud, A.; Normand, A.; Bisson, B. *Monographie de Notre-Dame de Paris et de la Nouvelle Sacristie de MM. Lassus et Viollet-Le-Duc*; A. Morel: Paris, France, 1853.
49. Mullins, S.S.; Canfield-Dafilou, E.K.; Katz, B.F.G. The development of the early acoustics of the chancel in Notre-Dame de Paris: 1160–1230. In Proceedings of the 2nd Symposium: The Acoustics of Ancient Theatres, Verona, Italy, 6–8 July 2022.
50. Lassus, J.B.A.; Viollet-le-Duc, E.E. *Projet de Restauration de Notre-Dame de Paris par MM. Lassus et Viollet-Leduc: Rapport…Annexé au Projet de Restauration, Remis le 31 Janvier 1843*; Imprimerie de Mme de Lacombe: Paris, France, 1843.
51. Travaux de Notre-Dame de Paris—Journal des Travaux 1844–1865. Available online: https://mediatheque-patrimoine.culture.gouv.fr/travaux-de-notre-dame-de-paris-1844-1865 (accessed on 1 June 2022).
52. Postma, B.N.; Katz, B.F.G. Acoustics of Notre-Dame Cathedral de Paris. In Proceedings of the 22nd International Congress on Acoustics (ICA), Buenos Aires, Argentina, 5–9 September 2016; pp. 5–9.
53. Katz, B.F.G.; Weber, A. An acoustic survey of the Cathédrale Notre-Dame de Paris before and after the fire of 2019. *Acoustics* **2020**, *2*, 791–802. [CrossRef]
54. Hoey, L.R. Pier Form and Vertical Wall Articulation in English Romanesque Architecture. *J. Soc. Archit. Hist.* **1989**, *48*, 258–283. [CrossRef]
55. Thurlby, M. Aspects of the architectural history of Kirkwall Cathedral. *Proc. Soc. Antiqu. Scotl.* **1998**, *127*, 855–888.
56. CNRS/MC. Chantier Scientifique Notre-Dame de Paris. Available online: https://www.notre-dame.science/ (accessed on 1 June 2022).
57. Klein, B. Naissance et formation de l'architecture gothique en France et dans les pays limitrophes. In *L'art Gothique, Architecture-Sculpture-Peinture*; Toman, R., Ed.; Könemann: Köln, Germany, 1999; Chapter 3, pp. 28–114.
58. Murray, S. Notre-Dame of Paris and the Anticipation of Gothic. *Art Bull.* **1998**, *80*, 229–253. [CrossRef]
59. Davis, M.T. Splendor and Peril: The Cathedral of Paris, 1290–1350. *Art Bull.* **1998**, *80*, 34–66. [CrossRef]
60. Viollet-le-Duc, E.E. *Dictionnaire Raisonné de l'Architecture Française du XIe au XVIe Siècle*; B. Bance: Paris, France, 1859.
61. Fernie, E. La fonction liturgique des piliers cantonnés dans la nef de la cathédrale de Laon. *Bull. Monum.* **1987**, *145*, 257–266. [CrossRef]
62. Olson, V. Colonnette Production and the Advent of the Gothic Aesthetic. *Gesta* **2004**, *43*, 17–29. [CrossRef]
63. Bony, J. French Influences on the Origins of English Gothic Architecture. *J. Warbg. Court. Inst.* **1949**, *12*, 1–15. [CrossRef]
64. Hoey, L. Piers versus Vault Shafts in Early English Gothic Architecture. *J. Soc. Archit. Hist.* **1987**, *46*, 241–264. [CrossRef]
65. Fabero, J.; Bautista, A.; Casasús, L. An explicit finite differences scheme over hexagonal tessellation. *Appl. Math. Lett.* **2001**, *14*, 593–598. [CrossRef]
66. Yamashita, O.; Tsuchiya, T.; Iwaya, Y.; Otani, M.; Inoguchi, Y. Reflective boundary condition with arbitrary boundary shape for compact-explicit finite-difference time-domain method. *Jpn. J. Appl. Phys.* **2015**, *54*, 07HC02. [CrossRef]
67. Tolan, J.G.; Schneider, J.B. Locally conformal method for acoustic finite-difference time-domain modeling of rigid surfaces. *J. Acoust. Soc. Am.* **2003**, *114*, 2575–2581. [CrossRef] [PubMed]

68. Bilbao, S.; Hamilton, B. Passive volumetric time domain simulation for room acoustics applications. *J. Acoust. Soc. Am.* **2019**, *145*, 2613–2624. [CrossRef] [PubMed]
69. Vázquez, P.; Menéndez, B.; Denecker, M.F.C.; Thomachot-Schneider, C. Comparison between petrophysical properties, durability and use of two limestones of the Paris region. In *Sustainable Use of Traditional Geomaterials in Construction Practice*; Geological Society of London: London, UK, 2016. [CrossRef]
70. Botteldooren, D. Acoustical finite-difference time-domain simulation in a quasi-Cartesian grid. *J. Acoust. Soc. Am.* **1994**, *95*, 2313–2319. [CrossRef]
71. Hamilton, B. Finite Volume Perspectives on Finite Difference Schemes and Boundary Formulations for Wave Simulation. In Proceedings of the 17th International Conference on Digital Audio Effects (DAFx-14), Erlangen, Germany, 1–5 September 2014.
72. Du, Q.; Faber, V.; Gunzburger, M. Centroidal Voronoi Tessellations: Applications and Algorithms. *SIAM Rev.* **1999**, *41*, 637–676. [CrossRef]
73. Geuzaine, C.; Remacle, J.F. Gmsh: A 3-D finite element mesh generator with built-in pre- and post-processing facilities. *Int. J. Numer. Methods Eng.* **2009**, *79*, 1309–1331. [CrossRef]
74. Rasmussen, K. *Calculation Methods for the Physical Properties of Air Used in the Calibration of Microphones*; Report PL-11b; Department of Acoustical Technology, Technical University of Denmark: Kongens Lyngby, Denmark, 1997.
75. Sheaffer, J.; van Walstijn, M.; Fazenda, B. Physical and numerical constraints in source modeling for finite difference simulation of room acoustics. *J. Acoust. Soc. Am.* **2014**, *135*, 251–261. [CrossRef] [PubMed]
76. Müller, S.; Massarani, P. Transfer-Function Measurement with Sweeps. *J. Audio Eng. Soc.* **2001**, *49*, 443–471.
77. Robinson, P.; Xiang, N. On the subtraction method for in-situ reflection and diffusion coefficient measurements. *J. Acoust. Soc. Am.* **2010**, *127*, EL99–EL104. [CrossRef]
78. Zhang, L.; Wu, X. On cross correlation based-discrete time delay estimation. In Proceedings of the IEEE International Conference on Acoustics, Speech, and Signal Processing (ICASSP'05), Philadelphia, PA, USA, 23 March 2005; Volume 4, pp. iv/981–iv/984. [CrossRef]
79. Krynkin, A.; Umnova, O.; Vicente Sánchez-Pérez, J.; Yung Boon Chong, A.; Taherzadeh, S.; Attenborough, K. Acoustic insertion loss due to two dimensional periodic arrays of circular cylinders parallel to a nearby surface. *J. Acoust. Soc. Am.* **2011**, *130*, 3736–3745. [CrossRef]
80. Markowskei, A.J.; Smith, P.D. Measuring the effect of rounding the corners of scattering structures. *Radio Sci.* **2017**, *52*, 693–708. [CrossRef]
81. Lilly, J.; Olhede, S. Higher-Order Properties of Analytic Wavelets. *IEEE Trans. Signal Process.* **2009**, *57*, 146–160. [CrossRef]
82. Robinson, P.W.; Xiang, N.; Braasch, J. Understanding the perceptual effects of diffuser application in rooms. *Proc. Meet. Acoust.* **2011**, *12*, 015002. [CrossRef]
83. Berry, D.L.; Taherzadeh, S.; Attenborough, K. Acoustic surface wave generation over rigid cylinder arrays on a rigid plane. *J. Acoust. Soc. Am.* **2019**, *146*, 2137–2144. [CrossRef]
84. Farhat, M.; Chen, P.Y.; Bağcı, H. Localized acoustic surface modes. *Appl. Phys. A* **2016**, *122*, 1–8. [CrossRef]
85. Weber, A.; Katz, B.F.G. An investigation of sound scattering by ancient and Gothic piers and columns. *Acta Acust.* 2022, *submitted*.
86. Zitron, N.R.; Davis, J. A note on scattering of cylindrical waves by a circular cylinder. *Can. J. Phys.* **1966**, *44*, 2941–2944. [CrossRef]

Article

Reviving the Low-Frequency Response of a Rupestrian Church by Means of FDTD Simulation

Francesco Martellotta *, Stefania Liuzzi and Chiara Rubino

Dipartimento di Architettura, Costruzione e Design, Politecnico di Bari, 70125 Bari, Italy
* Correspondence: francesco.martellotta@poliba.it; Tel.: +39-080-5963631

Abstract: Rupestrian churches are spaces obtained from excavation of soft rocks that are frequently found in many Mediterranean countries. In the present paper the church dedicated to Saints Andrew and Procopius, located close to the city of Monopoli in Apulia (Italy) is studied. On-site acoustical measures were made, obtaining a detailed description of the acoustics in the current state pointing out, thanks to a combination of analysis techniques, the presence of significant modal behavior in the low frequencies, causing reverberation time to be about 2 s, four times longer than in the other bands, as well as being strongly dependent on source and receiver position (with variations of about 1 s when source is moved outside the chancel). However, as the church is characterized by significant degradation of surfaces and large amounts of debris cover the floor, the original acoustic conditions can be expected to somewhat differ. Acoustical modelling can be very helpful in grasping the original conditions, but given the small dimensions of the space, conventional geometrical acoustic prediction methods cannot be applied to simulate the low-frequency behavior. Thus, the present paper proposes an application of finite-difference-time-domain (FDTD) computation to simulate the low-frequency behavior and analyze a possible reconstruction of the original state. Results showed that a very good agreement was obtained between predictions and measurements, both in terms of resonance frequencies and reverberation times that differed by less than 5%. Modal response strongly affected the acoustical conditions also in the hypothetical reconstruction of the original state, although the sound field proved to be more uniform than in the current state.

Keywords: acoustics; rupestrian church; acoustical simulation; FDTD modelling; heritage buildings

Citation: Martellotta, F.; Liuzzi, S.; Rubino, C. Reviving the Low-Frequency Response of a Rupestrian Church by Means of FDTD Simulation. *Acoustics* **2023**, *5*, 396–413. https://doi.org/10.3390/acoustics5020023

Academic Editors: Margarita Díaz-Andreu and Lidia Alvarez Morales

Received: 14 February 2023
Revised: 29 March 2023
Accepted: 10 April 2023
Published: 12 April 2023

1. Introduction

Rupestrian churches belong to the largest and most widespread group of artificial cavities of anthropic origin, often labelled as rocky or troglodytic settlements, which can be found all over the world [1]. However, in the Mediterranean area, they can be found in nearly every country, with some of them showing a significantly higher number, as outlined by a census of rocky sites in the Mediterranean area [2], from which Italy, Spain, and Turkey appear as the richest in settlements. Apulia (and the neighboring area of Matera), Sicily and Tuscany in Italy, Andalusia in Spain, and Cappadocia in Turkey showed the highest number of rocky sites, most frequently located in areas where soft stone in combination with meteorological agents already created natural caves that were subsequently enlarged and shaped according to the needs of the occupants. With reference to religious buildings, the sites in Southern Italy originated from the spreading of Greek monks following the iconoclastic prosecution in the Eastern regions but also from local communities that often found a safe place far from the frequent aggressions arriving from the sea. In Spain, many sites were built by the Mozarabic population during the Moorish occupation of the South. In Cappadocia the churches can be found in settlements of cenobite monks that developed from 6th and 7th century after the introduction of the Christian cult [3], but it is not unusual to also find mosques in some places.

The shape of the rupestrian churches usually recalled that of ordinary architectures of the time, with the obvious limitations related to the characteristics of the site and the nature of the stone, but it is not unusual to find spaces that may well compete in dimension and complexity with ordinary churches.

Cave acoustics has attracted a considerable interest in the last years, starting with the seminal studies that tried to demonstrate that acoustic effects might have influenced the choice of the locations. Infrasonic resonances [4], echoes, and long resonances might have influenced mural paintings [5,6], as well as the use of sites as burial places [7]. Some studies [8,9] seem to agree with these hypotheses, also suggesting a deliberate use of such effects in the underground galleries of Chavin de Huantar in Peru [8]. More recently the topic has been further investigated by several large interdisciplinary groups [10–13], and other studies of cave acoustics have been carried out involving natural caves mostly [14–16] or burial places like catacombs [17].

When dealing with rupestrian churches and sacred spaces, the number of studies is significantly reduced [18,19], suggesting that many specific features of such spaces still need to be better disclosed and understood. One key issue in such spaces is represented by the bad conservation state, often related to locations far from urban areas, lack of controls, and exposure to any type of weather agents. Thus, current acoustic conditions may significantly differ from their original state, in most cases. It is important to underline that rupestrian churches were, first of all, "worship" spaces although their usage was not simply ascribed to individual celebrations by monks practicing ascetism, but they were used by whole communities of the settlements. Consequently, as rupestrian churches were used for ordinary masses, they had to deal with problems of speech intelligibility and propagation of the typical Byzantine hymns, inducing us to hypothesize correlations between acoustics and the shape and size of the spaces. For this reason, a reconstruction of the original acoustics of such spaces is of the utmost importance to understand which were the conditions in which the church was used and whether such correlations might exist.

Given the small dimensions of such spaces, modal behavior is likely to appear, resulting in non-uniform sound distribution at low frequencies, sound coloration (i.e., amplification or attenuation of certain frequencies), and strong variations of reverberation time as a function of narrow band frequencies. Modal behavior originates from standing waves that appear whenever the distance between the walls is a multiple of half wavelengths [20]. Axial, tangential, or oblique modes may appear depending on the number of surfaces that are involved in such standing wave paths.

Consequently, it is important to underline that simulating the acoustics of such spaces may be particularly challenging because of their small dimension which prevents a straightforward application of conventional methods based on geometrical acoustics which would be valid only well above the Schroeder frequency. Therefore, in such cases, it is essential to take advantage of different wave-based computational tools that may better describe the low-frequency behavior of the spaces. To this purpose, in this paper an application of the finite-difference-time-domain (FDTD) method is proposed, by implementing a simple scheme with frequency-independent boundary conditions limited to the lowest frequencies. Even though this method has been already applied to larger buildings and covering a much broader spectrum [21,22] using parallel computation on several graphical processing units, the present implementation aims at providing a useful low-frequency complement to conventional geometrical acoustic-based simulations, without requiring a high computational load.

The paper, taking advantage of the results already published in Ref. [19], will be organized as follows: Section 2 outlines the church that was surveyed, the acoustic measurement methods, the modelling techniques, and a few notes about FDTD; Section 3 presents the results of the on-site measurements, the results of the FDTD simulation in the current conditions, and the reconstruction of the original state; Section 4 provides a brief discussion of the results in comparison with other existing studies; finally, Section 5 summarizes the major conclusions.

2. Materials and Methods

2.1. The Church Surveyed

The church of Saints Andrew and Procopius (Santi Andrea and Procopio) in Monopoli (Figures 1 and 2) is located in the Contrada L'Assunta, along the Via Traiana and was likely built after the city of Monopoli was destroyed in 1042 to fight the Normans. The dedication of the church confirms the strong Byzantine influence as Andrew the Apostle was the founder of the church of Constantinople, and Procopius, martyr of Caesarea, was the protector of the Byzantine armies. The church was at the center of a rock village which also served as a post station along the Via Traiana (later Francigena) made up of a large number of caves with two or more rooms, oil mills, and mills [23].

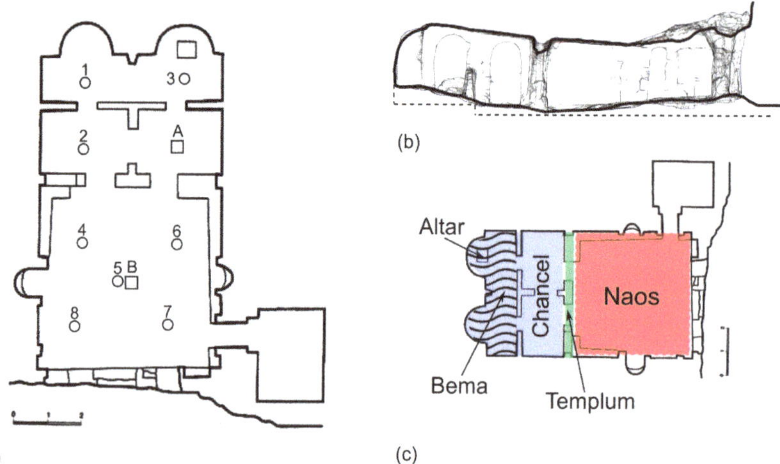

Figure 1. (**a**) Plan with source positions (A,B) and receiver positions (1–8), (**b**) cross section, and (**c**) schematic of the different areas of the surveyed church.

Figure 2. Photographs of the interior of the church taken during the survey.

The facade, just like in a masonry temple, has three arched entrances. The church is divided in two parts, the "naos" occupied by the congregation, having a simple rectangular shape (about 5.6 m by 5.2 m) and the chancel, divided by means of a "templum" (i.e., a stone iconostasis). The chancel was subdivided into four squares, two of them (the farthest from the entrance) define the "bema" where the altar was located. The ceiling is flat and its current height varies between 1.90 m and 2.40 m, but the original height is likely to be greater as a large amount of debris covers the floor. All the walls were originally covered by frescos but only a few fragments may be admired today.

Among the frescoes on the walls that are still visible, there are the apostles Peter and Paul, the saints Cosma, Damiano, Eligio (with the symbols of his patronage over the blacksmiths), Giorgio, Leonardo, a Virgin in throne and a scene of the Annunciation, datable between the 13th and 14th centuries. The painted scenes are all found in the bema and transept: Annunciation, Deesis, Trinity, Crucifixion, while the saints instead occupy the walls of the naos. The wall decoration is not contemporaneous with the excavation of the crypt but was constructed two centuries later. However, it reflects the continuity of worship from the Norman age, to which Saints Peter and Paul, Leonardo and Eligius belong, for example, up to the period of the Crusades (see, for example, the fresco of Saint George with the emblem of the cross on the saddle and on the shield). Unfortunately lost is the Byzantine iconography of the eleventh century relating to the eponymous saints of the sanctuary.

From the point of view of the liturgical functions, the church had the bema closed by a first iconostasis in stone opened by two doors in correspondence with the two apses. The transverse arm of the transept separates the presbytery from the naos, a common quadrangular room, not divided by pillars, with a second iconostasis, also with two doors. A summary of the main geometrical features is given in Table 1.

Table 1. Summary of geometrical data and surface finishing.

Volume	113 m^3
Plan surface	57 m^2
Total area	210 m^2
Average height	2.15 m
Surface openings	5 m^2
Floor	Dry ground with large cracks
Ceiling/vault	Extremely rough and peeling tuffaceous stone
Wall	Strongly degraded tuffaceous stone, with evident signs of erosion and, in some points, covered by frescoed plasters also with evident signs of superficial deterioration

2.2. Acoustic Measurement Methods

Given the location of the church in an open field, all the measurements were carried out with portable instruments powered with battery packs. An omni-directional sound source (Lookline D301) was located in two positions, one in front of the altar (Position A) and one in the congregation area (Position B), so as to simulate in this way the priest and the congregation, respectively. The source was fed by an equalized sine sweep played back by a portable music player and generated using MATLAB (v. 2021b) according to Müller and Massarani [24] so that the spectrum of the radiated sound was substantially flat from 50 Hz to 16 kHz. The duration of the sweep was kept short (about 8 s) in order to limit any potentially adverse effects due to lack of doors, which, determining significant air circulation, might compromise the linear and time-invariant hypothesis. In addition, given the conservation state of the surface finishing, longer sine sweeps might have induced excess mechanical stress, so it was preferred to keep signals short. Given the limited dimensions of the space, the signal-to-noise (S/N) ratio was sufficiently high to ensure perfectly usable impulse responses, even if a short sweep was used. Room responses were collected using a portable B-format microphone (Soundfield ST-350) connected to a multi-channel recorder (Tascam DR-680) and a pair of binaural microphones (Soundman OKM II) worn by one of the authors and connected to a second recorder (Tascam DR-07). The measurement chain was previously tested in the lab to ensure that the "open loop" settings did not create any sync problem.

All the measurements were carried out complying with ISO 3382-1 [25] standard and taking into account the guidelines for measurements in worship spaces [26], and, despite the small dimensions of the church, eight receiver positions were used to provide a detailed description of the point-by-point variations. Microphones were placed at 1.6 m from the floor, assuming that the congregation was standing during the celebration. Source and receiver locations were chosen to provide a description of acoustical conditions that could be produced by actual sound sources and heard by listeners distributed in the space. Given the small dimension of the churches, only one person stayed in the room during the measurements.

Impulse responses (IR) were calculated by deconvolving the signal used to feed the sound source and, despite a significant background noise due to birds and other natural sounds (resulting, on average, in an A-weighted sound pressure level of 45 dB), provided a minimum S/N ratio of about 55 dB over the whole spectrum of interest. The measured IRs were then processed in order to calculate the most important acoustic parameters and to investigate room resonances. In particular, in addition to monaural parameters based on the omni (W) response of the B-format microphone, lateral energy fraction was calculated using W and Y Ambisonic components (assuming X axis was aimed at the source), while inter-aural cross correlation was based on binaural responses.

2.3. Geometrical Modelling of the Space

Given the complex and irregular shape of the space, the only reasonable way to obtain a reliable geometrical model was to use 3D laser scanning. A point cloud of more than 150k elements was originally obtained using a Riegl VZ 400 scanner (with an original resolution of 5 cm), but for the purpose of the simulation, such level of detail was unnecessary and thus, after cleaning artifacts and imperfections using a specifically designed tool developed in MATLAB, the point cloud was further simplified using the open source software Meshlab through subsequent applications of the "Quadric Edge Collapse Decimation" algorithm (Figure 3).

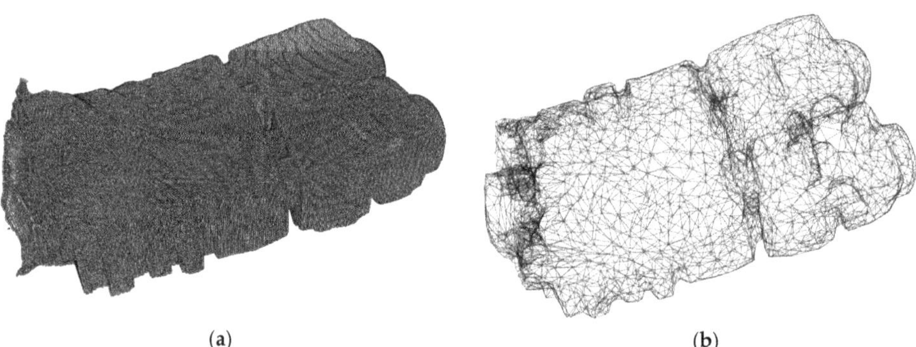

(a)   (b)

Figure 3. Comparison between the original point cloud (**a**) and the simplified mesh (**b**).

2.4. FDTD Simulation

FDTD acoustic modelling has been applied to solve acoustic problems for a long time [27] but given the computational load it has been mostly applied to low frequencies [28], while the availability of parallel computation distributed by several GPUs fostered a gradual extension to a much wider frequency range [21,22,28–30]. FDTD starts from the assumption that a generic derivation operator can be replaced by one of its finite difference forms:

$$\frac{df}{dx}(x_0) \approx \frac{f(x_0 + \Delta x) - f(x_0)}{\Delta x} \tag{1}$$

and hence the second-order derivative becomes:

$$\frac{d^2 f}{dx^2}(x_0) \approx \frac{f(x_0 + \Delta x) - 2f(x_0) + f(x_0 - \Delta x)}{\Delta x^2} \tag{2}$$

In acoustics, any propagation phenomenon is described by the wave equation expressed by:

$$\nabla^2 p = \frac{1}{c^2}\frac{\partial^2 p}{\partial t^2} \tag{3}$$

The equation is only expressed as a function of sound pressure p, but its finite-difference form would require three different time steps to be numerically solved, so in a 2D case in which T is the time step and X is the grid spacing, pressure at step $n + 1$ and node (x,y) will be given by:

$$p_{x,y}^{n+1} = \lambda^2 \left(p_{x+1,y}^n + p_{x-1,y}^n + p_{x,y+1}^n + p_{x,y-1}^n \right) + 2\left(1 - 2\lambda^2\right)p_{x,y}^n - p_{x,y}^{n-1} \tag{4}$$

where $\lambda = c\,T/X$, c being the speed of sound in air. However, considering that the wave equation is derived by two other fundamental equations, involving only first-order derivatives, but also including particle velocity, it is possible to find a perfectly equivalent [29,30] alternative formulation. This formulation is known as staggered Yee's grid, in which the grid of the pressure values is complemented by the grid of the particle velocity values (u). Thus, it is possible to first update particle velocity components (remembering that particle velocity is a vector quantity and, in 2D, it will have ux and uy components) and when this matrix is available, we can update the pressure at the next time step. Pressure at step $n + 1$ and node (x,y) will consequently be given by:

$$p_{x,y}^{n+1} = p_{x,y}^n - \frac{\rho c^2 T}{X}\left(ux_{x+0.5,y}^{n+0.5} - ux_{x-0.5,y}^{n+0.5} \right) - \frac{\rho c^2 T}{X}\left(uy_{x,y+0.5}^{n+0.5} - uy_{x,y-0.5}^{n+0.5} \right) \tag{5}$$

The above equations apply when non-boundary conditions are found, while at grid points close to walls, assuming a wall impedance Z, it is possible to find an update formula that, in case Z may be considered a real number independent of frequency, yields (in the 2D case) [31]:

$$ux_{x,y}^n = \frac{R_x - Z}{R_x + Z}ux_{x,y}^{n-1} + \frac{2}{R_x + Z}p_{x-0.5,y}^{n-0.5} \tag{6}$$

where $R_x = \rho\,T/X$, ρ being air density, and $Z = \rho \cdot c\,\{[1 + (1 - \alpha)^{0.5}]/[1 - (1 - \alpha)^{0.5}]\}$, with α being the absorption coefficient of the given surface. A number of alternative approaches have been proposed to account for frequency dependence [27,32], but in the present case, being the analysis limited to the low-frequency case, the proposed assumption was not considered a major limitation.

In order to obtain a reliable calculation using FDTD, it is essential to properly set grid spacing X and time step T. In fact, they are both related to the maximum frequency that can be analyzed, as a minimum of five points per wavelength is usually required to prevent errors. Therefore, this means that given the minimum wavelength of interest (λ), the grid spacing should be at least $\lambda/5$ or, better, $\lambda/10$. Thus, for a grid spacing of 0.1 m, the FDTD results will be accurate up to a maximum frequency of 340 Hz. The time step is not independent of the grid spacing if the stability condition given by:

$$T \leq \frac{X}{c\sqrt{D}}, \tag{7}$$

(where D is the number of dimensions of the problem) is satisfied. Thus, once the grid spacing and the problem dimension are defined, time step results consequently (in the present case corresponding to a sampling frequency of 5768 Hz). Proper limitation of the highest frequency of analysis is also useful to minimize dispersion errors that result in phase velocity being different from the actual value of the medium. In the present case, limiting to frequencies below 340 Hz ensures that dispersion errors will be below 2% [29].

The previously described FDTD framework was implemented in MATLAB (Figure 4), where the geometrical model was first voxelized using a 10 cm grid spacing, and then surface properties could be assigned together with source and receiver locations. Sound sources were modeled as simple point sources and emitted signals could vary between sine waves (for modal analysis) and short pulses.

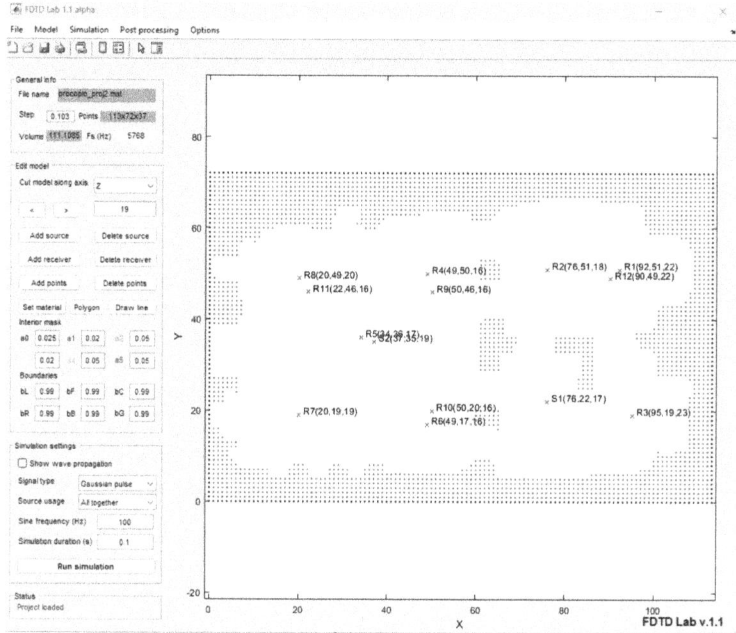

Figure 4. Graphical User Interface developed in MATLAB to perform FDTD simulations.

As per usual, surface properties were adjusted in order to have a suitable match between measured and predicted values of the reverberation time in the present condition at frequencies of 63 Hz and 125 Hz.

It is important to point out that for a space of such dimensions (113 m^3), it is essential to model sound propagation by means of wave-based models because geometrical acoustics can only be effective well above Schroeder's frequency that, being equal to $2000\sqrt{T/V}$, in this case was around 250 Hz. For this reason, modal response was also carefully compared to ensure that the model could realistically reproduce low-frequency propagation.

2.5. Material Characterization

In order to characterize sound absorption of the tuffaceous surface of the church, samples taken from quarries of geologically similar limestone were analyzed. Measurements of normal incidence sound absorption coefficient were carried out according to ISO 10534-2:1998 [33], using the transfer function method. As the objective of the study was to understand the low-frequency behavior of the materials, only the tube with an internal diameter of 10 cm was used, resulting in a maximum measurable frequency of 2 kHz and a low-frequency limit of 50 Hz. The emitting end consisted of an 11 cm loudspeaker sealed into a wooden case and suitably isolated from the tube structure by an elastic and protective layer. All the processing was performed by a MATLAB graphic user interface generating a linear sweep to feed the loudspeaker.

As shown in Table 2, the major difference between different types of tuffaceous stones appears above 80 Hz, where higher porosity may contribute to nearly double absorption compared to harder samples. At very low frequencies, no significant differences were observed in the measured values.

Table 2. Summary of measured normal incidence absorption coefficients of different limestone samples 5 cm thick.

Material	63 Hz	80 Hz	100 Hz	125 Hz	160 Hz
Soft limestone (Carparo)	0.010	0.020	0.030	0.050	0.080
Hard limestone	0.010	0.020	0.020	0.024	0.037

3. Results

3.1. On-Site Acoustic Measurements

The analysis of the reverberation times shows very interesting differences between the medium-high frequencies, where the values settle below 0.5 s, with negligible point-by-point variations, and the low frequencies, where the values are considerably longer and show some variability depending on the location of both source and receivers (Figure 5). In particular, when the source was close to the altar (Source A), T30 at 63 Hz and at 125 Hz assumed the longest values (particularly at receivers 1, 2, 4, and 7), smoothly decreasing when moving closer to the entrance. When the source was in position B, T30 dropped by about 0.5 s in the same position. This fact, combined with EDT values shorter than T30 whatever the source position, suggested that no evident reverberant coupling effects between the sub-volumes may explain the observed variations. Conversely, the analysis of the time decays at 63 Hz and 125 Hz shows (Figure 6) that decays are characterized by a staircased or "pulsating" trend, clearly evident at 63 Hz, but also appearing at 125 Hz, in particular for combinations A_01 and A_02. Such behavior is typically associated with repeated reflections (flutter echoes) or modal effects which could be better investigated by analyzing the narrow band spectra of the responses.

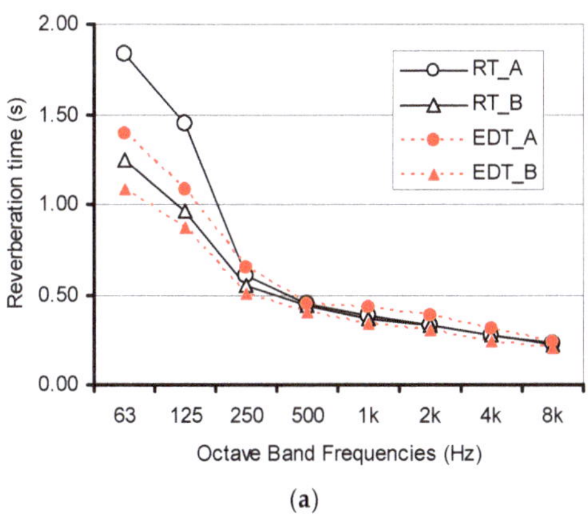
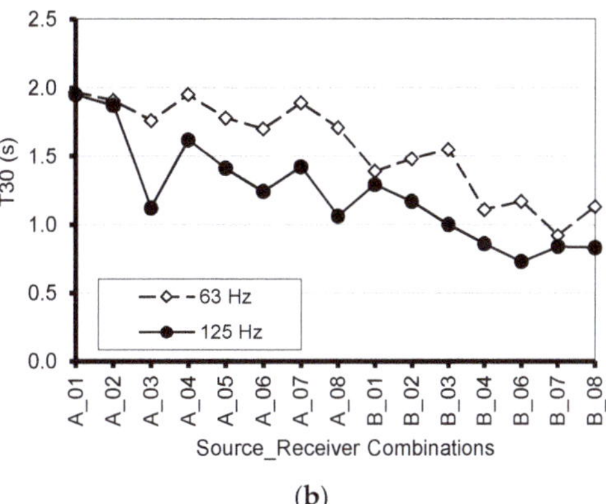

Figure 5. (**a**) Plot of spatially averaged reverberation time (T30) and early decay time (EDT) as a function of frequency and source position; (**b**) plot of T30 as a function of receiver position at 63 and 125 Hz.

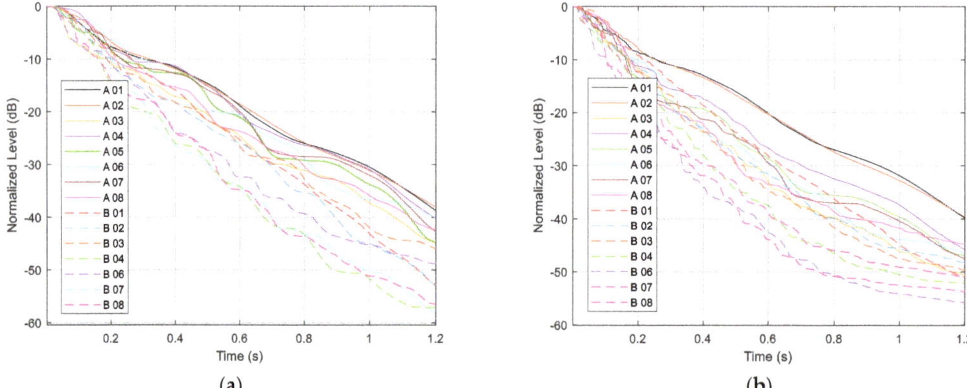

Figure 6. Plot of decay curves for all source–receiver combinations at (**a**) 63 Hz octave band and (**b**) 125 Hz octave band.

Before delving into the analysis of the spectra, it is worth noticing that the short reverberation time at medium-high frequencies was likely due to the presence of openings (about 4 m^2) and to the particular state of deterioration of the stone which appeared very pronounced, with frescoes occupying only a small part of the surfaces and the underlying layer characterized by greater porosity and tenderness. As shown in Table 2, a soft limestone is capable of absorbing a significantly higher amount of acoustic energy, which may easily explain the observed values. In addition, the floor was completely covered by waste soil overflowed during floods, characterized, at the time of measurement, by numerous cracks and having a thickness which could be estimated to vary between 10 cm and 50 cm in some points.

In order to better understand the observed low-frequency behavior, narrow band spectra were determined for all the source–receiver combinations (Figure 7). As expected, when source was in position A, receivers 1, 2, and 4 clearly showed a marked peak at 79 Hz, surrounded by many others appearing at 65 Hz, 73 Hz, and 87 Hz. In the other receivers, the same modes appeared but their energy content was significantly reduced. In fact, if the mid-frequency spectrum density is taken as a reference, the level variation around 79 Hz is about 30 dB between receivers 1 and 2 and the others.

When the source was moved into the "naos", the acoustic energy redistributed among modes. In fact, the new absolute peak appeared around 65 Hz and 73 Hz at receivers 1, 2, and 4, while in the others the mode energy was more evenly distributed but remained at least 10 dB higher than in the frequencies above 100 Hz.

This can clearly explain what was observed in terms of reverberation time because the peak appearing around 80 Hz due to its position and magnitude influenced both the octave bands of 63 Hz and 125 Hz, causing the slower decay as observed in Figure 6 to appear in receiver 1 and 2 also in the higher octave band. In the other receivers, as the energy in the modes decreases, the reverberation time becomes gradually shorter, remaining longer in the 63 Hz band due to persistence in time of modal behavior (Figure 8a). To this purpose, modal reverberation time [34,35] was calculated for combination A-01 at the frequency of 78.2 Hz, showing (Figure 8b) that the observed value basically coincided with the octave band value.

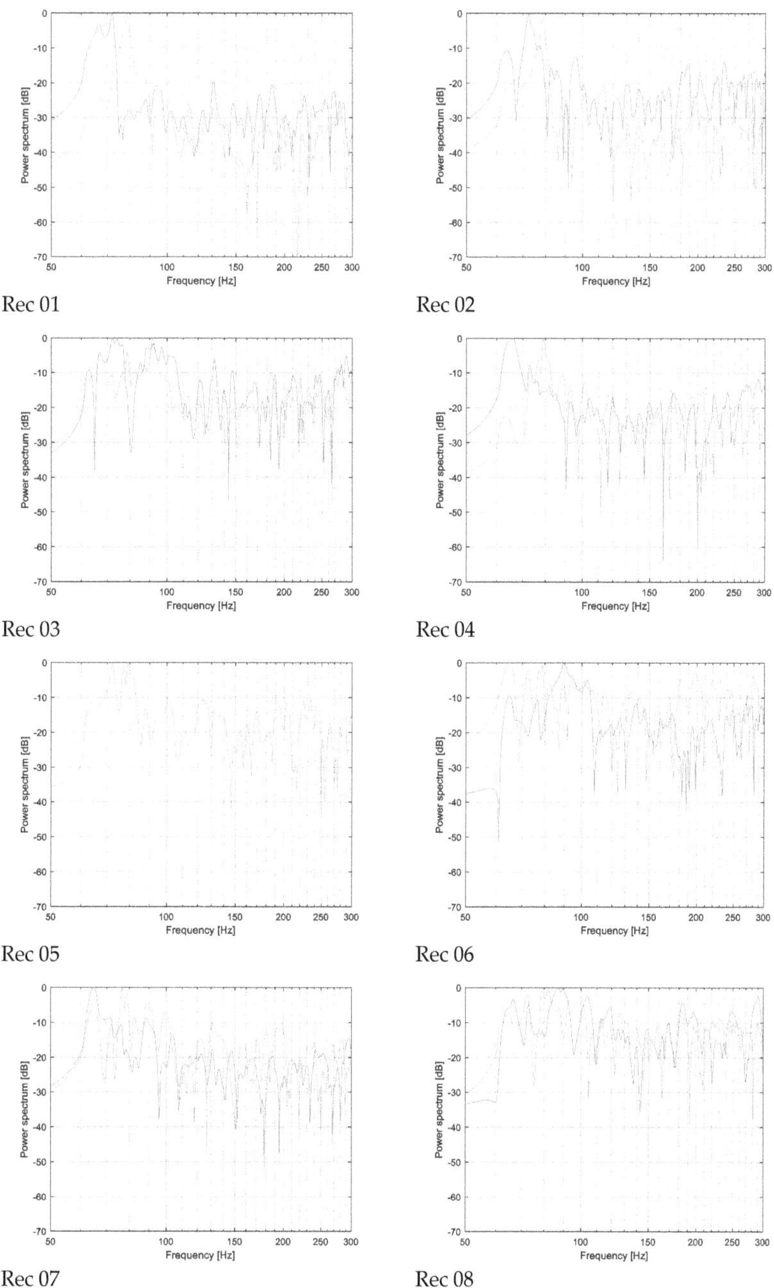

Figure 7. Plot of narrow band spectra measured in each receiver position (Rec. 01 to Rec. 08), having source located in position A (blue) and B (orange).

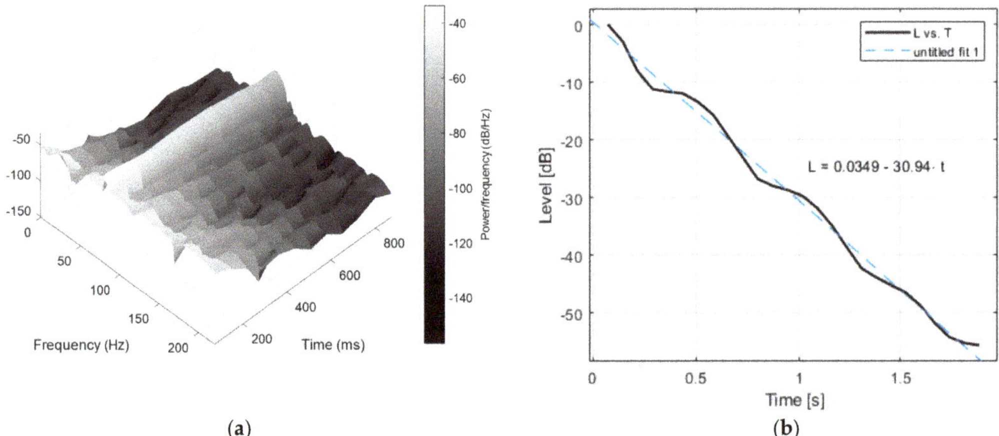

(a)   (b)

Figure 8. (**a**) Low-frequency spectrogram for combination A-01, and (**b**) modal decay curve for combination A-01, at frequency of 78.2 Hz, calculated according to Ref. [32].

The mathematics of the axial resonance modes [20] state that resonances appear at frequency $f_n = (c \cdot n)/(2L)$, where c is the speed of sound, L is the room dimension in the direction that is considered, and n is an integer index that defines the mode order. Thus, it can be easily found that 79 Hz corresponds to the first ($n = 1$) axial mode along the vertical direction assuming a height of 2.15 m (which is in good agreement with the average height of the space, although being far from a simple rectangular box), while 65 Hz corresponds well to the second ($n = 2$) axial mode along the two side walls (spaced by about 5.4 m), the only two large and continuous surfaces besides the floor and the ceiling.

Once the basic acoustical features have been explained by properly combining reverberation time and modal analysis, it is now possible to have a look at the other acoustical parameters (Figure 9). Given the short reverberation, very high speech intelligibility is obtained (STI = 0.82 on average) and strong frequency imbalance represented by a bass ratio (BR) equal to 2.1, although in the original conditions, with walls covered by frescos, it is likely that a more balanced condition was observed.

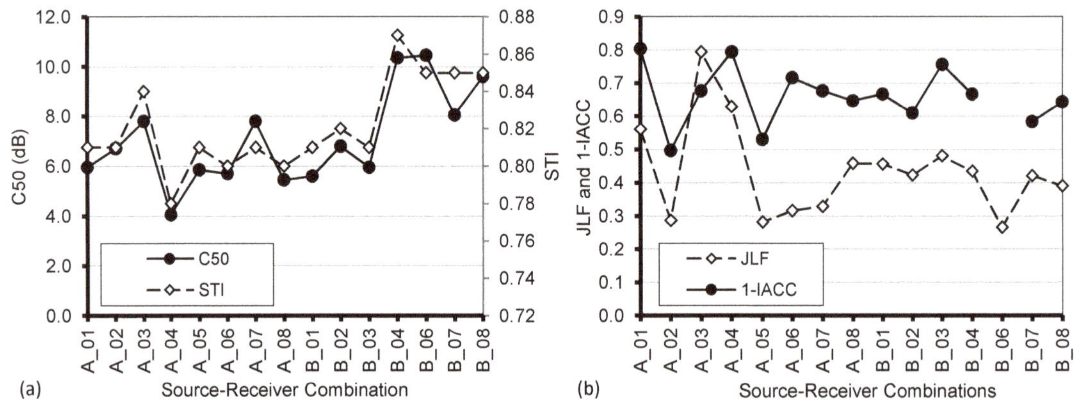

(a)   Source-Receiver Combination   (b)   Source-Receiver Combinations

Figure 9. Plot of multi-octave band averages (as defined in ISO 3382-1) of other acoustical parameters as a function of source–receiver combinations. (**a**) Clarity (C50) and Speech Transmission Index (STI). (**b**) Early lateral energy fraction (JLF) and early interaural cross-correlation (1-IACC).

The perception of sound spatiality and, more generally, the binaural impression were particularly interesting. JLF and 1-IACC applied to the initial part of the impulse response showed rather high values (1-IACC = 0.67 on average) which were reduced only (as expected) in correspondence with the points placed in direct proximity to the source (receivers 2 and 5). The large amount of surface irregularities distributed almost everywhere, together with the presence of obstacles (such as the iconostasis) all contribute to this effect, making the sound perception very "spacious" despite the small size. JLF showed extremely high values deriving from the strong contribution of lateral reflections combined with the frequent shielding of reflections coming from frontal directions.

3.2. FDTD Acoustical Simulation of Current State

FDTD simulations were carried out according to the procedures described in Section 2.4. As the finishing was relatively the same over all the surfaces, with the notable exception of the floor covered by hardened soil, and given the values measured for similar materials (Table 2), an initial value of 0.015 was used for the absorption coefficient (obtained by averaging among the 63 Hz and the 125 Hz values). As this value returned slightly longer T30 values than expected, while in acoustic modelling [22] a maximum difference of 5% (corresponding to one just noticeable difference [25]), is considered to be acceptable, step-by-step increases were applied until a value of 0.025 was reached. Under these conditions, the spatial average of T30 in one-third octave bands from 63 Hz to 80 Hz differed by only 2% from measurements, while at 100 Hz the error was 4%. At higher frequencies, the error was bigger, suggesting that a further increase in absorption coefficient was needed. By adopting an α value of 0.04, the mean error was finally reduced to 4% also in the 125 Hz band. Figure 10 shows the comparison between measured and predicted reverberation time in one-third octave bands at individual positions. It can be observed that in the lowest bands clear modal behavior appears at given receiver positions, resulting from the strong non-uniform distribution of the values. In addition, as discussed in the previous section, the strength of the first axial mode along the vertical direction clearly appears also in the simulated results, where T30 in the 80 Hz band is markedly higher than the corresponding values at 63 Hz despite the same absorption coefficient. At 100 Hz and 125 Hz the modal behavior is significantly attenuated in the simulations (although the measured values at combinations A-01 and A-02 still show some effects).

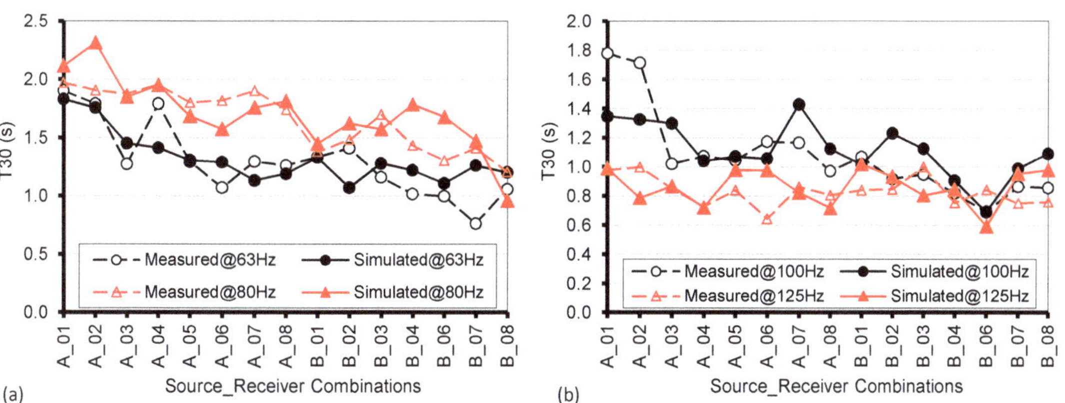

Figure 10. Plot of measured and simulated values of T30 as a function of source–receiver combination, at (**a**) 63 Hz and 80 Hz, and at (**b**) 100 Hz and 125 Hz.

Given the role played by modal response of the room, it was important to check the agreement between measured and predicted spectra. Considering that Figure 7 had already shown the combinations where a strong modal response was measured, the comparison was restricted to a subset of all the source–receiver combinations. Figure 11 shows that for combinations A-01 and A-02, the strong response around 79 Hz clearly appears, although the magnitude is attenuated compared to measurements assuming higher frequencies as a reference (which also explains why the 100 Hz T30 is shorter in the simulated response). Small shifts in the peak frequencies can be observed, but they rarely exceed 2 Hz. Combination A-05 is modelled pretty well, with the peak around 65 Hz well visible and a slightly less defined cluster around 79 Hz but a well comparable distribution of levels. Finally, combination B-04 is considered, as in the measured spectrum a clear peak appeared at 65 Hz. The simulated spectrum presents several peaks that are mostly aligned with measurements, but the stronger peak at 65 Hz seems more attenuated (by about 15 dB), and another peak below 60 Hz appears. However, considering that measured spectra were band-limited by the inherent frequency response of the sound source and by the signal used to feed the loudspeaker, this was considered a minor problem.

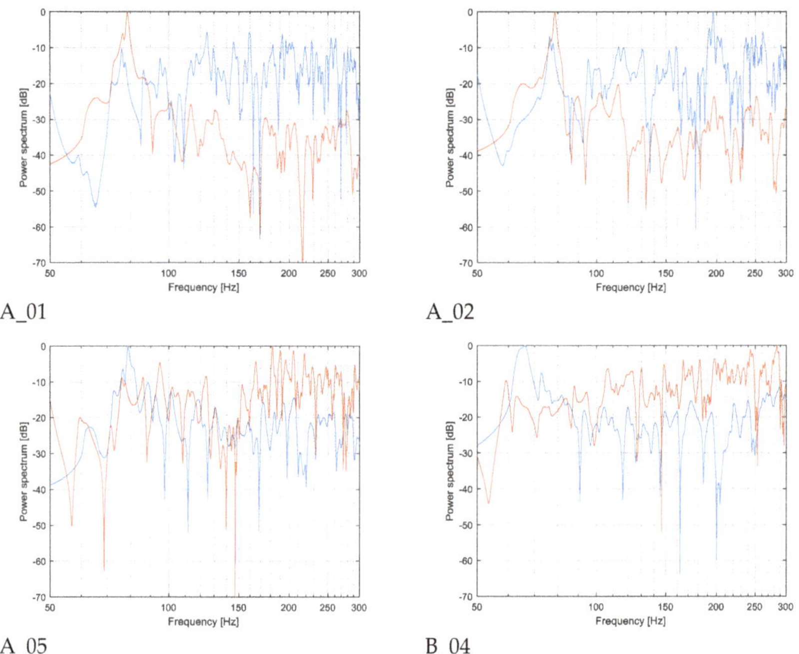

Figure 11. Plot of narrow band spectra measured (blue) and simulated (orange) in selected source and receiver combinations (A-01, A-02, A-05, and B-04).

A further element to be considered that might explain some of the small differences observed in the T30 values and in the responses is the strong point-by-point variations that appear in consequence of the modal behavior. Taking into account the FDTD model, the sound pressure level resulting from pure tones at 65 Hz and 79 Hz was calculated for the two source positions. As shown in Figure 12, dramatic sound pressure level variations take place by simply moving the receivers a few centimeters apart. Hence, it is not unlikely that small misplacement of the receivers (both during the measurements and in the simulation setup) might explain some of the observed differences.

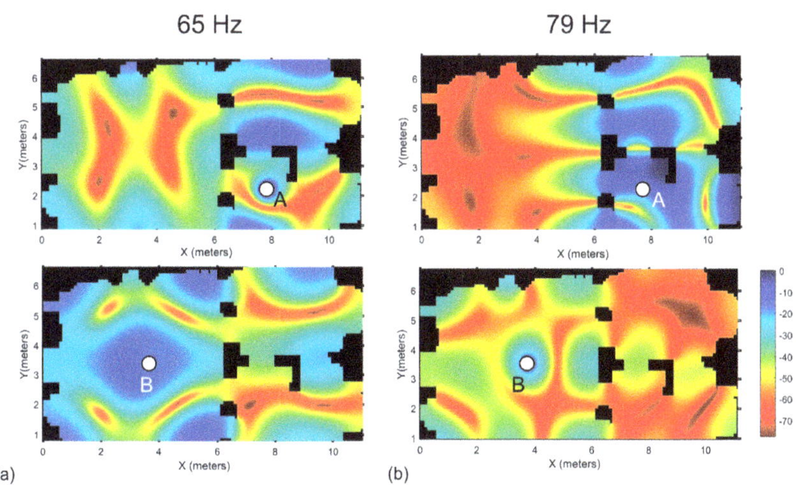

Figure 12. Sound pressure level distribution resulting FDTD simulation of pure tones emission at (**a**) 65 Hz and (**b**) 79 Hz, with sound source located in position A (top) and B (bottom).

However, Figure 12 is quite instructive because it shows which are the positions in the church where stronger resonances appear and, consequently, where longer reverberation might be experienced. Predictably, receivers close to the side walls in the "naos" could experience strong resonances whatever the source position was, particularly at 65 Hz (which was the second axial mode along the transverse direction), but also at the other frequencies.

3.3. FDTD Acoustical Reconstruction of Original State

Once the model was calibrated, it could be used to apply a few changes that could realistically correspond to the original state, allowing us to appreciate the implications for the acoustics. In the specific case, three simple changes were applied. First, the floor area was rectified assuming that the layers of debris that currently occupy the entrance and the chancel area, where the ceiling is unusually low (see dashed line in Figure 1b), could be removed. Surfaces were considered to be covered by plasters and frescos (like the limited portions still existing clearly suggest, and like it is found in many other rupestrian churches in the region), resulting in a reduced absorption coefficient of 0.015 at 63 Hz and 80 Hz, and 0.025 at 100 Hz. Finally, in order to simulate acoustics under occupied conditions, a seated audience was located along the two side walls of the "naos", according to the common practice of the time, as demonstrated by the usual presence of "subsellia" (carved benches) in other churches. This position was chosen in order to locate the absorption in the position where modal behavior might have more strongly affected the acoustic response in terms of damping. Absorption coefficients for this area were set to 0.2 [36] assuming the audience to be tightly distributed among the limited seats. All the openings were left in their actual state as no evidence of pre-existing doors or windows was found.

As shown in Figure 13, most of the main features observed in the "current" state also appear in the "reconstructed", but now reverberation is longer and the modal response at 79 Hz when the source is in A is stronger and extends well beyond the "templum". When the source is in B, the most evident variation is a drop appearing at receivers close to the entrance, possibly as a consequence of the change in room height (that is the part where the debris were thicker and after removal the ceiling height becomes about 2.7 m) and increase in the absorbing elements in the volume due to audience and increased opening surfaces (again as a consequence of debris removal).

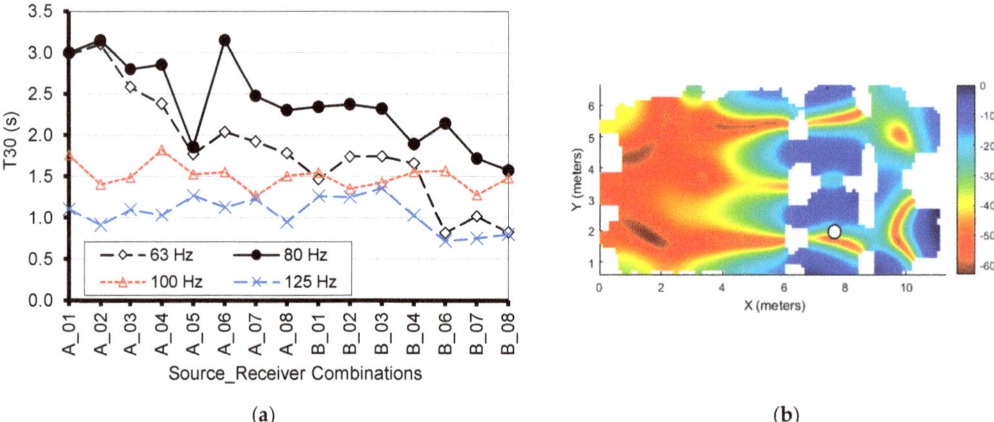

Figure 13. (**a**) Plot of T30 under simulated "reconstructed" state, as a function of receiver position at one-third octave bands from 63 Hz to 125 Hz. (**b**) Sound pressure level distribution resulting from FDTD simulation under simulated "reconstructed" state, of pure tones emission at 79 Hz, with sound source located in position A.

Such results seem to suggest that the resulting acoustic effects might have been exploited during liturgical celebrations and singing, in particular, considering that the specific features of many Byzantine hymns include many bass notes sustained for a long time, which could well excite the resonances of the space.

4. Discussion

After presenting the results of the measurements and of the simulations, a brief discussion to put the results in the context of the existing literature can be developed. Among the spaces that have been surveyed by other authors, the highest similarity can be found with the rock-cut structures in Cappadocia [18]. Although the only space having a comparable volume was the Avanos dining room (114 m^3), and the observed reverberation times spanned over a much broader range (up to 5 s for the Hallaç Church and main hall and for Açıksaray Hall), in all of the cases a significant low-frequency imbalance was observed, with reverberation time being up to 3 times longer than at mid-frequencies. This was likely caused by the characteristics of the stone that due to its porosity caused increased absorption as frequency grew. However, no specific low-frequency measures were made, and the frequency range of analysis was limited to 125 Hz.

Similarly, among the catacombs in Southern Italy [17], it was difficult to find spaces that are entirely comparable, but the two cubicles in the "San Callisto" catacombs have several similarities in terms of room dimensions and volumes. Materials are also similar, the walls being made of tufa stone, resulting in short reverberation times usually well below 1 s. In this case, the frequency range of the measurements extends into the 63 Hz band, showing only a moderate increase in reverberation time compared to mid-frequencies, with the notable exception of the so called "double cubicle", where reverberation is longer (up to 1.5 s at 63 Hz) and characterized by point-by-point variations. Regardless, even in this study, although the "resonant" behavior in the low frequencies is supposed, no additional investigations were made.

Low-frequency resonances and modal behavior were investigated in many stone chambers and cairns in the British Isles [7]. In this case, experimental measurements were carried out using sine sweeps to find resonant frequencies and the distribution of nodes and antinodes in the space. A theoretical justification was consequently found based on the dimensions of the space, showing that in most of the cases, resonance frequencies varied between 95 Hz and 120 Hz, slightly above the resonances observed in the present work, likely as a consequence of the smaller dimension.

Finally, one interesting example where the modal behavior of a space was investigated also by means of wave-based acoustic simulation is the Hal Saflieni Hypogeum [11]. In this Neolithic structure, the authors investigated the possibility that the resonance frequencies observed in different chambers have been "tuned" in some way. Based on the on-site measurements, a numerical model was developed to test whether different dimensions of the chambers might have altered this tuning. However, as in the present study, this reconstruction reflects the "current" conditions which might have been different from the original conditions which, as the authors clearly state, may be difficult to imagine.

5. Conclusions

In this paper, the case of the rupestrian church of Saints Andrew and Procopius in Monopoli (Apulia, Italy) was considered. The church has a small volume and is in a very bad conservation state, but it was acoustically analyzed in its current state by means of measurements carried out according to ISO 3382-1 standard. A combination of different analysis methods, including not only simple room acoustical parameters, as in most previous studies, but also spectrograms, modal reverberation time, and detailed maps based on FDTD modelling, were used to interpret the results. Measurements pointed out a singular behavior characterized by short reverberation times (around 0.5 s) at frequencies above 250 Hz, and much longer values in the lowest bands reaching up to 2 s, also in combination with strong resonances due to the modal response of the room. The analysis of the narrow band spectra and of modal reverberation time confirmed such dependance and resonant frequencies, especially those appearing at 65 Hz and 79 Hz, which were explained as a function of the room dimensions (although with some approximations due to the strongly irregular shape). In order to acoustically simulate the space and analyze the distribution of sound both in the current and in any possible reconstructions of the original state, given the small dimensions that make it impossible to use geometrical acoustic models in the low-frequency range, the numerical FDTD method was applied by using a proprietary code developed in MATLAB. Thanks to a laser scanner survey, the geometry was simplified and voxelized so that FDTD could be applied. Results showed a very good agreement both in terms of predicted reverberation time and modal response, implying that FDTD method, despite its simplified implementation, is the most reliable approach to simulate such small spaces and obtain time-dependent responses. Consequently, a simulation of a possible reconstructed condition, characterized by plastered surfaces, debris removed from the floor, and some occupants in the space, was carried out. Results showed that the strong modal response still appeared, suggesting that liturgical singing might have taken advantage of such specific acoustic features.

Author Contributions: Conceptualization, F.M.; methodology, F.M.; software, F.M.; validation, F.M., S.L. and C.R.; investigation, F.M., S.L. and C.R.; data curation, F.M., S.L. and C.R.; writing—original draft preparation, F.M.; writing—review and editing, F.M., S.L. and C.R.; visualization, F.M. All authors have read and agreed to the published version of the manuscript.

Funding: This research was partly funded by Fondazione Cassa di Risparmio di Puglia, within the project "Suoni di Pietra".

Institutional Review Board Statement: Not applicable.

Informed Consent Statement: Not applicable.

Data Availability Statement: Data are available from the authors on request.

Conflicts of Interest: The authors declare no conflict of interest.

References

1. UNESCO World Heritage Centre. Available online: http://whc.unesco.org/ (accessed on 28 March 2023).
2. Bixio, R.; De Pascale, A.; Mainetti, M. Census of Rocky Sites in the Mediterranean Area. In *The Rupestrian Settlements in the Circum-Mediterranean Area*; Crescenzi, C., Caprara, R., Eds.; Publisher DAdsp–UniFi: Florence, Italy, 2012.
3. Crescenzi, C. Typology of Rupestrian Churches in Cappadocia. In *The Rupestrian Settlements in the Circum-Mediterranean Area*; Crescenzi, C., Caprara, R., Eds.; Publisher DAdsp–UniFi: Florence, Italy, 2012.
4. Plummer, W.T. Infrasonic Resonances in Natural Underground Cavities. *J. Acoust. Soc. Am.* **1969**, *46*, 1074–1080. [CrossRef]
5. Dayton, L. Rock art evokes beastly echoes of the past. *New Sci.* **1992**, *1849*, 14.
6. Waller, S.J. Sound and rock art. *Nature* **1993**, *363*, 501. [CrossRef]
7. Jahn, R.; Devereux, P.; Ibison, M. Acoustical resonances of assorted ancient structures. *J. Acoust. Soc. Am.* **1996**, *99*, 649–658. [CrossRef]
8. Abel, J.S.; Rick, J.W.; Huang, P.P.; Kolar, M.A.; Smith, J.O.; Chawning, J.M. On the acoustics of the underground galleries of Ancient Chavin de Huantar. In Proceedings of the Acoustics 08, Paris, France, 29 June 29–4 July 2008; pp. 4165–4170.
9. Reznikoff, I. Sound resonance in prehistoric times: A study of Paleolithic painted caves and rocks. In Proceedings of the Acoustics 08, Paris, France, 29 June–4 July 2008; pp. 4135–4139.
10. Fazenda, B.; Scarre, C.; Till, R.; Pasalodos, R.J.; Guerra, M.R.; Tejedor, C.; Peredo, R.O.; Watson, A.; Wyatt, S.; Benito, C.G.; et al. Cave acoustics in prehistory: Exploring the association of Palaeolithic visual motifs and acoustic response. *J. Acoust. Soc. Am.* **2017**, *142*, 1332. [CrossRef] [PubMed]
11. Wolfe, K.; Swanson, D.; Till, R. The frequency spectrum and geometry of the Hal Saflieni Hypogeum appear tuned. *J. Arch. Sci. Rep.* **2020**, *34*, 102623. [CrossRef]
12. Santos da Rosa, N.; Alvarez Morales, L.; Martorell Briz, X.; Fernández Macías, L.; Díaz-Andreu García, M. The acoustics of aggregation sites: Listening to the rock art landscape of Cuevas de la Araña (Spain). *J. Field Archaeol.* **2023**, *48*, 130–143. [CrossRef]
13. Mattioli, T.; Farina, A.; Armelloni, E.; Hameau, P.; Díaz-Andreu, M. Echoing landscapes: Echolocation and the placement of rock art in the Central Mediterranean. *J. Archaeol.* **2017**, *83*, 12–25. [CrossRef]
14. Till, R. Sound Archaeology: A Study of the Acoustics of Three World Heritage Sites, Spanish Prehistoric Painted Caves, Stonehenge, and Paphos Theatre. *Acoustics* **2019**, *1*, 661–692. [CrossRef]
15. Iannace, G.; Trematerra, A. The acoustics of the caves. *Appl. Acoust.* **2014**, *86*, 42–46. [CrossRef]
16. Małecki, P.; Lipecki, T.; Czopek, D.; Piechowicz, J.; Wiciak, J. Acoustics of Icelandic lava caves. *Appl. Acoust.* **2022**, *197*, 108929. [CrossRef]
17. Iannace, G.; Trematerra, A.; Qandil, A. The acoustics of the catacombs. *Arch. Acoust.* **2014**, *39*, 583–590. [CrossRef]
18. Adeeb, A.H.; Sü, Z.; Henry, A. Characterizing the Indoor Acoustical Climate of the Religious and Secular Rock-Cut Structures of Cappadocia. *Int. J. Archit. Herit.* **2021**, 1–22. [CrossRef]
19. Martellotta, F.; Cirillo, E. Acoustic characterization of Apulian Rupestrian churches. In Proceedings of the Forum Acusticum 2011, Aalborg, Denmark, 27 June–1 July 2011. Paper 00052.
20. Kuttruff, H. *Room Acoustics*, 5th ed.; Spon Press: London, UK, 2009.
21. Fratoni, G.; Hamilton, B.; D'Orazio, D. Feasibility of a finite-difference time-domain model in large-scale acoustic simulations. *J. Acoust. Soc. Am.* **2022**, *152*, 330–341. [CrossRef] [PubMed]
22. D'Orazio, D.; Fratoni, G.; Rovigatti, A.; Hamilton, B. Numerical simulations of Italian opera houses using geometrical and wave-based acoustics methods. In Proceedings of the 23rd International Congress on Acoustics, Aachen, Germany, 9–13 September 2019; pp. 5994–5996.
23. Dell'Aquila, F.; Messina, A. *Le chiese rupestri di Puglia e Basilicata*; Adda Editore: Bari, Italy, 1998.
24. Müller, S.; Massarani, P. Transfer-function measurement with sweeps. *J. Audio Eng. Soc.* **2001**, *49*, 443–471. Available online: http://www.aes.org/e-lib/browse.cfm?elib=10189 (accessed on 28 March 2023).
25. *ISO 3382-2009*; Acoustics–Measurement of Room Acoustic Parameters–Part 1: Performance Spaces. ISO: Geneva, Switzerland, 2009.
26. Martellotta, F.; Cirillo, E.; Carbonari, A.; Ricciardi, P. Guidelines for acoustical measurements in churches. *Appl. Acoust.* **2008**, *70*, 378–388. [CrossRef]
27. Botteldooren, D. Finite-difference time-domain simulation of low-frequency room acoustic problems. *J. Acoust. Soc. Am.* **1995**, *98*, 3302–3308. [CrossRef]
28. Southern, A.; Siltanen, S.; Murphy, D.T.; Savioja, L. Room Impulse Response Synthesis and Validation Using a Hybrid Acoustic Model. *IEEE Trans Audio Speech Lang Process* **2013**, *21*, 1940–1952. [CrossRef]
29. Kowalczyk, K. Boundary and Medium Modelling Using Compact Finite Difference Schemes in Simulations of Room Acoustics for Audio and Architectural Design Applications. Ph.D. Dissertation, Queen's University Belfast, Belfast, UK, 2008.
30. Kowalczyk, K.; van Walstijn, M. Room acoustics simulation using 3-D compact explicit FDTD schemes. *IEEE Trans. Audio Speech Lang. Process.* **2011**, *19*, 34–46. [CrossRef]
31. Hill, A.J.; Hawksford, M.O.J. Visualization and Analysis Tools for Low-Frequency Propagation in a Generalized 3D Acoustic Space. *J. Audio Eng. Soc.* **2011**, *59*, 321–337. Available online: http://www.aes.org/e-lib/browse.cfm?elib=15932 (accessed on 28 March 2023).

32. Kowalczyk, K.; van Walstijn, M. Formulation of locally reacting surfaces in FDTD/K-DWM modelling of acoustic spaces. *Acta Acust. United Acust.* **2008**, *94*, 891–906. [CrossRef]
33. *ISO 10534-2*; Acoustics–Determination of Sound Absorption Coefficient and Impedance in Impedance Tubes-Part 2: Transfer-function Method. ISO: Geneva, Switzerland, 1998.
34. Rindel, J.H. Modal energy analysis of nearly rectangular rooms at low frequencies. *Acta Acust. United Acust.* **2015**, *101*, 1211–1221. [CrossRef]
35. Prato, A.; Casassa, F.; Schiavi, A. Reverberation time measurements in non-diffuse acoustic field by the modal reverberation time. *Appl. Acoust.* **2016**, *110*, 160–169. [CrossRef]
36. Martellotta, F.; D'Alba, M.; Della Crociata, S. Laboratory measurement of sound absorption of occupied pews and standing audiences. *Appl. Acoust.* **2011**, *72*, 341–349. [CrossRef]

Article

Investigation of a Tuff Stone Church in Cappadocia via Acoustical Reconstruction

Ali Haider Adeeb and Zühre Sü Gül *

Department of Architecture, Bilkent University, Ankara 06800, Turkey; haider.adeeb@bilkent.edu.tr
* Correspondence: zuhre@bilkent.edu.tr

Abstract: This study investigates the indoor acoustical characteristics of a Middle Byzantine masonry church in Cappadocia. The Bell Church is in partial ruins; therefore, archival data and the church's remains are used for its acoustical reconstruction. The study aims to formulate a methodology for a realistic simulation of the church by testing the applicability of different approaches, including field and laboratory tests. By conducting qualitative and quantitative material tests, different tuff stone samples are examined from the region. Impedance tube tests are performed on the samples from Göreme and Ürgüp to document their sound absorption performances. Previous field tests on two sites in Cappadocia are also used to compare the sound absorption performance of tuff stones, supported by acoustical simulations. The texture, physical and chemical characteristics of the stones together with the measured sound absorption coefficient values are comparatively evaluated for selecting the most suitable material to be applied in the Bell Church simulations. The church was constructed in phases and underwent architectural modifications and additions over time. The indoor acoustical environment of the church is analyzed over objective acoustical parameters of EDT, T30, C50, C80, D50, and STI for its different phases with different architectural features and functional patterns.

Keywords: acoustical reconstruction; archaeoacoustics; church acoustics; tuff stone; Cappadocia

Citation: Adeeb, A.H.; Sü Gül, Z. Investigation of a Tuff Stone Church in Cappadocia via Acoustical Reconstruction. *Acoustics* **2022**, *4*, 419–440. https://doi.org/10.3390/acoustics4020026

Academic Editors: Margarita Díaz-Andreu and Lidia Alvarez Morales

Received: 23 March 2022
Revised: 5 May 2022
Accepted: 11 May 2022
Published: 16 May 2022

Publisher's Note: MDPI stays neutral with regard to jurisdictional claims in published maps and institutional affiliations.

1. Introduction

Archaeoacoustics is an interdisciplinary field that focuses on the understanding of acoustical qualities of archaeological sites. Developments in technology have enabled researchers to document and discuss various soundscapes: acoustic environments based on functional context and perception by humans [1]. Scholars who are interested in archaeoacoustics try to gather non-material pieces of evidence from the human past, as a contextual acoustical study of a space can reveal a number of important features regarding the function of the space and the people that inhabited it. Acoustics can bring us invaluable information about intangible parts of a culture such as music, ritual and religion [2]. Consequently, it becomes relevant to enrich the academic discourse on the use of Middle Byzantine spaces in Cappadocia, a region in Central Anatolia, through an archaeoacoustic study.

Cappadocia is a region in Turkey. The region is known for its idiosyncratic volcanic landscape, and is marked by the present-day cities of Aksaray, Nevşehir, Kayseri, and Niğde. Most of the surviving rock-cut and masonry structures in the region correspond to the Middle Byzantine period, especially to the ninth to eleventh centuries [3]. The rock-cut structures of Cappadocia are listed UNESCO World Heritage Sites.

A number of historical structures have been acoustically reconstructed by means of simulations in order to evaluate the spaces' acoustical characteristics for those locations that cannot be field-tested due to restorations, limited time permissions or for in-depth sound field analysis [4–9]. For instance, the acoustics of the buildings from Greek antiquity, such as the Temple of Zeus, have been explored using computer-aided simulations and the results indicate that due to the high reverberance as well as low clarity and intelligibility of the space, the temple is more apt for the performance of hymns and ritual songs rather than speech communication [10]. The acoustic properties of Benevento Roman theatre from A.D. 2 [11] and the little theatre of Pompeii, also known as the "Odeon" [12], have been predicted using ray-tracing simulations by evaluating some of the major acoustical parameters such as EDT, T30, C80, etc. Furthermore, the Roman theatre of Posillipo in Naples has excellent speech comprehension, which has been tested through both field measurements and a virtual model [13].

Similarly, the acoustic make-up of churches has also been virtually recreated; these studies include a domus and basilica [14], the Romanesque cathedral of Santiago de Compostela [15], and the Maior Ecclesia in Cluny [16]. All these studies analyze the acoustic performance of these spaces in regard to liturgical celebrations and other religious music. Virtual aural reconstruction is also used to analyze the three-choir configurations that have been employed in the Cathedral of Granada, Spain [17]. A similar technique is used to document the soundscape of a 14th century Spanish church, which was abandoned in 1836 and is currently in ruins [18]. The effect of occupancy and festive decorations on the acoustics of a church from the Baroque Period (San Petronio Basilica) reveals that during festive seasons of the era, the church would have significantly reduced reverberation times [19].

The study is one of the first archaeoacoustic studies pursued in Cappadocia, which is a region that has always been a topic of academic discourse. With archaeoacoustic studies that would bring forward intangible aspects of the Middle Byzantine Cappadocian life, the region, its people and their lifestyles can be better understood and documented. In order to contribute to the acoustical archive of historical sites, this study aims to analyze the indoor acoustical climate of the remains of a Middle Byzantine masonry church (Bell Church) in Cappadocia by means of virtual reconstruction of its aural environment. The methodology covers ray-tracing room acoustic simulations, which are supported by previous field tests conducted in different structures of the nearby region, and impedance tube tests for determining the sound absorption characteristics of tuff rocks. The study compares the changes in the acoustical performance of the church over its two phases of construction by using available drawings of the original states. Gathering the sound absorption coefficient data of the material, which is predominantly tuff rock, is the most challenging part of this research. In this regard, field tests of other structures from the same region (Hallaç Church and Avanos Dining Hall) [20] and their simulations, as well as impedance test results of tuff stone samples from different areas of the region are compared for their further application in the church's simulations. This paper is structured as follows. Section II sets out the historical and architectural features of Cappadocia and the Bell Church. Section III gives details of the methodologies used for collecting and analyzing the data, ray-tracing model implementation and calibration, and impedance tube tests. In Section IV, the results are discussed in detail. Section V concludes the paper by emphasizing the major findings.

2. Historical and Architectural Description of Cappadocia and Bell Church (Çanlı Kilise)

Cappadocia is registered as a mixed natural and cultural heritage site by UNESCO and is located in Central Anatolia. The Cappadocian region is composed of a number of towns that are known for their characteristic rock-cut structures (Figure 1). The majority of the structures date back to the 9th and 11th centuries, when there was a peak in construction activities especially between the end of the Arab invasion in the 10th century and the gradual takeover by the Seljuks by the end of the 11th century [21] when Cappadocia remained a provincial settlement under the rule of the Byzantine Empire.

There exist a number of problems that hinder comprehensive understanding of the settlement patterns of Byzantine Cappadocia and archaeological fieldwork conducted in the region remains limited. One of the main problems is the lack of any textual evidence that comes from the region [22]. However, there is a plethora of physical study material in the form of Cappadocia's signature rock-cut structures which are studied in an interpretative way to understand their functions and the people of the time. Therefore, studies concentrating on different perspectives would contribute to a better understanding of such spaces of the Byzantine Cappadocia. One such physical character is the indoor acoustical climate of these spaces. This study is an attempt to understand the context of the use of a Middle Byzantine church, the Bell Church (Çanlı Kilise), in Cappadocia through acoustical reconstruction of the partially demolished structure based on the available data.

Figure 1. A map of the towns in Cappadocia taken from Google Maps. A similar map has been used by Öztürk [3], p. 137.

The Bell Church settlement is located on the outskirts of present-day Aksaray, Turkey. The church stands aloof from the rest of the settlement (Figure 2). According to Ousterhout, as opposed to the perspective of previous scholarship, the Bell Church region was not a monastic center; instead, the region had a significant residential and military character [23]. The church is located on a hilltop, at the edge of the settlement; and was the main church of the settlement. The church is considered unique because it is one of the few churches in the Cappadocian region that is a masonry construction as opposed to being carved out of the volcanic tuff, as is the case of the Hallaç Church [21]. Furthermore, the Bell Church is known to have undergone alterations and additions in the time span of its use; hence, it is significant to compare the indoor acoustical changes of the church over time.

The church was built in three main phases [23]. Phase I of the church only included the main structure, the naos, which is approximately 9.2 m^2, and was constructed in the early eleventh century [22]. The second phase of the church included a narthex to the south and the north, while during the third phase of the construction, a parekklesion (side chapel) was added to the church [23]. Figure 3 shows the two main phases of construction on which this study focuses.

The naos, with three apses on the east, has four central piers that supported the dome until the twentieth century. The dome holds immense importance as it was the only Middle Byzantine dome in Cappadocia that survived and could be documented in the 20th century. Although the dome does not exist today, it is known to have been approximately 16 m in height. While the drum of the dome was decorated with slit windows, it was smooth on the inside [24]. The plan of the cross-in-square church remains compact, a characteristic that is in line with the local practice. Almost all Cappadocian churches are more compact versions of examples from Constantinople [22,25].

Figure 2. (a) An exterior view of Bell Church; (b) an interior view of the main church space (naos) with three apses; (c) the entrance hallway (narthex) of the church.

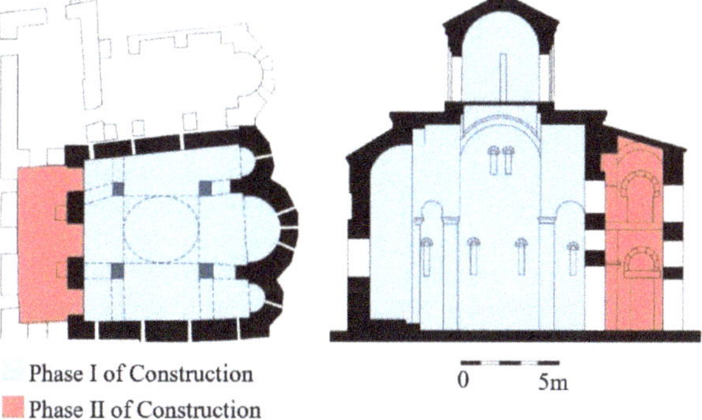

Figure 3. Plan and section drawings of the Bell Church showing its two phases of construction. Drawn by the authors based on the drawings from Ousterhout [22], pp. 92, 94.

The church's underground structures, the finely detailed facades, the faceted apses, and thick mortar beds all affirm the notion that the design of the church comes from Constantinople, and a few builders must have received their training in the capital, while the rest of the work was carried out by a local workforce [23,26]. There are some striking similarities in the design of the church with the churches of Constantinople, although there are a few aspects that make Bell Church different. Instead of using brick for the entire thickness of the walls, stone and brick have been utilized as revetments for a rubble core. The vaulting of the church is also made of stone, which reflects the local construction technique of the region [22]. Therefore, the limited use of bricks in the façade indicates the possibility of the bricks being imported in order to create a prominent landmark in Cappadocia, as a reminder of the center, Constantinople. The limited use of bricks in the arches of the slit windows on the interior wall surfaces has a trivial effect on the indoor sound field of the church.

The architectural typology of the Bell Church is a simple cross-in-square composition that was developed from a single nave after the addition of two lateral naves. This architectural composition was common in Middle Byzantine churches [27]. Similarly to the St Nicholas Monastery of Mesopotamia which was constructed in 1224 [28], the Bell Church is built on top of a hill and consists of one interior volume that has been divided by columns and arches. As opposed to four domes of the St Nicholas Monastery of Mesopotamia, the Bell Church has only one dome with a tall drum and slit windows.

The naos of a Middle Byzantine church was utilized during liturgical practice for congregational purposes. The practice would involve a congregation of people and clergy in the church [29]; some parts of the Divine Liturgy were composed of a dialogue between the two groups. Singers used to be a distinguishing part of the congregation and they had fixed places in the church; the choir would either be on an ambo, an elevated platform in the naos, or divided into two groups on either side of the center of the naos. Other than the Divine Liturgy, other regular and intermittent services were also celebrated; these services would range from baptism and marriage to simple prayers that would address a specific need of the people such as curing a sickness. The Byzantine liturgy was composed of a variety of services to praise God and saints. The rites would be expressed in speech, text, and song. Furthermore, movement was also an important part of a few rituals such as the Divine Liturgy, where the congregation would meet in the church, go around the town, and gather at the naos of the church again [29]. In order to decipher the auditory environment during liturgical practices in terms of the effects of the physical features of the space, the sound field of the main church is searched in detail within this study.

Instead of a permanent separation between the sanctuary and the naos, it is possible that a wooden screen (templon) was used during times of liturgy in the Bell Church [22], which would block any visible connection while maintaining an aural connection between the clergy and the congregation [30]. The purpose of the templon was to reinforce the sacredness of the holy space by separating the clergy from others who would just attend the services from the naos [29]. Therefore, in this study, the screen is included in the naos before carrying out acoustical simulation tests on the church in order to mimic and/or generate a more realistic reconstruction of the church's acoustic environment.

This study also documents the acoustics of the church with the additional features of Phase II, which mainly included a double-storied southern narthex with a wooden roof [23] (Figure 3). According to Ousterhout, based on the iconography of the paintings in the naos, a painter from the center of Constantinople must have been hired by the patron of the church to do the job. It is possible that the painter would have reached the area well after the completion of Phase II [22]. Therefore, based on this theory, the study includes a virtual and aural study of the Phase I construction (naos) with and without the paintings found on the interior surfaces. These paintings were made with the application of a very thin layer of plaster on the interior walls, which could be the reason behind the present-day deterioration of the paintings [23]. The study documents the acoustics of the church from both Phase I and Phase II, to evaluate the aural performance of the space over time.

3. Materials and Methods

The data collection of the study has been conducted in two main parts. The Bell Church has been virtually and aurally reconstructed using computational 3D-modeling tools based on the information from available archival data (drawings, etc.) [22–24] and on-site measurements of window slits and wall thickness. Impedance tube tests have also been performed on two rock samples from two different towns of Cappadocia. By dint of ray tracing, the church's acoustic environment has been documented and analyzed.

In order to have a more reliable acoustical analysis of the church, it is important to document the sound absorption performances of the building material. Due to the archaeological significance of the masonry blocks at the Bell Church, the blocks could not be taken from the nearby site for a material analysis. However, rock samples from three different regions of Cappadocia, Göreme, Ürgüp, and Hallaç (Figure 1), were collected. These samples are all tuff stone, which is the characteristic rock type of the region. The sample collection areas house a number of rock-cut churches. For instance, the Hallaç Complex is a mid-eleventh century complex that has almost entirely been carved out of living rock and Hallaç Church is a cross-in-square church on the east of the main courtyard of the complex [21]. Apart from material tests of the rocks from these sites, previously held field tests held in Hallaç Church and Avanos [20] are used to tune the acoustical models of selected structures. The field test tuned acoustical simulations aid to identify the sound absorption performance of tuff rocks.

3.1. Impedance Measurements

For the samples from Göreme and Ürgüp, impedance tube (S.C.S. Kundt Impedance Tubes) tests are performed (Figure 4). Kundt / TL Tubes (S.C.S) measure absorption coefficient, standard impedance, characteristic impedance and transmission loss as a function of frequency. A two-microphone-transfer function method is applied, according to ISO 10534-2:1998 [31]. The double-tube setup of the sound absorption measurements includes small and large tubes. The small-tube setup is composed of small-sized (Ø28 mm) devices for measurements of acoustic properties in the high frequency range (800 Hz to 6300 Hz), while the large-tube setup is made up of larger tubes (Ø100 mm) and is used for measurements in the low frequency range (50 Hz to 1200 Hz).

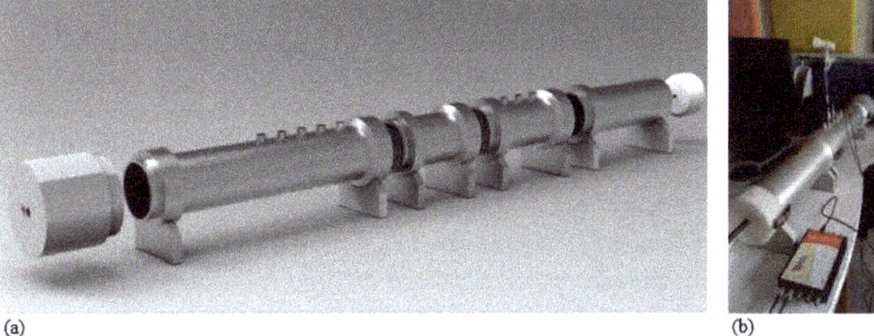

(a)   (b)

Figure 4. (**a**) A diagrammatic view of Kundt Impedance Tubes; (**b**) the Kundt Tube setup used in the study.

The Kundt Tube method has been proven to be an efficient and faster method to calculate absorption coefficients of sound-absorbing materials than both Standing-Wave-Ratio (SWR) and reverberant room methods [32,33]. Another reason to use the Kundt Tube method is the fact that large samples cannot be brought into the reverberation chambers for sound absorption coefficient measurements.

For this study, two types of samples are prepared for each rock, collected from three different areas of the region. For the low-frequency measurements (100 mm samples), rings of cardboard have been stacked to create a mold to hold the tuff stones. The corners of the mold are filled with the same tuff stone to reduce air gaps. For the high-frequency samples (28 mm samples), a steel mold has been filled in with the rock tuff. Since the rock samples are porous, an acoustically transparent piece of cloth has been used on both ends of the mold to ensure that the material stays compact (Figure 5). For the low-frequency setup, the rock samples have been cut to fit in the mold, just like a masonry block. Therefore, the tuff is primarily in the shape that was used in the construction of the rock-cut spaces. However, for the high-frequency setup, fragments of the tuff stone have been put together to fill the mold. The method is considered appropriate due to the presence of air spaces in the porous rock materials. The inability to clearly cut the rock samples for the high-frequency setup is a limitation of this study. Each sample, for both low- and high-frequency setups, has been tested in the tube 10 times to ensure the repeatability and accuracy of the results.

Figure 5. Göreme rock sample (**a**) for low-frequency setup; (**b**) for high-frequency setup; Ürgüp rock sample (**c**) for low-frequency setup; (**d**) for high-frequency setup.

The rock sample from Hallaç Complex has also been brought to study its acoustical properties (Figure 6). The sample cannot be used for impedance tests as it could not be cut in the proper shape to fit into the tube due to the hardness of the material. As Figure 6 shows, the rock sample cannot be cut using the same cutter that has been used for the rock samples from Göreme and Ürgüp. Accordingly, there are basically two types of tuff stones which are different due to their changing porosity and stiffness. This may be due to their exposure to environmental factors as well as different chemical compositions. Even based on these cutting and shaping experience of the rock samples, it can be said that the tuff stone samples from different areas of Cappadocia have different physical properties due to their different compositions or exposure to environmental factors. Hence, it would be misleading to assume that all the tuff rocks from Cappadocia have similar acoustical properties, and thus, it is important to document different rock samples and their absorption coefficients from different locations. In order to better characterize the samples, Qualitative and Quantitative Mineral Tests were performed on all three samples. These tests were carried out at the General Directorate of Mineral Research and Exploration (MTA) in Ankara. The test codes are 35-30-MP-19 (Standard Qualitative Mineral Tests under XRD Analysis) and 35-30-AJ-31 (XRF analysis) in accordance with TS EN ISO/IEC 17025 Standard. The results of the tests are presented in tables under Section 4.1.1.

3.2. Acoustical Simulations

With the current advancements in computer-aided tools to create virtual 3D spaces, it has become more convenient to study the acoustics of spaces that are otherwise difficult or impossible to study. The Bell Church is one such site; the remains of the Middle Byzantine masonry church include only the four walls that encapsulate the main hall of the church. Initially, an acoustical model with all the main volumetric arrangements (Figure 5) of the church was generated.

The acoustical simulations are held by ODEON Room Acoustics Software version 16.08. The program uses a hybrid calculation system that makes use of both the image-source method and ray tracing [34].

Figure 6. Rock sample from Hallaç Complex, which could not be cut to be used for impedance tube testing.

The interior surfaces of the church are dominantly tuff stone, according to the information gathered from both archival data and on-site observations. For estimating the sound absorption performance of tuff rocks in order to support or compare impedance test results, acoustical simulations are utilized. Field test results from two sites in Cappadocia have been included in this study. As part of a previous study, field measurements have already been held in the Hallaç Church and Avanos Dining Hall [20]. The full set of measurements are held by using a B&K (Type 4292-L) standard dodecahedron omni-power sound source, a B&K (Type 2734-A) power amplifier, a B&K (Type 4190ZC-0032) microphone combined with a B&K (Type 2250-A) hand-held analyzer, and a portable PC. The sampling frequency of the recorded multi-spectrum impulse is 48 kHz, covering the interval of interest between 100 and 8000 Hz. The impulse response length is kept at 10 s. Noise signals are generated using DIRAC Room Acoustics Software Type 7841 v.4.1. The same software is also used for post-processing of the measured impulse responses for all receiver positions. Three types of signals are tested: E-sweep, MLS, and balloon pop.

3.2.1. Simulations Based on Field Tests

A simplified virtual model of the Hallaç Church (estimated volume: 595 m^3) has been created based on the plan drawing by Ousterhout [22] and the geometrical measurements taken on the site. ODEON Room Acoustics Software version 16.08 has been used for acoustic analysis. The transition order is set at 2, the room impulse response length is kept at 5000 ms, with a resolution of 5.0 ms, and the number of late rays is set at 7736. Figure 7 shows images of the church's ray tracing model and a 3D Open GL view. The same process is repeated for the Avanos Dining Hall (Figure 8). A simplified acoustical model of the hall is generated based on the three-dimensional measurements taken on the site (estimated volume: 123 m^3). The calculation setup of the acoustical model is as follows: the transition order is kept at 2, the room impulse response length is kept at 2000 ms, with a resolution of 2.0 ms, and the number of late rays is set at 1898.

Acoustical parameters and specifically T30 values from the Hallaç Church and Avanos Dining Hall have been tuned with the field test results in their respective simulations with a maximum deviation of 1 just-noticeable-difference (JND), which corresponds to 5% of the value, in order to obtain sound absorption coefficients of the surrounding tuff stone surfaces [35]. JND is a measure of the smallest perceivable difference of a given parameter, and is used to accurately correlate subjective sound perception to objective measurements [36,37]. After tuning, sound absorption coefficients of the rock materials are noted over octave bands for the Hallaç Church and Avanos Dining Hall. The tuff stone is the dominating material in these spaces; the other materials, such as marble for flooring and wood for the roof of the narthex, occupy comparatively less surface area.

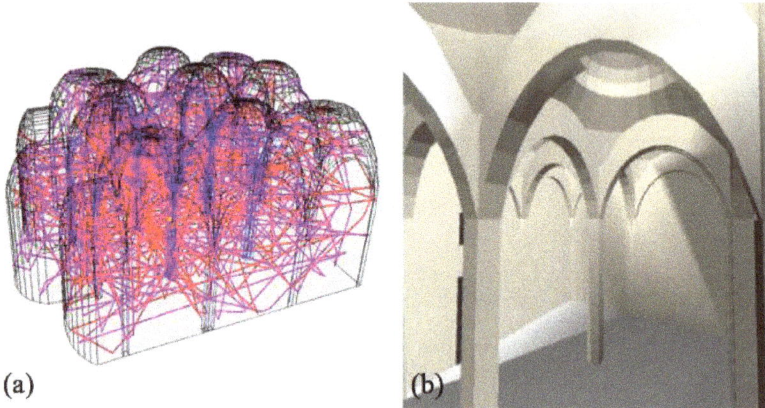

Figure 7. Hallaç Church (**a**) ray tracing model with source (red) and receiver (blue) positions; (**b**) 3D Open GL of an interior view.

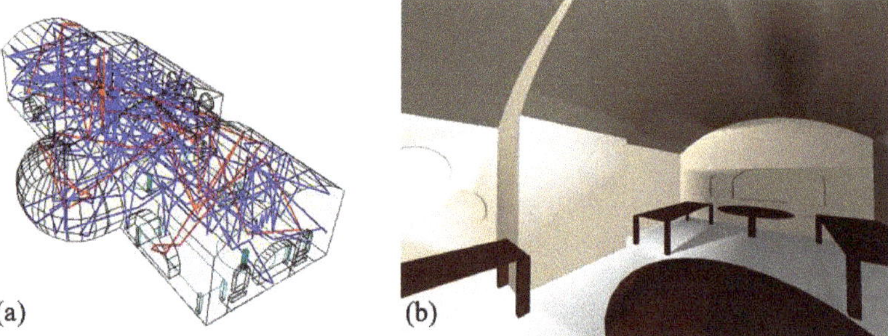

Figure 8. Avanos Dining Hall (**a**) ray tracing model with source (red) and receiver (blue) positions; (**b**) 3D Open GL of an interior view.

3.2.2. Acoustical Simulation Setup of Bell Church

The architectural drawings (Figure 3) and dimensions recorded at the site have been used to create simplified computational 3D models of the Bell Church. The simulations are run on three different models of the Bell Church; Phase I (without frescoes), Phase I (with frescoes), and Phase II (with frescoes and narthex). In the simulations of the church, the transition order is set at 2, the room impulse response length is kept at 4000 ms, with a resolution of 5.0 ms, and the number of late rays is set at 2000.

For the Bell Church, the same materials are used for the floor as in the model of the Hallaç Church. Furthermore, the Hallaç Church material (tuff stone type I) is used for the walls and the dome of Bell Church. Based on the impedance tests and material tuning through simulation, the materials presented a significantly large difference between their absorption coefficients. Therefore, the sound absorption properties of the tuff stone type I were used because of the material's physical texture, which is similar to the Bell Church's masonry blocks. Moreover, the rock samples from Göreme and Ürgüp are more porous than the masonry blocks of the Bell Church, which is why they have not been employed to the church's acoustical model. The choice of the material used for the Bell Church in the simulation is further discussed in the next section.

In the simulations of Bell Church Phase I (estimated volume: 975 m^3), a wooden screen is added between the apses and the naos to represent an iconostasis [23]. Therefore, the sound source is placed behind the screen to recreate the acoustical environment during liturgical practices. Figures 9a and 10a show a ray tracing model and a 3D OpenGL view of the church from Phase I.

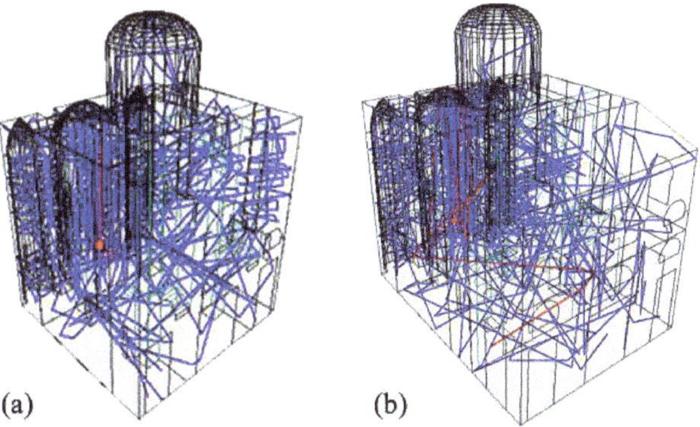

Figure 9. Ray tracing model of Bell Church from (**a**) Phase I; (**b**) Phase II.

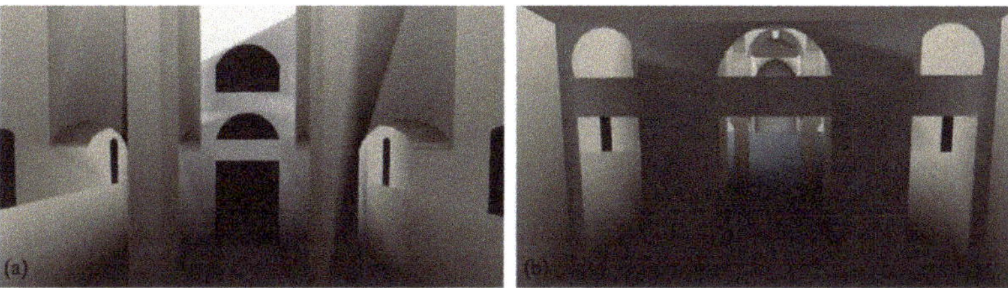

Figure 10. 3D Open GL of an interior view of Bell Church from (**a**) Phase I; (**b**) Phase II.

For Phase II of the model, a two-storied narthex, with a wooden roof, is added to the main church space. The estimated volume of the Bell Church Phase II model is 1200 m^3. Figures 9b and 10b present a ray tracing model and a 3D OpenGL view of the church from Phase II.

Distribution maps are visual representations of the quantitative probability of an acoustical parameter in grid form. Therefore, distribution maps of the church are also analyzed (Section 4.2) in order to see the overall distribution of acoustic quality in the space. In order to obtain distribution maps for Bell Church simulations, the floor of the naos and narthex (if present) are divided into a grid of 80 cm to 80 cm, with a height of 1.5 m to correspond to the average height of human ears.

4. Results and Discussions
4.1. Acoustical Material Analysis
4.1.1. Impedance Test Results

As detailed in Section 3.1, tuff stone samples from Göreme and Ürgüp have been collected for further analysis of their sound absorption performance. These samples can be different from other samples collected from the same region based on their different composition. However, as part of this study, it is an attempt to document the characteristic tuff stone examples from the Cappadocia region. Most of the tuff stone samples from Cappadocia are comparatively soft and porous, but the sample from Hallaç (Figure 6) is stiff and could not be even cut properly to be tested in the impedance tube setup. Therefore, qualitative and quantitative mineral tests are conducted on the three samples under study to have a better understanding of the physical makeup of the samples. Table 1 shows the results of the qualitative test, while the results of the quantitative test are summarized in Table 2.

Table 1. Results of the Qualitative Mineral Test on the tuff stone samples from Göreme, Ürgüp, and Hallaç.

Sample	Mineral Composition
Göreme	Gypsum, kaolinite, clay, calcite, quartz, plagioclase, cristobalite, alkali feldspar, mica, zeolite, amphibole
Ürgüp	Plagioclase, clay, quartz, cristobalite, dolomite, zeolite, alkali feldspar, mica, calcite
Hallaç	Quartz, plagioclase, clay, mica, alkali feldspar, cristobalite, tridymite, amphibole (little traces)

Table 2. Results of the Quantitative Mineral Test on the tuff stone samples from Göreme, Ürgüp, and Hallaç.

Sample	Chemical Compounds (%)										
	A.Za	Al_2O_3	CaO	Fe_2O_3	K_2O	MgO	MnO	Na_2O	P_2O_5	SiO_2	TiO_2
Göreme	8.9	10.0	2.2	1.2	3.2	0.2	<0.1	0.7	0.1	70.9	0.1
Ürgüp	6.7	11.3	1.3	3.0	3.9	0.5	<0.1	1.0	0.1	71.7	0.1
Hallaç	4.2	10.9	2.0	2.8	3.5	0.3	0.1	1.6	<0.1	74.2	0.2

Based on the results of the qualitative mineral test, the three tuff stone samples are mainly composed of the same minerals such as plagioclase, quartz, mica, etc. However, there are some differences in these tuff stones. For example, tridymite is only found in the sample of Hallaç. The difference in the physical properties of the samples could be a result of the quantitative mineral makeup of the tuff stones. From a chemical point of view, the samples show comparable differences in the percentages of chemical compounds such as Na_2O, SiO_2, and A.Za in the sample from Hallaç (Table 2). The quantitative differences of chemical compounds in the samples can account for the different textures, porosity, and hardness of the tuff stone samples.

As a result, the two samples from Göreme and Ürgüp are much more porous and softer than the tuff stone from Hallaç, and hence, as mentioned in Section 3.1, impedance tube tests could be performed on the tuff stone samples from Göreme and Ürgüp. The results of the test are discussed as follows.

Figure 11 shows the sound absorption coefficients of the tested two rock samples over the octave bands from 125 Hz to 4000 Hz. According to the results, the absorption coefficient values of the samples are very close to each other, except for 500 Hz, where the absorption coefficient of Ürgüp's rock sample (0.48) is slightly higher than that of Göreme (0.39). Maximum sound absorption (0.49) is achieved by the sample from Göreme at 4000 Hz. These results indicate that the collected and measured rock samples (not the hard sample that cannot be measured) are highly absorptive (especially with absorption coefficient values reaching up to 0.50) and can be categorized as absorptive materials under today's diverse sound-absorptive material umbrella, which include both natural and man-made materials. This information is shared for future research on tuff stone and its potential application in architectural acoustics. However, the impedance tube results could not be applied to the Bell Church, as the porous samples do not reflect the current tuff in the Bell Church. For this reason, rather than the tube results, field-tuned acoustical simulation absorption data of the hard tuff stone (tuff stone type I), obtained from the Hallaç Church, are applied to the acoustical models of Bell Church.

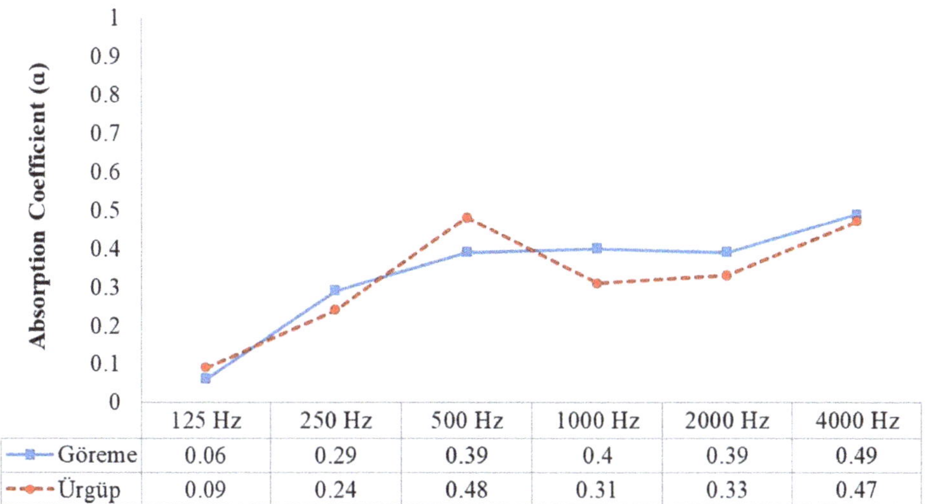

	125 Hz	250 Hz	500 Hz	1000 Hz	2000 Hz	4000 Hz
Göreme	0.06	0.29	0.39	0.4	0.39	0.49
Ürgüp	0.09	0.24	0.48	0.31	0.33	0.47

Figure 11. Absorption coefficient (α) values over 1/1 octave bands of the rocks from Göreme and Ürgüp. Maximum sound absorption (0.49) is achieved by the sample from Göreme at 4000 Hz.

4.1.2. Acoustical Simulation Results of the Hallaç Church and Avanos Dining Hall

The main reason for generating acoustical models of the Hallaç Church and Avanos Dining Hall is to gather information regarding the tuff rocks' acoustical performance. It is important because even though all the spaces are composed of volcanic tuff, the make-up of Cappadocian rocks from different regions is generally different based on varying rock ingredients and physical conditions. In this regard, the test results obtained from MTA (Tables 1 and 2) prove that hypothesis. The tube results of the tested group of rocks, which are much softer and more porous in comparison to the Hallaç stone sample, have shown very high values of sound absorption. On the other hand, the current rocks of Hallaç Church (Figure 6), in texture and hardness, are much similar to the existing stones of Bell Church. Thus, a group of simulation studies is necessary for the calibration of the Bell Church's acoustical model.

The tuning procedure is explained under Section 3.2. After adjusting the absorption coefficients of the walls and the ceiling of the church, T30 values close to (with less than 5% difference) the in-situ measurements are obtained (Figure 12). Accordingly, the attained sound absorption coefficients of the Hallaç Church's interior tuff rock surfaces are listed in Table 3.

A similar tuning process is held for the Avanos Dining Hall. The sound absorption coefficient values over octave bands for the rock-cut tuff stone walls and ceiling are tuned according to the T30 results of the site measurements. After the absorption coefficients are adjusted to obtain T30 values for the model close to 1 JND (with less than 5% difference) to the field measurements, the tuning process is completed. The results are shown in Figure 12. The figure shows that the T30 values of both the field test and the simulated model are close to each other along the frequency spectrum, except for values at 125 Hz. Even when the absorption coefficient is kept at a minimum (0.01), the T30 value of the simulated model remains much shorter than the field test result. The additional energy obtained at 125 Hz in the field test could be due to the presence of neighboring galleries around the dining hall. The estimated absorption coefficients of the tuff material (tuff stone type II) around the walls and the ceiling vault of the Avanos Dining Hall are shown in Table 4.

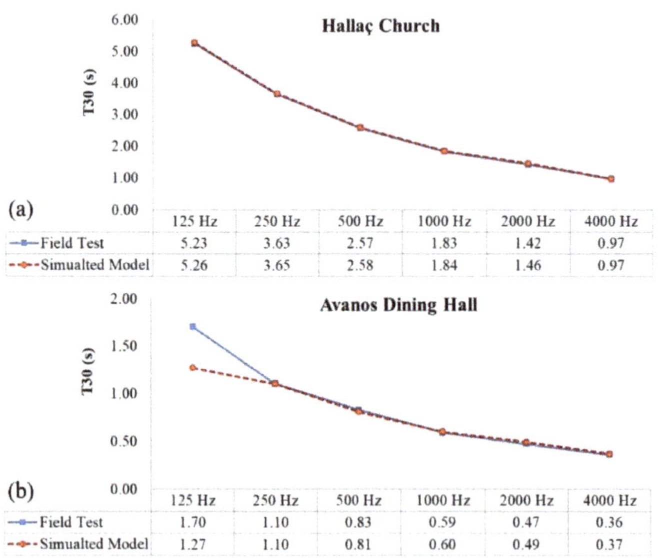

	125 Hz	250 Hz	500 Hz	1000 Hz	2000 Hz	4000 Hz
Field Test	5.23	3.63	2.57	1.83	1.42	0.97
Simualted Model	5.26	3.65	2.58	1.84	1.46	0.97

	125 Hz	250 Hz	500 Hz	1000 Hz	2000 Hz	4000 Hz
Field Test	1.70	1.10	0.83	0.59	0.47	0.36
Simualted Model	1.27	1.10	0.81	0.60	0.49	0.37

Figure 12. T30 values over 1/1 octave bands of the field test and simulated model of (**a**) the Hallaç Church; (**b**) the Avanos Dining Hall. T30 values of both the field test and the simulated model are close to each other along the frequency spectrum.

Table 3. Absorption coefficients (α) over 1/1 octave bands of the materials applied to the Hallaç Church acoustical model.

Materials/Locations	Frequency (Hz)					
	125	250	500	1000	2000	4000
Stone (marble)/Floor	0.01	0.01	0.01	0.02	0.02	0.02
100% open to outdoor	1.00	1.00	1.00	1.00	1.00	1.00
Hallaç church (tuff stone type I)/Walls & ceiling	0.03	0.05	0.07	0.09	0.11	0.15

Table 4. Absorption coefficients (α) over 1/1 octave bands of the materials applied to the Avanos Dining Hall acoustical model.

Materials/Locations	Frequency (Hz)					
	125	250	500	1000	2000	4000
Stone (marble)/Floor	0.01	0.01	0.01	0.02	0.02	0.02
50% Absorbent (open door to kitchen)	0.50	0.50	0.50	0.50	0.50	0.50
Avanos dining hall (tuff stone type II)/Walls & ceiling vault	0.01	0.05	0.07	0.13	0.17	0.24

4.1.3. Comparison of Sound Absorption Coefficients Obtained by Simulations and Impedance Tests

By means of impedance measurements and acoustical simulations, four materials from different parts of Cappadocia (Göreme, Ürgüp, Hallaç Church and Avanos Dining Hall) and their absorption coefficients have been documented. It is essential to compare the data regarding these materials in order to see the similarities and differences between their sound absorption performances, as shown in Figure 13. It should also be noted that impedance tube results are not random incidence measurements, but instead rely on plane wave propagation; plane wave reflection coefficients can only be estimated. Although this is a limitation, the tube method is still one of the few techniques to be used in experimental analysis when it is not possible to collect the large samples for further reverberation chamber tests.

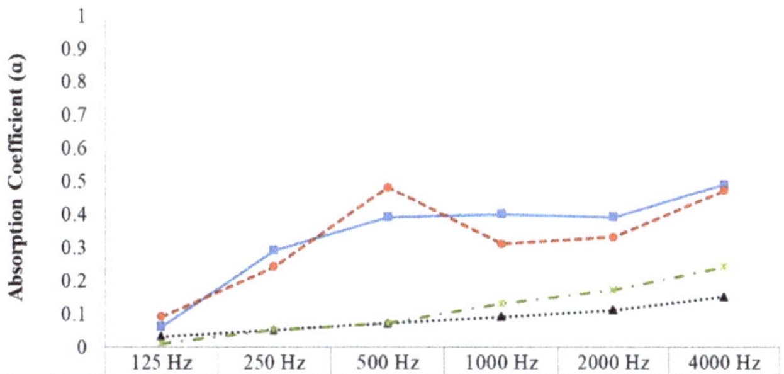

	125 Hz	250 Hz	500 Hz	1000 Hz	2000 Hz	4000 Hz
Göreme-tube	0.06	0.29	0.39	0.4	0.39	0.49
Ürgüp-tube	0.09	0.24	0.48	0.31	0.33	0.47
Hallaç-simulation	0.03	0.05	0.07	0.09	0.11	0.15
Avanos-simulation	0.01	0.05	0.07	0.13	0.17	0.24

Figure 13. Absorption coefficients (α) of the materials over 1/1 octave bands from the Hallaç Church, Avanos Dining Hall, Göreme, and Ürgüp tuff stones. Göreme has the highest sound absorption, while the Hallaç Church seems to be the most reflective sample.

According to Figure 13, among all the materials under study, Göreme has the highest sound absorption, while the Hallaç Church seems to be the most reflective. The sound absorption values of the samples studied with the impedance tests (Göreme and Ürgüp) have an overall higher sound absorption than the materials obtained from the tuning of virtual models (Hallaç Church and Avanos Dining Hall).

The higher sound absorption performance of the rock samples from Göreme and Ürgüp is initially thought to be due to the fact that they are untouched rocks with higher porosity. On the other hand, the tuff stones of the Hallaç Church and Avanos Dining Hall are in their current state after the tuff around them had been carved off and the final surface smoothened. Moreover, over time, the rock materials in these spaces have been in contact with air, which could have led to the thickening of the surfaces, and hence resulted in their increased reflectivity.

Originally, the masonry church was constructed of hard gray tuff, set above an ash layer, which is different than the softer purple tuff which was carved to make the rock-cut spaces around the neighboring settlement [38]. There are basically two types of rock samples that have been evaluated: soft and hard. The difference in the hardness of these rock samples is due to their different mineral compositions. Therefore, due to a lack of data regarding the hard gray tuff found around the Bell Church settlement, the tuff stone type I (from the Hallaç Church) is used for the surrounding walls and the dome of the masonry church for its aural reconstruction. The tuff stone type I is chosen for its similar texture and hardness to the masonry blocks of the Bell Church.

4.2. Acoustical Parameter Results and Comparisons of the Different Phases of Bell Church

The Bell Church, which in its current condition cannot even be assessed by field tests, can best be understood aurally through the utilization of virtual reconstruction and acoustical simulations. The common acoustical parameters, as previously utilized in other archaeoacoustic studies, are also used in this study to assess the acoustics of the Bell Church.

As mentioned before, the Bell Church was constructed in phases. This study documents the acoustics of three likely states of the church: Phase I (without frescoes), Phase I (with frescoes), and Phase II (with frescoes and narthex).

For Phase I (without frescoes), the tuff stone type I has been used (Table 3). For Phase I (with frescoes), the same material has been used for the walls without any traces of frescoes. In order to account for the frescoes, the sound absorption coefficients of plaster of the Hagia

Sophia [8] are applied in the simulations. Although the Hagia Sophia and the Bell Church are from different time periods, the building stands as the closest example from Anatolia. The main frescoes in the Bell Church are found over the dome of the church, on the three apses, and the eastern wall. For the iconostasis of the church, a wooden material has been used. The same material is also used for the roof of the narthex, which is added in Phase II of the church. The sound absorption coefficient values of the applied materials in the acoustical models of the Bell Church are summarized in Table 5.

Table 5. Absorption coefficients (α) over 1/1 octave bands of the materials applied to Bell Church acoustical model.

Materials/Locations	Frequency Band (Hz)					
	125	250	500	1000	2000	4000
Stone (marble)/Floor	0.01	0.01	0.01	0.02	0.02	0.02
100% open to outdoor	1.00	1.00	1.00	1.00	1.00	1.00
Tuff stone type I/Walls & ceiling	0.03	0.05	0.07	0.09	0.11	0.15
Fresco paintings	0.13	0.09	0.07	0.05	0.03	0.04
Wooden screen & narthex roof	0.15	0.20	0.10	0.10	0.10	0.10

Early decay time (EDT), reverberation time (T30), clarity (C80), definition (D50), and speech intelligibility (STI) values are estimated from the room impulse responses collected from the simulations run on the three models representing three potential phases of the Bell Church.

Figure 14 displays EDT and T30 values of the three states of the church. For Bell Church Phase I (estimated volume: 975 m^3), an optimum range for liturgical music has been estimated by applying RT = 0.55 × log10(Vol.) − 0.14 [39]. Accordingly, T30 of 1.5 s over the mid-range frequencies is optimal for liturgical performances.

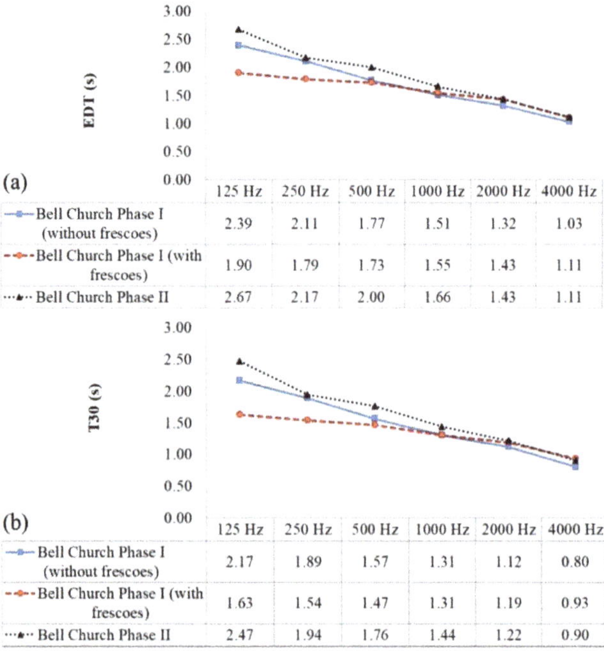

(a)

	125 Hz	250 Hz	500 Hz	1000 Hz	2000 Hz	4000 Hz
Bell Church Phase I (without frescoes)	2.39	2.11	1.77	1.51	1.32	1.03
Bell Church Phase I (with frescoes)	1.90	1.79	1.73	1.55	1.43	1.11
Bell Church Phase II	2.67	2.17	2.00	1.66	1.43	1.11

(b)

	125 Hz	250 Hz	500 Hz	1000 Hz	2000 Hz	4000 Hz
Bell Church Phase I (without frescoes)	2.17	1.89	1.57	1.31	1.12	0.80
Bell Church Phase I (with frescoes)	1.63	1.54	1.47	1.31	1.19	0.93
Bell Church Phase II	2.47	1.94	1.76	1.44	1.22	0.90

Figure 14. (**a**) EDT; (**b**) T30 over 1/1 octave bands for Bell Church acoustical models. The highest EDT of 2.42 s and a T30 of 1.60 s at the mid-range frequencies is observed in Phase II of Bell Church.

The T30 value for Bell Church Phase I (without frescoes) is around 1.44 s over the mid-range frequencies (Figure 14). Based on this result, the church satisfies the average value for liturgical music (1.50 s) which would include both speech and music. For the same phase of the church, an EDT of 1.64 s for the mid frequency range is observed. The bass ratio (BR) a ratio of the reverberation times of low frequencies over middling frequencies. The BR for the church is around 1.4. The high BR, as in this case, is a means of augmenting male voices [9], which should have been an important part of the liturgical practices at a Medieval Byzantine church.

The second condition of the Phase I Bell Church (with frescoes) Is found to have an EDT of approximately 1.64 s, and a T30 of 1.39 s over the mid-range frequencies (Figure 14). The church remains suitable for practices related to liturgy with the addition of plastered frescoes in the naos. The BR of the space drops to 1.1, which is still a high BR indicating the bass frequency augmentation.

The Phase II of Bell Church includes the addition of a narthex and the frescoes in both the naos and the narthex. The estimated volume of the Phase II church is approximately 1200 m^3. For this volume, an optimum RT for liturgical music is 1.55 s for mid frequencies. Figure 14 shows that the church has an EDT of 2.42 s and a T30 of 1.60 s at the mid-range frequencies. The BR of the church is approximately 1.4 s. Although the church in its Phase II favors liturgical practices, it has a much higher EDT than Phase I, which means that the perceived reverberance of the church has increased due to the addition of the narthex. The Phase II version of the church has become more alive.

Overall, the EDT, T20, and T30 values are close to each other over the whole frequency range. In most cases, the difference between EDT and RT values is less than 10%. This signifies that there is an even distribution of sound within the church space in all three stages of construction and fresco decoration. The T30 of Bell Church Phase II (1.60 s) is higher than the other two phases (Figure 14). This is due to the addition of the frescoes. The plastered surfaces in the church are more reflective compared to the tuff stone type I applied to the church over the mid- and high-range frequencies. Furthermore, the increase in T30 is also due to the additional volume of the narthex. With the new additions to the Bell Church the interior space has become more reverberant, but in all cases the decay rates are high enough and consistent with the optimums defined in the literature of church acoustics.

The EDT value of St Nicholas's church is around 2.2 s at 1000 Hz [27]. Bell Church Phase II has a very similar EDT with 2.4 s over the mid-range frequencies. The higher EDT values are a result of larger volumes and the presence of frescoes in these churches.

The T30 at 1000 Hz distribution map (Figure 15) of the three states of the Bell Church shows that there is a fairly uniform distribution of the T30 values that are predicted over the entire floor of the church as presented by the uniform distribution of the acoustical parameter value. Bell Church Phase I (with frescoes) has the lowest T30 over the mid-range frequencies (around 1.39 s). On the other hand, Bell Church Phase II shows the highest value of T30 in the whole congregational space, with a value of approximately 1.60 s.

C50, C80 and D50 are ratios of early sound energy to late sound energy (50, 80, and 50 ms respectively), where C80 is directly correlated with music while C50 and D50 are used to analyze speech [40]. The parameters are dependent on the location of source and receiver. The optimum C50 values for speech-related activities are in the range of -2dB and $+2$ dB [41]. The optimum C80 values for liturgical music should be in between -2 dB and $+3$ dB for locations in the space that are closer to the source, whereas for larger distances, the C80 values can be in the optimum range of 0 to $+5$ dB [42]. The Bell Church is a comparatively small space; consequently, the optimum range of C80 for nearby locations (-2 dB and $+3$ dB) can be utilized. On the other hand, for speech-related activities, the optimum value of D50 should be higher than 0.15 [43].

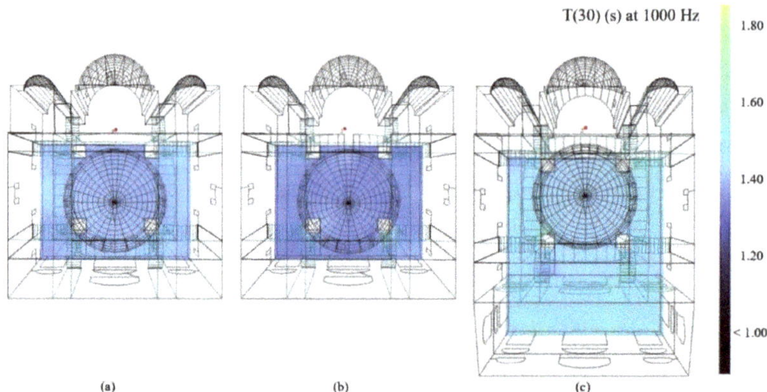

Figure 15. T30 (1000 Hz) distribution maps of (**a**) Bell Church Phase I (without frescoes); (**b**) Bell Church Phase I (with frescoes); (**c**) Bell Church Phase II. There is a fairly uniform distribution of the T30 values over the entire floor of the church.

Negative values of C50 are observed during all three states of the church (Figure 16). Bell Church Phase I (without frescoes) and Phase I (with frescoes) have the same values of C50 over the whole frequency range, which signifies that the addition of frescoes has no or minimal effect on the clarity of speech in the space. With the addition of volume, the church seems to experience lower C50 values over the entire spectrum. However, the C50 values are not in the optimum range of −2 dB to +2dB over the whole frequency range. This signifies poor condition of verbal clarity. A similar trend of negative C50 values is observed in other Byzantine churches due to late reflections, which are caused by architectural features such as vaults and domes [27]. The same reasoning could be applied for the negative C50 values of Bell Church in all its three different states. Furthermore, the presence of a wooden screen as a barrier between the naos and the apse also results in negative values of C50.

For Bell Church Phase I (without frescoes), the simulation results present negative values of C80 over the whole frequency range. The negative values of C80 indicate the absence of crisp acoustics in the space. The church is found to have an approximate C80 of −4.81 in the mid frequency range (Figure 16). The values of C80 increase with increasing octaves; however, they do not fall in the optimum range (−2 dB and +3 dB), except at 4000 Hz where C80 is −1.20 dB. For D50, the results indicate values lower than 0.15. This could be accounted by the presence of the wooden screen, which blocks early energy. Moreover, the low D50 values are also due to the high reverberation in the space. The D50 has a mean value of 0.08 at mid-range octave bands.

For Bell Church Phase I (with frescoes), negative values of C80 over the whole frequency range are observed. With the addition of frescoes, the church is found to have an approximate C80 of −4.94 dB (Figure 16). The same effect of increasing C80 values with increasing octaves is also observed. Figure 15 also shows low values of D50, with a mean value of 0.09 at mid-range frequencies.

For Bell Church Phase II, the mean C80 value is around −5.40 dB. The decrease in clarity over the different phases of the church is due to the increase in reverberation times. Similarly to the results from the earlier phases of the church, the D50 values also remain lower than 0.15 for the assessed octave bands (Figure 16). At the mid-range octave bands, the mean D50 value is 0.07. D50 values decrease with an increase in reverberation in the three versions of the church.

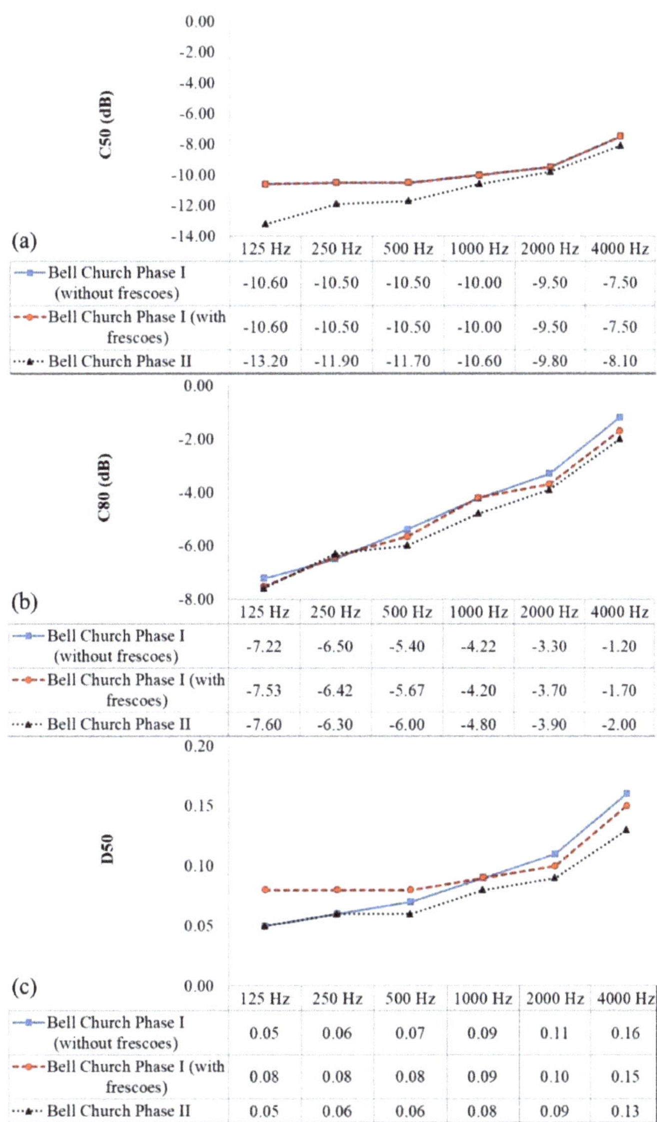

(a)

	125 Hz	250 Hz	500 Hz	1000 Hz	2000 Hz	4000 Hz
Bell Church Phase I (without frescoes)	-10.60	-10.50	-10.50	-10.00	-9.50	-7.50
Bell Church Phase I (with frescoes)	-10.60	-10.50	-10.50	-10.00	-9.50	-7.50
Bell Church Phase II	-13.20	-11.90	-11.70	-10.60	-9.80	-8.10

(b)

	125 Hz	250 Hz	500 Hz	1000 Hz	2000 Hz	4000 Hz
Bell Church Phase I (without frescoes)	-7.22	-6.50	-5.40	-4.22	-3.30	-1.20
Bell Church Phase I (with frescoes)	-7.53	-6.42	-5.67	-4.20	-3.70	-1.70
Bell Church Phase II	-7.60	-6.30	-6.00	-4.80	-3.90	-2.00

(c)

	125 Hz	250 Hz	500 Hz	1000 Hz	2000 Hz	4000 Hz
Bell Church Phase I (without frescoes)	0.05	0.06	0.07	0.09	0.11	0.16
Bell Church Phase I (with frescoes)	0.08	0.08	0.08	0.09	0.10	0.15
Bell Church Phase II	0.05	0.06	0.06	0.08	0.09	0.13

Figure 16. (**a**) C50; (**b**) C80; (**c**) D50 obtained over 1/1 octave bands for Bell Church Phase I & II. Negative C50 values are observed in all phases of the church. The values of C80 increase with increasing octaves, with Phase II showing a mean C80 value of around −5.40 dB. The low D50 values observed are due to high reverberation in the space.

Generally, the energy ratio results show C50, C80 and D50 values lower than the optimum range in all phases of the Bell Church, mainly due to the blockage (barrier effect) provided by a wooden screen between the sound source and the receiver, as would be the practice during a liturgical ceremony in the church. Generally, optimum values of C80 indicate the ease of noticing differences in musical details of different instruments. However, it appears that the distinctness of sound is less audible in the Bell Church. It may be possible that the blurriness of sound could have created a more spiritual environment for the people visiting the church. It is certain that the invisible sound source and its audible effect is much different than the effect in the Hallaç Church [20] or any other church that does not have a templon separating the priest from the congregation. All the phases of the church have very similar D50 values. The low values of D50 indicate that the church is not the most suitable space for speech-related activities, which is reasonable as the church would mainly be used for congregational prayers and ceremonies where both speech and music must have been employed. The quality of speech in the church is further discussed in regard to STI scores.

Speech intelligibility can be evaluated by assessment with the Speech Transmission Index (STI), which is a metric that denotes the quality of speech while considering the loss of speech articulation caused by reverberation [44]. Figure 17 displays average male and female STI results of the three states of Bell Church under study. The differences between the male and female STI values are considerably small in all three states of the church. The STI values seem to drop with the addition of plaster in the church space, as the reverberation in the space increases. The lowest STI value is 0.49 (STI male, Bell Church Phase II), whereas the highest value is 0.53 (STI female, Bell Church Phase I, without frescoes). STI values between 0.45 and 0.60 are fair for speech intelligibility [39]. Hence, Bell Church Phase I and Phase II results indicate that the space is only fairly good for speech intelligibility, which is also supported by the church's D50 values. Similar results in relation to speech intelligibility are obtained in other Byzantine churches due to the distortion of the speech signal in high-reverberation areas [27].

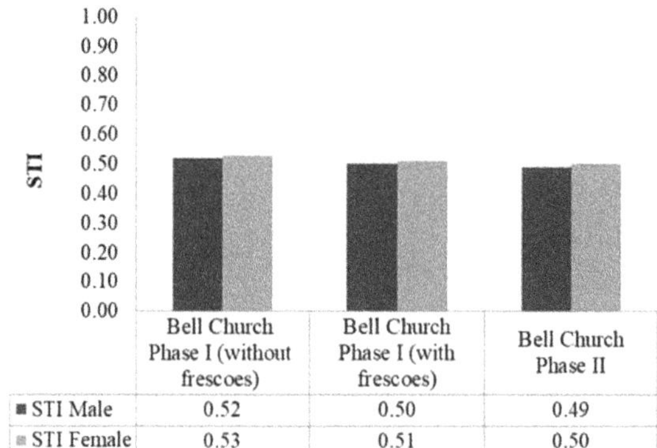

	Bell Church Phase I (without frescoes)	Bell Church Phase I (with frescoes)	Bell Church Phase II
STI Male	0.52	0.50	0.49
STI Female	0.53	0.51	0.50

Figure 17. Male and female STIs obtained for Phase I & II of Bell Church. The lowest STI value is 0.49 (STI male, Bell Church Phase II), whereas the highest value is 0.53 (STI female, Bell Church Phase I, without frescoes).

5. Conclusions

Table 3 values to field measurements. After the process is completed, sound absorption coefficients for the tuff stones in the Hallaç Church and the Avanos Dining Hall are obtained. The absorption coefficients of the material from these spaces are found to be much more reflective than those of the tuff stone samples collected from Göreme and Ürgüp, which have been tested in the tube. After comparing the results of tube tests and tuning of materials through simulations, the tuff stone of the Hallaç Church seems to be the most plausible option to be used in the simulation of Bell Church because of the material's physical similarity to the masonry blocks of the church in terms of texture and hardness. Hence, in ray-tracing simulations, the absorption coefficient data obtained from the Hallaç Church (tuff stone type I) is applied to the acoustical models of the Bell Church.

Once the acoustical simulations are performed, the results of three states of Bell Church, Phase I (without frescoes), Phase I (with frescoes), and Phase II, are obtained and compared with each other to see the effect of the architectural changes on the acoustics of the church over time. First of all, the decay rate results from all three sates of the church show that the space is suitable for liturgical practices. As expected, T30 values show an increase with the addition of plastered frescoes and volume (narthex). Due to a screen between the source and the receiver, negative C50 and C80 values are obtained over the whole frequency range, indicating that the distinctness of musical and speech tones in a liturgical ceremony might not be clear to the listeners. The blurriness of sound in the space may have created a spiritual atmosphere. STI results denote that the church space, in all its phases, is fair for speech intelligibility. In general, the results obtained from the simulation run on the church present a different set of results (low C80 and D50 values) for different states of the church in its timespan, but these results present the characteristic indoor acoustical environment of the Bell Church, which, in combination with the unconventional architecture (e.g., wooden templon) of the church, would have made it a unique space offering a distinct aural experience.

The main motivation behind the study is to document the soundscape of a Middle Byzantine built environment in Cappadocia. With the academic discourse on the identification and functions of the spaces in Cappadocia, an acoustical approach can provide invaluable information that can help broaden the current literature. The study of the acoustical environments of Middle Byzantine churches and other rock-cut structures can be useful in identifying functional patterns of these spaces. Future archaeoacoustic studies can use the model of this research to study the acoustical performance of remote and/or ruined spaces. The Byzantine liturgy rites can also be further studied in regard to different churches and different ceremonial occasions using a similar computational reconstruction method.

To conclude, this study primarily focuses on creating the most proximate representation of the acoustical climate of a Middle Byzantine masonry church over its different phases of construction, and it has only been possible by taking into account a couple of typical methodological approaches and combining them to make a more informed decision. The challenges of aural and virtual reconstruction of a ruined church are discussed hoping that the research steps will guide similar future archaeoacoustic studies.

Author Contributions: Conceptualization, A.H.A. and Z.S.G.; methodology, Z.S.G.; software, A.H.A.; validation, A.H.A.; formal analysis, A.H.A.; investigation, A.H.A. and Z.S.G.; resources, Z.S.G.; data curation, A.H.A.; writing—original draft preparation, A.H.A.; writing—review and editing, Z.S.G.; visualization, A.H.A.; supervision, Z.S.G.; project administration, Z.S.G.; funding acquisition, Z.S.G. All authors have read and agreed to the published version of the manuscript.

Funding: This research received no external funding.

Data Availability Statement: Not applicable.

Acknowledgments: The authors would like to acknowledge Mezzo Stüdyo Ltd. for providing the necessary tools for simulations, laboratory conditions and equipment for impedance tube tests. The authors are also grateful to Ayşe Henry for her valuable contributions on the site and case selections, and her inspiration for this study.

Conflicts of Interest: The authors declare no conflict of interest.

References

1. Brown, A.L.; Gjestland, T.; Dubois, D. Acoustic environments and soundscapes. In *Soundscape and the Built Environment*, 1st ed.; Kang, J., Schulte-Fortkamp, B., Eds.; Taylor & Francis Group: Oxfordshire, UK, 2017; pp. 1–16.
2. Gibson, J.J.; Carmichael, L. *The Senses Considered as Perceptual Systems*; Houghton Mifflin: Boston, MA, USA, 1966; Volume 2.
3. Öztürk, F.G. Transformation of the 'Sacred' Image of a Byzantine Cappadocian Settlement. In *Architecture and Landscape in Medieval Anatolia, 1100–1150*, 1st ed.; Blessing, P., Goshgarian, R., Eds.; Edinburgh University Press: Edinburgh, UK, 2017; pp. 135–154.
4. Sü Gül, Z.; Xiang, N.; Çalışkan, M. Investigations on sound energy decays and flows in a monumental mosque. *J. Acoust. Soc. Am.* **2016**, *140*, 344–355. [CrossRef] [PubMed]
5. Sü Gül, Z.; Xiang, N.; Çalışkan, M. Diffusion Equation-Based Finite Element Modeling of a Monumental Worship Space. *J. Comput. Acoust.* **2017**, *25*, 1750029. [CrossRef]
6. Sü Gül, Z.; Çalıskan, M.; Tavukçuoğlu, A.; Xiang, N. Assessment of acoustical indicators in multi-domed historic structures by non-exponential energy decay analysis. *Acoust. Aust.* **2018**, *46*, 181–192. [CrossRef]

7. Sü Gül, Z.; Odabaş, E.; Xiang, N.; Çalışkan, M. Diffusion equation modeling for sound energy flow analysis in multi domain structures. *J. Acoust. Soc. Am.* **2019**, *145*, 2703–2717. [CrossRef] [PubMed]

8. Sü Gül, Z. Acoustical impact of architectonics and material features in the lifespan of two monumental sacred structures. *Acoustics* **2019**, *1*, 493–516. [CrossRef]

9. Sü Gül, Z. Exploration of room acoustics coupling in Hagia Sophia of Istanbul for its different states. *J. Acoust. Soc. Am.* **2021**, *149*, 320–339. [CrossRef]

10. Vassilantonopoulos, S.L.; Mourjopoulos, J. Virtual acoustic reconstruction of ritual and public spaces of ancient Greece. *Acta Acust. United Acust.* **2001**, *87*, 604–609.

11. Sukaj, S.; Ciaburro, G.; Iannace, G.; Lombardi, I.; Trematerra, A. The Acoustics of the Benevento Roman Theatre. *Buildings* **2021**, *11*, 212. [CrossRef]

12. Giuseppe, C.; Iannace, G.; Lombardi, I.; Trematerra, A. Acoustic design of ancient buildings: The odea of Pompeii and Posillipo. *Buildings* **2020**, *10*, 224.

13. Gino, I.; Berardi, U. Acoustic virtual reconstruction of the Roman theater of Posillipo, Naples. Proceedings of Meetings on Acoustics 173EAA. *Acoust. Soc. Am.* **2017**, *30*, 1–9.

14. Suarez, R.; Sendra, J.J.; Alonso, A. Acoustics, Liturgy and Architecture in the Early Christian Church. From the domus ecclesiae to the basilica. *Acta Acust. United Acust.* **2013**, *99*, 292–301. [CrossRef]

15. Suárez, R.; Alonso, A.; Sendra, J.J. Intangible cultural heritage: The sound of the Romanesque cathedral of Santiago de Compostela. *J. Cult. Herit.* **2015**, *16*, 239–243. [CrossRef]

16. Suárez, R.; Alonso, A.; Sendra, J.J. Archaeoacoustics of intangible cultural heritage: The sound of the Maior Ecclesia of Cluny. *J. Cult. Herit.* **2016**, *19*, 567–572. [CrossRef]

17. Alonso, A.; Suarez, R.; Sendra, J.J. Virtual reconstruction of indoor acoustics in cathedrals: The case of the Cathedral of Granada. *Build. Simul.* **2016**, *4*, 431–446. [CrossRef]

18. Sender, M.C.; Planells, A.; Perelló, R.R.; Segura, J.G.; Giménez, A. Virtual acoustic reconstruction of a lost church: Application to an Order of Saint Jerome monastery in Alzira, Spain. *J. Build. Perform. Simul.* **2018**, *11*, 369–390. [CrossRef]

19. Francesco, M.; Cirillo, E.; Della Crociata, S.; Gasparini, E.; Preziuso, D. Acoustical reconstruction of San Petronio Basilica in Bologna during the Baroque period: The effect of festive decorations. *J. Acoust. Soc. Am.* **2008**, *123*, 3607.

20. Adeeb, A.H.; Sü-Gül, Z.; Henry, A.B. Characterizing the Indoor Acoustical Climate of the Religious and Secular Rock-Cut Structures of Cappadocia. *Int. J. Archit. Herit.* **2021**, 1–22. [CrossRef]

21. Mathews, T.F.; Daskalakis Mathews, A.C. Islamic-style mansions in Byzantine Cappadocia and the development of the inverted T-plan. *J. Soc. Archit. Hist.* **1997**, *56*, 294–315. [CrossRef]

22. Ousterhout, R.G. *Visualizing Community: Art, Material Culture, and Settlement in Byzantine Cappadocia*; Dumbarton Oaks Research Library and Collection: Washington, DC, USA, 2017.

23. Ousterhout, R.G. *A Byzantine Settlement in Cappadocia*; Dumbarton Oaks: Washington, DC, USA, 2005.

24. Ramsay, W.M.; Bell, G.L. *The Thousand and One Churches (London, 1909)*; Reprint, with a new foreword; Outsterhout, R.G., Jackson, M.P.C., Eds.; University of Pennsylvania Press: Philadelphia, PA, USA, 2008.

25. Rodley, L. *Cave Monasteries of Byzantine Cappadocia*; Cambridge University Press: Cambridge, UK, 2010.

26. Krautheimer, R.; Ćurčić, S. *Early Christian and Byzantine Architecture*; Yale University Press: New Haven, CT, USA, 1992; Volume 24.

27. Sukaj, S.; Bevilacqua, A.; Iannace, G.; Lombardi, I.; Parente, R.; Trematerra, A. Byzantine Churches in Albania: How Geometry and Architectural Composition Influence the Acoustics. *Buildings* **2022**, *12*, 280. [CrossRef]

28. St. Nicholas (Mesopotam) Monastery. Available online: https://www.intoalbania.com/attraction/st-nicholas-mesopotam-monastery (accessed on 26 April 2022).

29. Vasileios, M. *Architecture and Ritual in the Churches of Constantinople: Ninth to Fifteenth Centuries*; Cambridge University Press: Cambridge, UK, 2014.

30. Teteriatnikov, N.B. The Liturgical Planning of Byzantine Churches in Cappadocia. In *Orientalia Christiana Analecta*; Pontificio Istituto Orientale: Rome, Italy, 1996; p. 252.

31. International Organization for Standardization (ISO). *ISO 10534-2*; Acoustics–Determination of Sound Absorption Coefficient and Impedance in Impedance Tubes-Part 2: Transfer-Function Method.; International Organization for Standardization (ISO): Geneva, Switzerland, 1998.

32. McGrory, M.; Castro Cirac, D.; Gaussen, O.; Cabrera, D. Sound Absorption Coefficient Measurement: Re-Examining the Relationship between Impedance Tube and Reverberant Room Methods. In Proceedings of the Acoustics, Fremantle, Australia, 21–23 November 2012; 2012.

33. Chung, J.Y.; Blaser, D.A. Transfer function method of measuring in-duct acoustic properties. II. Experiment. *J. Acoust. Soc. Am.* **1980**, *68*, 914–921. [CrossRef]

34. Rindel, J.H. The use of computer modeling in room acoustics. *J. Vibroengineering* **2000**, *3*, 219–224.

35. International Organization for Standardization (ISO). *ISO. ISO 3382-1*; Acoustics: Measurement of the Reverberation Time of Rooms with Reference to Other Acoustical Parameters; International Organization for Standardization (ISO): Geneva, Switzerland, 2009.

36. Martellotta, F. The just noticeable difference of center time and clarity index in large reverberant spaces. *J. Acoust. Soc. Am.* **2010**, *128*, 654–663. [CrossRef] [PubMed]

37. Bork, I. A comparison of room simulation software-the 2nd round robin on room acoustical computer simulation. *Acta Acust. United Acust.* **2000**, *86*, 943–956.
38. Ousterhout, R. *Survey of the Byzantine Settlement at Çanlı Kilise in Cappadocia: Results of the 1995 and 1996 Seasons*; Dumbarton Oaks Papers: Wahington, DC, USA, 1997; pp. 301–306.
39. Beranek, L.L. *Acoustical Measurements*; Acoustical Society of America: New York, NY, USA, 1988.
40. Barron, M. *Auditorium Acoustics and Architectural Design*; Routledge: Oxfordshire, UK, 2009.
41. Giron, S.; Alvarez-Morales, L.; Zamarreno, T. Church acoustics: A state-of-the-art review after several decades of research. *J. Sound Vib.* **2017**, *411*, 378–408. [CrossRef]
42. Makrinenko, L.I. *Acoustics of Auditoriums in Public Buildings*; American Institute of Physics: College Park, MD, USA, 1994.
43. Templeton, D. *Acoustics in the Built Environment: Advice for the Design Team*; Butterworth-Heinemann: Oxford, UK, 1998.
44. Steeneken, H.J.; Houtgast, T. A physical method for measuring speech-transmission quality. *J. Acoust. Soc. Am.* **1980**, *67*, 318–326. [CrossRef]

acoustics

Article

An Acoustic Reconstruction of the House of Commons, c. 1820–1834

Aglaia Foteinou [1,*], Damian Murphy [1] and J. P. D. Cooper [2]

[1] School of Physics, Engineering and Technology: Audio Lab, University of York, York YO10 5DD, UK; damian.murphy@york.ac.uk

[2] Department of History, University of York, York YO10 5DD, UK; j.p.d.cooper@york.ac.uk

[*] Correspondence: aglaia.foteinou@york.ac.uk

Abstract: This paper presents an acoustic reconstruction of the UK House of Commons between c. 1820 and 1834. Focusing on a historically important site where political decisions were debated over the centuries, we aim to simulate and present the intangible principles of the acoustic properties and sounds heard within the space. The acoustic model was created based on available historical evidence with the aid of commercial acoustic simulation software. We discuss the decisions made for this reconstruction based on further experimentation with the acoustic characteristics of the constituent materials and settings of the available software. An additional comparison of the achieved acoustic results with spaces of similar historical importance and layout is presented, as a calibration of the model with in situ measurements was not possible in this case study. The values of T30, EDT, C50 and Ts are presented, while auralization examples are also available for a subjective evaluation of the results.

Keywords: room acoustics; intangible heritage; UK House of Commons; St Stephen's Chapel Westminster; geometrical acoustics

1. Introduction

In an effort to investigate some of the intangible aspects of history, including the acoustics of heritage spaces and the audio experiences perceived within them, new technologies and tools have been used in recent studies. For a more complete study of heritage sites, historians, archaeologists, musicologists and acousticians are just some of the specialists that should collaborate and exchange knowledge, methods and experience. The current paper focuses on the virtual acoustic reconstruction of the UK House of Commons as it appeared between c. 1820 and 1834. The interior of the Commons chamber had undergone multiple changes since its conversion from St Stephen's Chapel in the mid-sixteenth century; we focus on the last stage before it was severely damaged by a fire in Westminster Palace in 1834 and demolished in 1837. Considering the importance of this building in shaping the history of the UK, and the political speeches and decisions that took place within it, we aim to study the acoustic characteristics of the space and understand more about its impact on history. An acoustic model has been created based on the historical evidence that is available to us. In this paper, we discuss the process of creating this model and the decisions we have made to arrive at a plausible acoustic result for the simulated virtual space. We also compare the acoustic results with spaces that appear to have some similarities with the history, layout, and fabric of the lost Commons chamber. Despite the uncertainties and limitations of not having in situ measurements of the space for further calibration of our model, we believe that our estimations contribute to a more holistic history of this historic site.

Citation: Foteinou, A.; Murphy, D.; Cooper, J.P.D. An Acoustic Reconstruction of the House of Commons, c. 1820–1834. *Acoustics* **2023**, *5*, 193–215. https://doi.org/10.3390/acoustics5010012

Academic Editors: Margarita Díaz-Andreu, Lidia Alvarez Morales

Received: 23 January 2023
Revised: 17 February 2023
Accepted: 21 February 2023
Published: 24 February 2023

2. History of the House of Commons Chamber in the Palace of Westminster

The modern debating chamber of the UK House of Commons is not the same space that elected Members of Parliament historically occupied, nor is it in precisely the same location. The original Commons chamber was located in the former medieval chapel of St Stephen within the Palace of Westminster until 1834, when a major fire destroyed the space, and a replacement chamber was built nearby in the Gothic style. This House of Commons, in turn, was destroyed during the Second World War, following which the chamber was rebuilt on the same footprint in a similar but modernised style.

The first dedicated House of Commons chamber was located in the former royal chapel of St Stephen in the Palace of Westminster. The chapel itself was built during the reigns of Edward I, Edward II and Edward III, and was completed in the 1360s. The building was then refurbished and restructured in 1548–1550 during the Reformation of Edward VI, when the former college of St Stephen was dissolved and the upper chapel became the Commons chamber: the first time that the House of Commons had a permanent home of its own. The choir stalls were replaced with tiered benches arranged in parallel rows across the length of the room facing each other, with further seating wrapped around the east end of the space. As the purpose and the needs of Parliament were different from those for which the space had originally been designed, several alterations had to be made over the years. In 1692, the architect Sir Christopher Wren was called in to stabilise the building and modernise the interior. Wren lowered the ceiling of the medieval chapel, according to some in an effort to improve the acoustics of the space or to make the chamber warmer [1]. The galleries were extended to the north and south sides, and more benches were added to the existing west gallery. The medieval decorated stone walls were also covered with wainscot panelling, leaving a gap between the original walls and the panels and reducing the dimensions by 3 feet (0.91 m) at both the north and south sides. After the Acts of Union (1706–1707), additional seats were needed to accommodate the new Scottish Members of Parliament. The solution offered by Wren was to widen the galleries by one more row on each side.

In 1801, with the entry of Ireland into the United Kingdom, the architect James Wyatt was given the task of creating more space for the one hundred new members [2]. While his initial intention was to extend the west side of the chamber by merging it with the lobby located in the former antechapel, his work was limited to removing parts of the wainscot introduced by Wren and (more controversially) any remaining stonework and wall-paintings that had survived from the medieval St Stephen's chapel. Wyatt replaced the wainscot but without leaving any gap between the original stone walls and the panels. This created extra space for additional rows on the main floor of the chamber and space underneath all the windows and between the piers for additional benches. Additional columns were added underneath the galleries for extra support. After Wyatt's alterations, the final dimensions of the space were 18.6 m in length, 11.4 m in width and 8.6 m in height, based on the detailed architectural plans (as in Figure 1) that are available for all three levels (main floor, galleries, and ventilation space) [3] with a volume approximately 1823.5 m^3 without furnishing and about 1402 m^3 with furniture.

The capacity of the House of Commons, however, was always problematic, leading to overcrowding and uncomfortable sessions for the Members. The conditions in the chamber were poor; there was a constant issue with ventilation and room capacity while hosting the debates [2]. The heat, the smoke from candles and fires, and the background noise were reported on numerous occasions. Manuscript evidence implies that the architects involved in the reconstruction and renovation of the House of Commons between the 17th and 19th centuries included acoustical considerations in their planning and design [1].

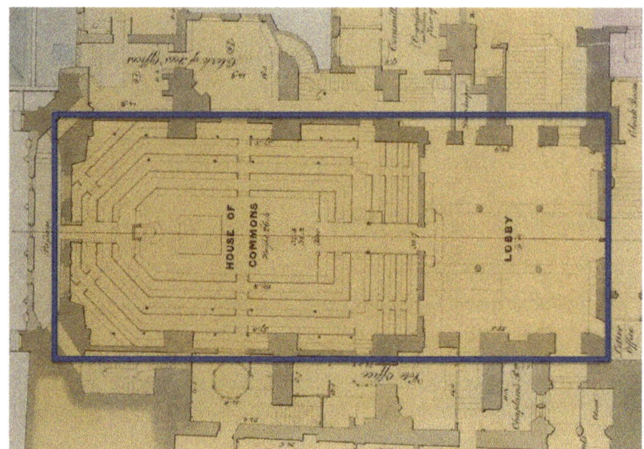

Figure 1. Detail from the architectural plans of the pre-fire House of Commons, as seen at [3]. The blue frame highlights the original boundaries of the medieval chapel of St Stephen, which was then separated into the House of Commons (east) and the Lobby (west).

The period of our investigation is focused between c. 1820 and 1834. During that time, the Commons chamber was in its very last stage of development before the fire. Additionally, we aim to include another historical element, which will be an interesting subject of future work. The ventilation space above the Commons ceiling, represented mainly as an octagon funnel in the middle of the House of Commons chamber, opened out into the original upper section of the medieval St Stephen's chapel (as appears in Figure 2). Evidence from the 1820s and early 1830s reveals that this space was used by women, formally excluded from the chamber itself, to gather and listen to the parliamentary debates going on below. An initial and partial study of this space was conducted previously [4], and we aim to follow this up with the current acoustic model as a basis for future work.

Figure 2. Section from west to east, showing the storeys of the House of Commons spaces, August 1834 [3].

3. Establishing the Geometry for the Acoustic Model

While focusing on the period from c. 1820 until the Palace fire of 1834, it was necessary also to retrace architectural changes over the centuries since a lot of surfaces and materials had remained the same or had been reused during the renovations of the space. Visual sources such as engravings, paintings [5–9], editorial cartoons [10–13], and architectural plans (such as Figures 1 and 2 shown above) have been studied for different periods of the House of Commons. Relevant literature, including modern publications and reports of committee meetings in the chamber before or after the fire, have helped to create a better understanding of the space over these years.

The House of Commons and St Stephen's Chapel have been the focus of two previous projects, which have proven very useful sources for this current work. Figure 3 is a virtual reconstruction of the House of Commons in 1692–1707 from [14], giving information on the seating layout and its materials, wall materials and decorations, as well as floor and ceiling as they appeared after Wren's alteration.

Figure 3. The House of Commons Remodelled, 1692–1707, St Stephen's Chapel Westminster Project 2017, image from [14].

An acoustic reconstruction of the House of Commons in c. 1789 was created in [4] using CATT-Acoustic [15], as presented in Figure 4a,b. This model was used as a starting point for this current study. Several changes had to be applied to this original model to reflect the state of the chamber after the significant alterations by Wyatt in 1801.

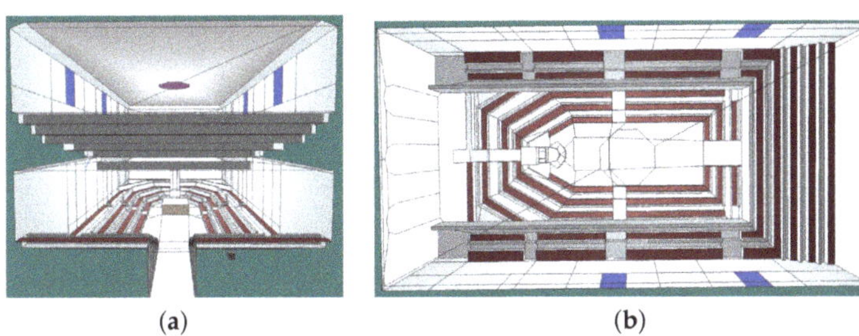

(a)   (b)

Figure 4. The original version of the model in CATT-Acoustic by [4], representing the House of Commons in c. 1789: (**a**) Section from north to south where the main entrance of the House of Commons was. (**b**) An elevation view of the model.

For the new version of the model, the wall surfaces on the north and south side of the room were set back between the piers, with additional seating areas underneath the windows. Additional ventilators have been added to the ceiling, working from new evidence on the ventilation systems of the period [2], and information about the use of the space above the ceiling by those well-connected women who were able to access it. The positioning of several doors and windows has been reconsidered for this model, following architectural plans (Figures 1 and 2) and discussions with historians [16–19].

It appears that the construction of adjacent buildings (as can be seen in Figure 1) had led to the blocking of some windows and others being covered with wooden panels, leaving the space with two windows on each side of the galleries, one window per side on the north and south walls of the ground floor and the three principal windows to the east. Evidence from engravings suggests that some windows at the ground floor level were covered with heavy green curtains, as were external passageways on the other sides of those windows, as also shown in Figure 5. All of these details have been simulated in the latest version of the space.

Figure 5 was also a great source of evidence for the additional seating areas between the columns at the north and south walls, and the additional fifth row of benches om the east side of the chamber and behind the Speaker's chair. Discussions with historians and researchers [16–19] specialising in the history of House of Commons and House of Lords gave assurance of the final decisions made in the creation of the geometric model and the materials used for the latest design.

Figure 5. The House of Commons, 1833 by Sir George Hayter, oil on canvas, 1833–1843, 136 1/4 in. × 213 3/8 in. (3460 mm × 5420 mm) overall, © National Portrait Gallery, London [20].

In this updated geometric model, the new seats have been added following the design of the previous version; however, it was considered to be an unnecessarily detailed design of the benches, as there were many surfaces smaller than 0.5 m^2 (as shown in Figure 4a,b). Literature on acoustic simulations and the comparison of detailed and simple versions of the geometric models on which they are based have indicated that by applying different scattering coefficients and overall settings in the acoustic modelling software, the results of the two versions could be sufficiently close [21,22]. ODEON [23,24] and CATT-Acoustic [15] guidance, however, suggest that this detailed approach can significantly increase the calculation time, while for the authors of this study, it was considered as an additional challenge to adjust the significant number of surfaces with different materials during the calibration process of this model.

It was also considered necessary to merge geometric planes (such as several planes forming part of the same wall or sides of the floor) and hence define them with the same materials. The updated and simplified version has 361 surfaces in total, which is reduced by half from the original detailed model, and a total surface area of 1175.68 m^2 with adjusted absorption and scattering coefficient for the seating area, as will be discussed in Sections 4.1 and 4.2. The model was also imported into the ODEON 14.00 Auditorium with some small changes to the subdivided surfaces as recommended for the use of this software. Figure 6 in CATT-Acoustic and Figure 7 in ODEON show the versions of the model based on the different detailed levels of the seating areas.

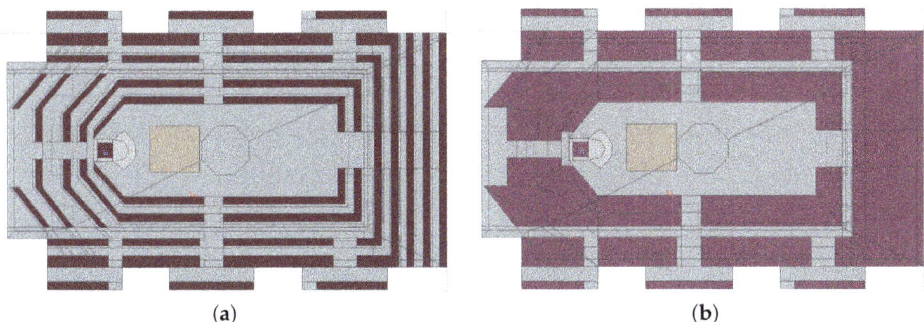

Figure 6. The updated version of the model extended on the north and south walls by the window areas in CATT-Acoustic. (**a**) A detailed simulation of the benches is represented. (**b**) The benches' surfaces have been simplified.

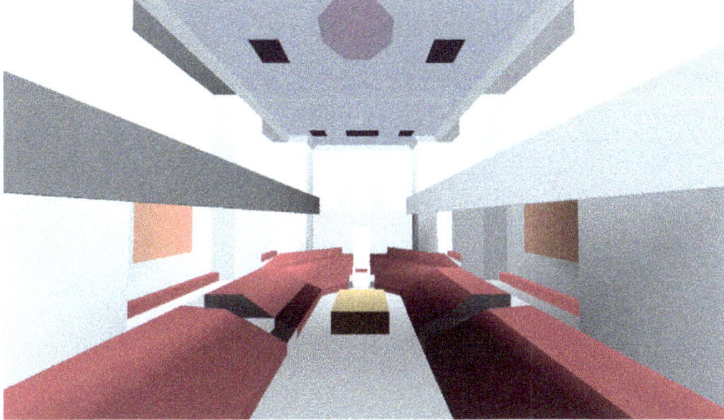

Figure 7. A virtualisation of the final model in ODEON with the simplified representation of the seating areas.

4. Materials and Methods

Examples in previous studies [22,25,26] where the acoustics of heritage sites have been reconstructed and studied using the two commercial acoustic software applications also used here, CATT-Acoustic and ODEON, have been considered, and useful information was taken regarding the estimation of the materials used, their absorption and scattering coefficients, and the calculation settings specific to each software.

For both ODEON and CATT-Acoustic models, the conditions in the acoustic environment were set to 20 °C for air temperature and 50% for the relative humidity. The results were obtained from an omnidirectional source at position P2, on the right side of the Speaker's chair (presented in red in Figure 8) and from eleven additional receiver positions (presented in blue), excluding receiver 2, which represents the seated speaker for future scenarios. The Speaker's position was not used as a candidate source position, as the speaker chair with the high marble back would have had a strong impact on the distribution of the early sound.

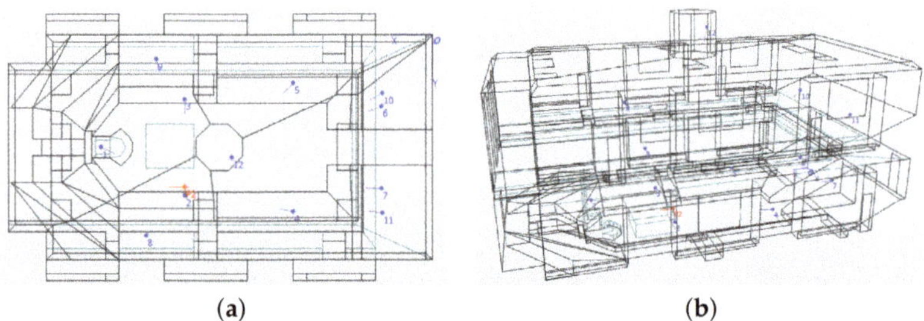

(a)   (b)

Figure 8. (**a**) An elevation view of the final ODEON model and (**b**) an east-to-west view of the space demonstrating the chosen source as a Member of the Parliament (MP) speaker on the right side of the Speaker's chair (presented in red), and twelve receiver positions (presented in blue) in several positions on the main floor, galleries and ventilation space, all facing the source.

4.1. Absorption Coefficients

The materials below (Table 1) have been identified and defined by following information from historical references and images of the chamber and in combination with the available materials' library properties from CATT-Acoustic and ODEON software.

Table 1. The chosen materials with their absorption coefficient across the frequency bands, and the overall surface area of each of them, as used in the early versions of the model.

Materials	125 Hz	250 Hz	500 Hz	1 kHz	2 kHz	4 kHz	Surface Area
Woodenpanel	0.20	0.15	0.10	0.08	0.04	0.02	630.4 m^2
Audience	0.52	0.68	0.85	0.97	0.93	0.85	222.9 m^2
Plaster	0.20	0.15	0.10	0.08	0.04	0.02	148.0 m^2
Wooden floor	0.15	0.11	0.10	0.07	0.06	0.07	112.0 m^2
Glass	0.18	0.06	0.04	0.03	0.02	0.02	24.7 m^2
Light velour	0.03	0.04	0.11	0.17	0.24	0.35	9.6 m^2
Marble	0.01	0.01	0.01	0.01	0.02	0.02	10.8 m^2
Metal perf	0.76	0.76	0.90	0.99	0.85	0.70	11.5 m^2
Curtains	0.04	0.23	0.14	0.57	0.53	0.40	5.8 m^2

The next step was to experiment with different possible materials and replace those surfaces of a considerable size, that is, the wooden panels on the walls and the seating area, and the wooden floor. Information on the absorption coefficient values for the materials used in the simulated model is limited in the literature, and even less for the scattering coefficient values. However, absorption and scattering coefficient values from church, cathedral and ancient theatre simulations have been taken into consideration, and the model was tested under a variety of different settings [22,25–31]. It was observed that different definitions for wooden panels (e.g. *16 mm wood on 40 mm studs* or *wooden floor on joists* or *wood, 25 mm with air space*) will only just slightly change the obtained results as the absorption coefficient of the possible options documented does not differ significantly (as also discussed in [24]). Thus, the challenge for this project was mainly in the estimation of an appropriate absorption coefficient for the seating area.

For the purpose of this study, it was considered best to redefine the space as unoccupied in order to help the calibration process and comparison with similar physical spaces, which have been measured unoccupied (see Section 5). Our evidence for the benches in the House of Commons c. 1820–1834 is limited to the images of the period where some indication of upholstering can be observed when they are occupied. References from an earlier period suggest that the green fabrics that covered the seating area during meetings of Parliament were then restored at the end of each session. The seating area can play a significant role in the overall absorption and reverberation of a space, especially in large rooms that further lack additional absorbing surfaces. The prediction of absorption coefficients for chairs and pews varies significantly with the type and design of the seats and any upholstered material that might partly cover these seats. Table 2 shows some of the different materials with their absorption coefficient across the frequency bands, as found in previous studies [32–36], which have been tested in the calibration of this model, with some being more plausible than others. Scattering coefficients are also provided for some of these materials (as the second values across the frequency bands) and were also used for the acoustic experimentation.

Table 2. Frequency-dependent absorption coefficient of the materials found in previous studies and tested for the seating surfaces in the model. The material *wooden pews* (WP) is taken from [37], and its scattering coefficient values are listed as well (figures on the right in the corresponding row). The material *unoccupied moderately upholstered chairs* (UC) is from [38]. The material *empty chairs, upholstered with leather* (LC) is taken from [23]. The material *empty chairs upholstered with cloth cover* (CC) is from [23]. The material *wooden pews (A/S) with seat cushions (without persons)* (WPC) is from [36].

Materials	125 Hz	250 Hz	500 Hz	1 kHz	2 kHz	4 kHz
WP	0.10/0.30	0.15/0.40	0.18/0.50	0.20/0.60	0.20/0.70	0.20/0.80
UC	0.44	0.56	0.67	0.74	0.83	0.87
LC	0.40	0.50	0.58	0.61	0.58	0.50
CC	0.44	0.60	0.77	0.89	0.82	0.70
WPC	0.21	0.53	0.77	0.69	0.58	0.59

Figure 9 presents the T30 results of the model from the different variations of the materials used on the seats as listed above. For this analysis, and what follows, the Aurora tools v.12.2.23-alpha [39] in Audacity 20.0 were used to obtain the relevant ISO3382 acoustic parameters from impulse responses generated as an output from the acoustic model. The observations from the previous studies highlighting the importance of the unoccupied seats in a space with the dimensions, furnishing, and materials as the current model have also been confirmed here. Significant changes in the average T30 values are obtained with the different materials on the seating area.

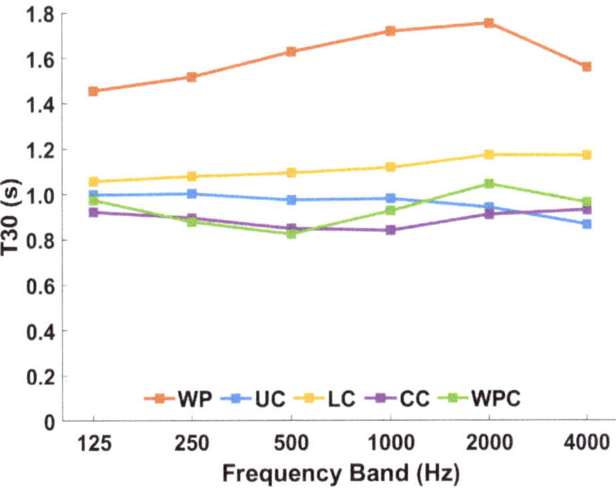

Figure 9. Average values of T30 obtained from experimental versions in ODEON by using different materials to define the surfaces of the seating area.

4.2. Scattering Coefficients

Scattering coefficients were also assigned to the experimental versions listed above. Initially, scattering coefficients were defined as 0.05 (as the default values in ODEON). By assigning detailed scattering coefficients, for the *wooden pews* (WP), as shown in Table 2, and for the rest of the surfaces, as shown in Table 3, the results of global parameter T30 were slightly changed. The results of C50, however, evidenced a significant increase, as shown in Figure 10.

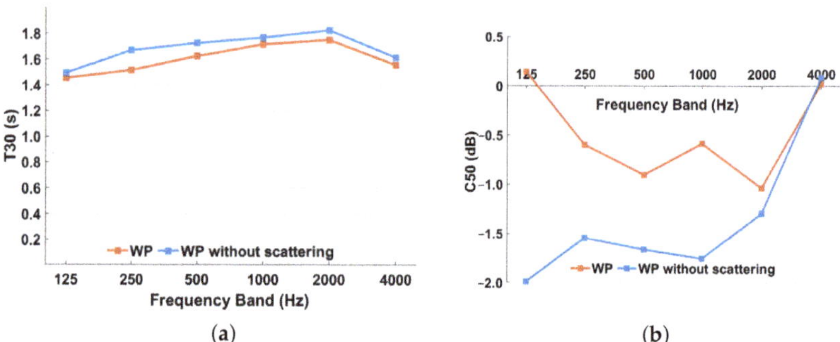

Figure 10. (**a**) Average T30 values and (**b**) C50 values with and without scattering coefficient information applied to all surfaces.

Table 3. Scattering coefficients applied to the used materials in ODEON and CATT-Acoustic.

Materials	Scattering Coefficient in ODEON	Scattering Coefficient in CATT-Acoustic (in %)					
		125 Hz	250 Hz	500 Hz	1 kHz	2 kHz	4 kHz
Wooden panel	0.2–0.5 for coffered panels and ceilings [40]	2	5	25	45	70	90
Wooden pews	0.7 [40]	9	35	60	80	90	90
Plaster	0.1 for large smooth surfaces [40]	1	2	5	30	60	85
Wooden floor	0.1	1	2	5	30	60	85
Glass	0.15	1	3	5	16	54	85

The *actual surface scattering* has been used on the settings in ODEON in order to assign scattering coefficients to each surface, and the *screen diffraction* option was selected due to the non-visible source from various locations/receivers in the space as recommended by the ODEON manual [23]. These are the receivers R8, R10 and R11, as shown in Figure 8, which do not have visibility of the source S2. Using this software, scattering coefficients are defined for the middle frequency of 707 Hz, while for the rest of the frequencies, the software takes into consideration the *Sets of Scattering Coefficient curves* as suggested in the manual.

For further confirmation of the results by the use of geometric acoustic models, the final model from ODEON was imported and adjusted in CATT-Acoustic v9.1b. The same absorption coefficients were used for each individual surface (as shown in Tables 1 and 2); however, scattering coefficients are not applied to individual surfaces as in ODEON but to each group of surfaces with the same material. Materials with the same name are defined with the same absorption and scattering coefficient. This can result in differences in the estimated parameters between the two pieces of software. The setting of the *2D Lambert scattering function* was chosen for the materials in CATT-Acoustic, and the values of the scattering coefficients across the six different frequency bands were estimated by the same scattering coefficient curves provided by ODEON, all listed in Table 3.

4.3. Software Settings

For the final settings in ODEON, 30,000 rays were used, a *Transition Order* of 2, an impulse response resolution of 3 ms and an impulse response length of 3000 ms. These settings were the result of experimentation with a variety of recommendations from relevant literature [24,41], providing no unusual issues with the estimation of the acoustic parameters, or suspicious reflections/spikes in the obtained impulse responses, as explained in [36], while the auralisation results also sounded reasonable at the informal subjective evaluation.

For the calculation in CATT-Acoustics, *Algorithm 2* was used in TUCT (see e.g. [15]), and similarly, several recommendations from the literature [22,25,26,37] were taken into consideration. For the number of rays/cones and impulse response length, the suggested values from the software were used, which were 470,688 rays/cones and 731.4 ms, correspondingly.

5. Model Validation Based on a Survey of Comparable Spaces

When developing an acoustic model, whether for an existing space or one that does not yet exist, it is important to build in calibration processes to justify the decisions made at key stages in the model definition and execution and to verify that the results obtained are plausible within a range of possible outcomes. This is particularly important in cases where it is not possible to obtain in situ measurements from the acoustic space being studied. In such circumstances, it is common to calibrate the model based on a comparison between predicted results obtained from the model and acoustic measurements from sites with similar characteristics or purposes. There are a significant number of studies that consider the acoustic properties of concert halls, ancient theatres, classrooms, and performance spaces in general [25,26,42]. The data from these studies, either based on acoustic measurements or acoustic reconstructions using computer modelling software, is of great value in informing the development and improvement of the acoustic experience in future similar designs [27]. In the acoustics of heritage sites, however, there is minimal supporting information available to better understand and compare the acoustic characteristics of spaces that no longer exist, and this information is usually limited to the study of other spaces with a comparable construction, design, or purpose. Thus, the design and the prediction of any acoustic result using computer modelling is based on a researcher's best experience and knowledge of acoustics, materials used in the construction of a place, and its characterisation.

In terms of this study, the architectural style and acoustic characteristics of the c. 1820–1834 House of Commons have not been discussed much in the literature on the acoustics of heritage buildings or performance spaces. As a result, we embarked upon a study of historical sites around the UK that demonstrate some comparison with the House of Commons. The five chosen spaces have been surveyed and acoustically measured by various research teams from the AudioLab, University of York from 2017–2018, as independent studies, and some of their results are presented in [43]. In this section, we summarise these results and present additional acoustic information for the comparison and verification of the acoustic results of the final House of Commons model.

Two of the spaces, the Convocation House and the Divinity School in Oxford, were used as meeting places for the House of Commons away from the Palace of Westminster in the 17th century. A third space, the modern House of Commons, has largely retained the layout and architectural style of previous chambers; however, the current space is significantly larger and with contemporary materials and fabric. The last two spaces, the Holywell Music Room in Oxford and the York Guildhall Council Chamber, were not used for Parliamentary meetings; however, their layout and volume are similar to the c. 1820–1834 House of Commons chamber. The Holywell Music Room has been mainly used for music concerts, while the York Guildhall Council Chamber is the one space that is perhaps closest to the purpose and period of the model in this study. It was built in the 1880s and is still used as a place of political debate and discussion for current local government.

For the acoustic measurements of the Convocation House, Divinity School, House of Commons, and Holywell Music Room, a Genelec 8030 (Genelec Oy, Iisalmi, Finland) was used as the sound source and a SoundField ST450 microphone (SoundField, Silverwater, Australia) as the receiver, using a 20 Hz–20 kHz exponential sine sweep of length 15 s. In the Guildhall Council Chamber, in York, the research team used a Genelec 8130A for the sound source, a SoundField ST450 MkII microphone for the receiver and a 30s 20 Hz–22 kHz sine sweep for the excitation signal. For the analysis, MATLAB R2021a was used to post-process the recorded files, while the Aurora tools [39] in Audacity were used to obtain the ISO3382 acoustic parameters T30, C50, EDT and Ts from the resulting impulse responses. Overall, the methodology established by Farina [44] has been followed in all cases, such that the data obtained can later be used in a variety of auralization applications.

5.1. Convocation House, Oxford

The Convocation House (CH) at the University of Oxford is part of the Bodleian Library and was formerly used for meetings of University committees. During the English Civil War and again in 1665 and 1681, it was also used for meetings of the House of Commons when Parliament was summoned to Oxford, which makes it an especially interesting space to compare with the designed model of the House of Commons at Westminster. It is a rectangular shape with dimensions of 18.55 m in length, 8.4 m wide and 7.63 m in height, with wooden stalls and wainscot decorating half of the height of the walls. During the measurements, three source locations were chosen and six receiver positions, resulting in fifteen source/receiver combinations, as shown in Figure 11. For the obtained average acoustic results in this paper, the measured position S2-R4 was excluded. The results of energy-based acoustic parameters (such as C50, C80 and D50) obtained from this position were quite different from the rest of the measurements in the space, as presented in [43]. Our explanation is that at location R4, the microphone was placed close to a wooden stand, which will have unduly influenced the acoustic behaviour at this location that was unrepresentative of the space more generally.

While the size and the architectural layout of the seating area are comparable with the House of Commons in the 1820s, the tiled floor and the vaulted ceiling make the space more reverberant, as is also found with examples of different ceiling designs in mosques [45], and as presented in T30 and EDT values shown in Figure 12. C50 and Ts values show high energy in the late reflections, which also indicate the perception of a reverberant space.

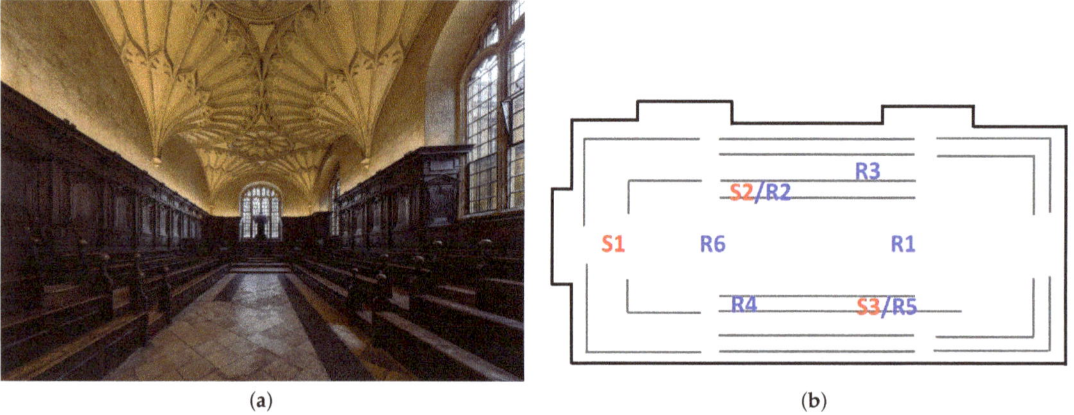

(**a**)   (**b**)

Figure 11. (**a**) Convocation House, University of Oxford, photo by DAVID ILIFF. Licence: CC BY-SA 3.0 via Wikimedia Commons [46] and (**b**) floor plan of the measured positions.

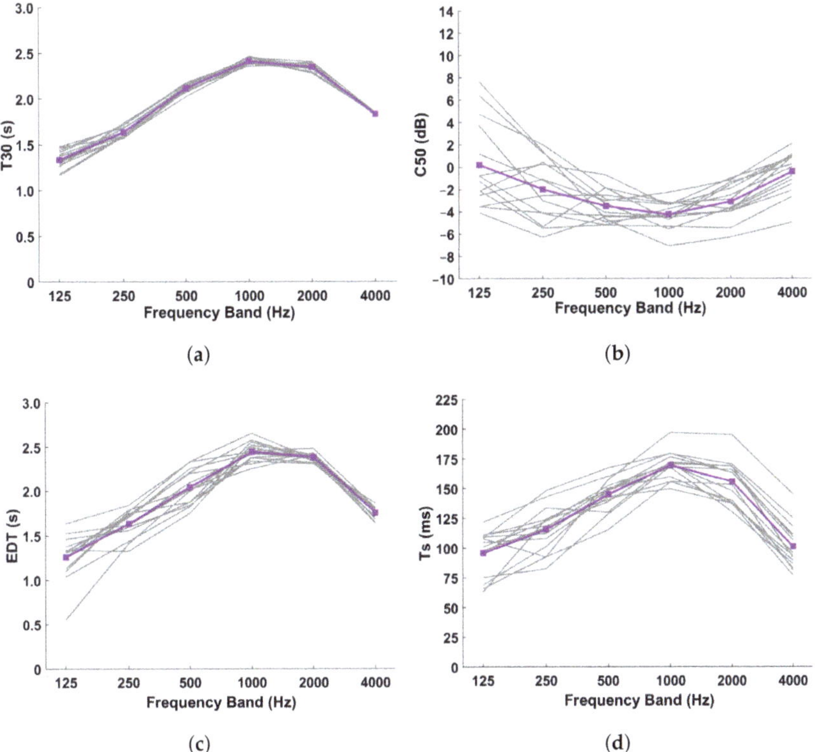

Figure 12. Results of the fourteen individual measured positions in the Convocation House, and the average values across frequency bands for (**a**) T30, (**b**) C50, (**c**) EDT and (**d**) Ts.

5.2. Divinity School, Oxford

The Divinity School (DS) is adjacent to the Convocation House at the University of Oxford. During the English Civil War and in 1625, it too was used for meetings of the House of Commons. It has dimensions of 27.6 m in length, 10.3 m in width and 7 m in height and is constructed from stone, marble, and large panelled windows. A single source/receiver position measurement was obtained for this space, as shown in Figure 13.

Its architectural characteristics and materials, as well as the lack of additional furnishing, indicate a much longer reverberation time compared to the c. 1820–1834 House of Commons chamber. Similar to the Convocation House, T30, EDT and Ts values (as shown in Figure 14) are quite high, with very poor clarity for speech (C50) as well. The highly reflective materials on the walls, as well as the long rectangular shape of the space, also cause high reverberation of low frequencies in comparison with the wooden furnishing in the Convocation House.

Figure 13. (**a**) Divinity School, University of Oxford, photo by DAVID ILIFF. Licence: CC BY-SA 3.0 via Wikimedia Commons [47] and (**b**) floor plan of the measured position.

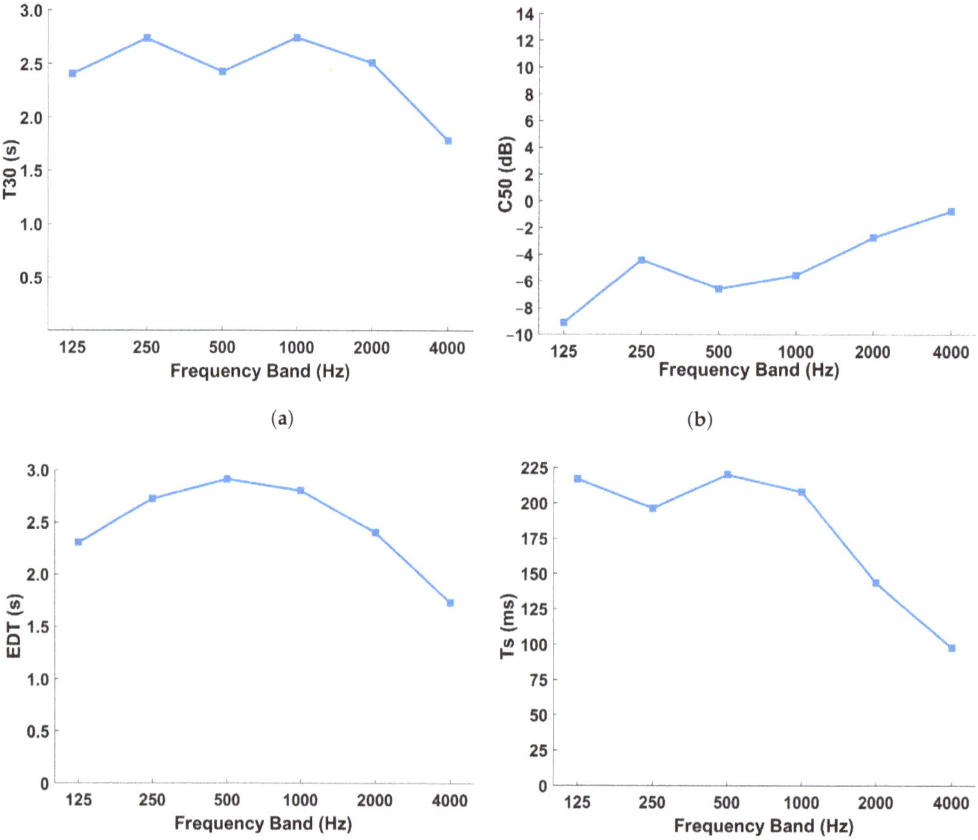

Figure 14. Results of the measured positions in Divinity School across frequency bands for (**a**) T30, (**b**) C50, (**c**) EDT and (**d**) Ts.

5.3. Modern House of Commons Chamber, Palace of Westminster

The modern House of Commons chamber (HoC), being built in a somewhat different style, is not directly comparable with the House of Commons as it was in the period 1820–1834. The use and purpose of the space for speech and debate were taken into consideration when the space was designed. This represents a significant difference between the modern House of Commons and the model under study, as St Stephen's Chapel had originally been intended for the sung liturgy. The study of the acoustics of the modern House of Commons, however, offers useful information on the possible acoustic behaviour of the 1820–1834 model.

Designed by Sir Giles Gilbert Scott in 1950, the current House of commons deliberately echoes the same architectural style as the previous chambers, with the benches laid out in parallel rows across the length of the room. Its dimensions are 21 m in length, 16 m in width and 14.7 m in height, with a volume of approximately 4839 m^3 without furnishings. Two-thirds of the walls are covered with wooden coffered benches and galleries, which work as acoustic diffusers, while the tops of the walls are covered with stone columns and large windows. The roof beam ends on a long roof window, the floor is covered with carpet, while the wooden benches are upholstered with green leather at the back and on the seats. Two source locations and six receiver positions were used during the measurement process, resulting in eight source/receiver combinations, as shown in Figure 15. The measured position S2-R5 was excluded from the averaged results obtained here. It was noted that the receiver position was in the near field of the sound source, and this unduly influenced the acoustic results (e.g., T30) obtained at this location that, upon inspection, could be seen to be unrepresentative of the space more generally [43].

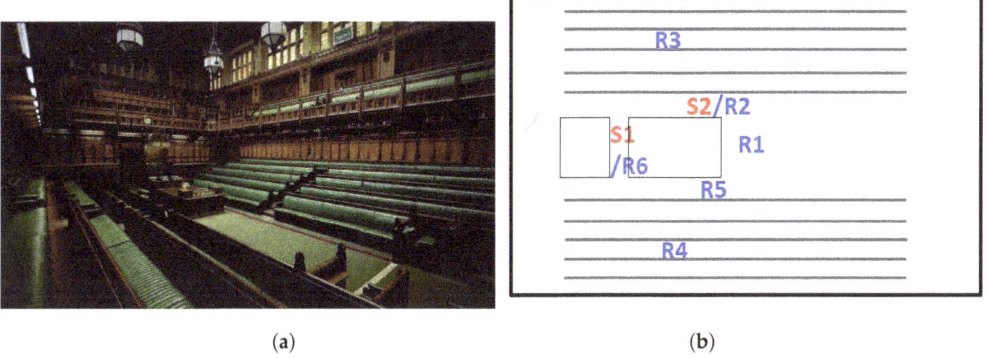

(a) (b)

Figure 15. (a) Chamber of House of Commons, Palace of Westminster, in its current condition. Licence: CC BY-SA 3.0 via Wikimedia Commons [48] and (b) floor plan of the measured positions.

Despite the large volume, the materials in the space absorb a significant amount of sound energy, which impacts the values of the acoustic parameters obtained from the measurements, as shown in Figure 16.

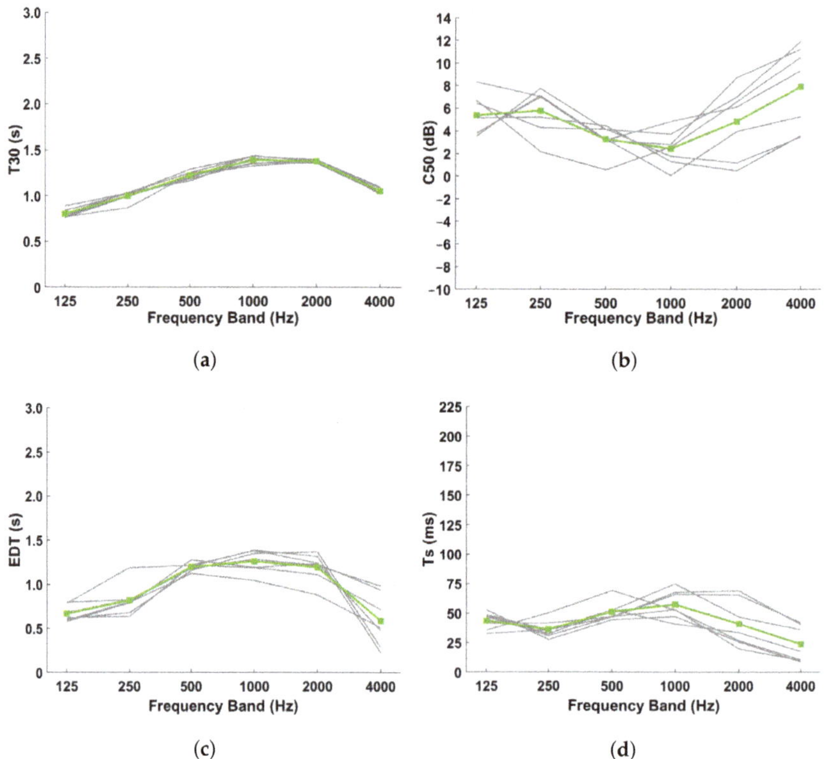

Figure 16. Results of the seven individual measured positions in the modern House of Commons and the average values across frequency bands for (**a**) T30, (**b**) C50, (**c**) EDT and (**d**) Ts.

5.4. Holywell Music Room, Oxford

The Holywell Music Room (HMR) was built in 1748 and is considered to be the first custom-built concert hall in Europe [49] (Figure 17), with dimensions of 19.7 m in length, 9.77 m in width and height of 9 m, and a volume of approximately 1732.22 m³. While its acoustics are suited for chamber music, and it is particularly well-known for performances of Haydn, there are no direct links with parliamentary history or meetings of the House of Commons within this space. However, its layout with the parallel benches across the room and a gallery at the back, together with the wooden floor and benches covered with cushions, makes the space comparable with the previous spaces and the model under study. It is worth noting, however, that the ceiling and walls are covered with plastered brick, making the space more reverberant than the c. 1820–1834 House of Commons model with its wooden coffered walls, and this was noted after a comparison of absorption coefficients for these materials in [15,23,38]. There is also a curtain installed across the west wall of the space, which should reduce the reverberation time further, as well as help prevent flutter echoes between the two parallel walls. Twenty-seven impulse responses were obtained during the in situ measurements from a combination of three sound source locations and nine receiver points. The measured position S1-R9 was also excluded from the averaged results presented in Figure 18. Upon inspection, the results of energy-based parameters (C50, C80, D50) were quite different from the rest of the measurements. A closer look at the time domain impulse response revealed that strong early reflections could be evidenced, potentially due to the concave nature of the stage and its focal point coinciding with position

R9. This measurement was, therefore, considered to be unrepresentative of the space more generally and so unsuitable for this acoustic analysis.

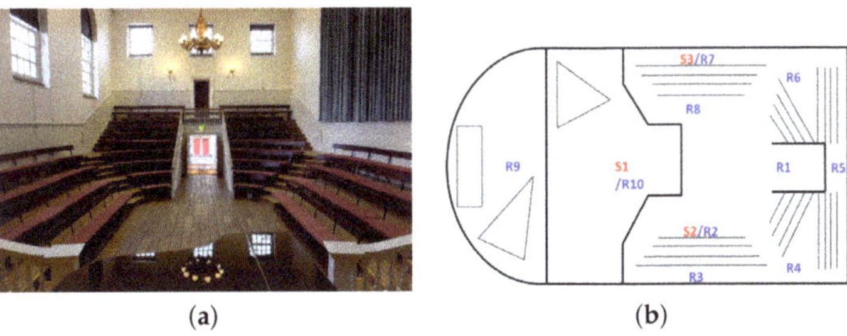

(a)

(b)

Figure 17. (**a**) Holywell Music Room, in Oxford in its current condition [49] and (**b**) floor plan of the measured positions.

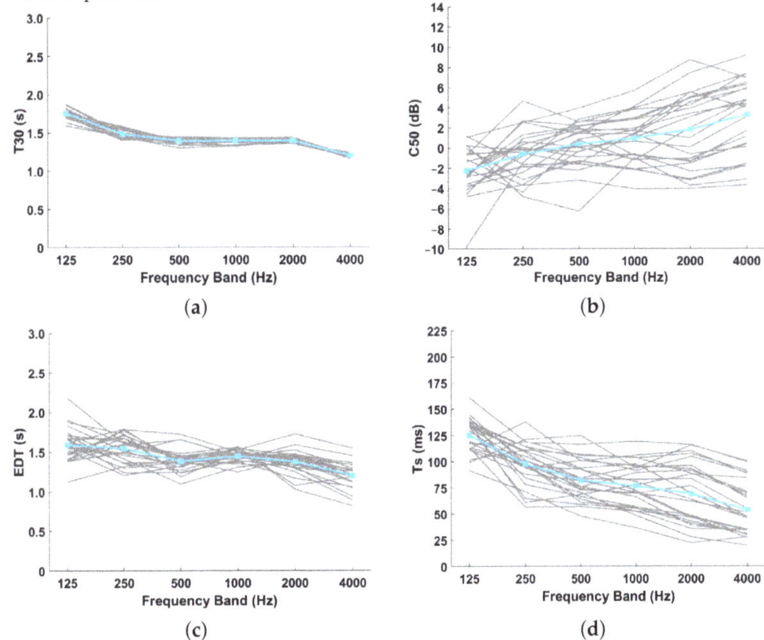

(a)

(b)

(c)

(d)

Figure 18. Results of the twenty-six individual measured positions in Holywell Music Room and the average values across frequency bands for (**a**) T30, (**b**) C50, (**c**) EDT and (**d**) Ts.

5.5. The Guildhall Council Chamber, York

The Council Chamber of York Guildhall (YGCC), built in 1888–1891, has significant architectural similarities with the historic House of Commons chamber. As Figure 19 shows, the members of the council are seated in an oval layout of wooden benches, surrounded by wooden panels on the walls from the floor to approximately 2.5 m in height. The rest of the height of the walls is wallpapered plaster up to the ceiling, which is arched and timbered. In addition to the comparable layout of the seated area, where members face each other during debates, it is worth noting that this particular space shares very similar dimensions and construction materials to the House of Commons chamber of c. 1820–1834. The York Council Chamber measures 15.23 m in length, 10.04 m in width, and 8.68 m in height and is approximately 1327.24 m³ without furnishings. Twelve combinations of source/receiver had been obtained from three source and four receiver positions. The acoustic results of all these source/receiver combinations for T30, C50, EDT and Ts, were averaged and presented in Figure 20. It is noticeable that all the studied parameters have a flat curve across all the frequency bands, overall presenting very good acoustic characteristics for speech.

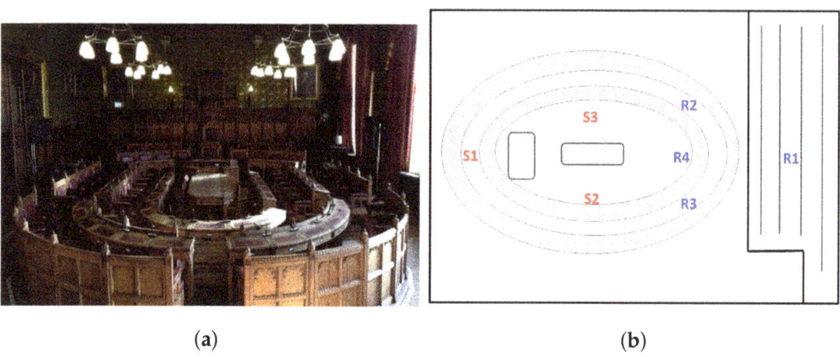

(**a**)　　　　　　　　　(**b**)

Figure 19. (**a**) Photograph of the York Guildhall Council Chamber (2023) and (**b**) floor plan of the measured positions.

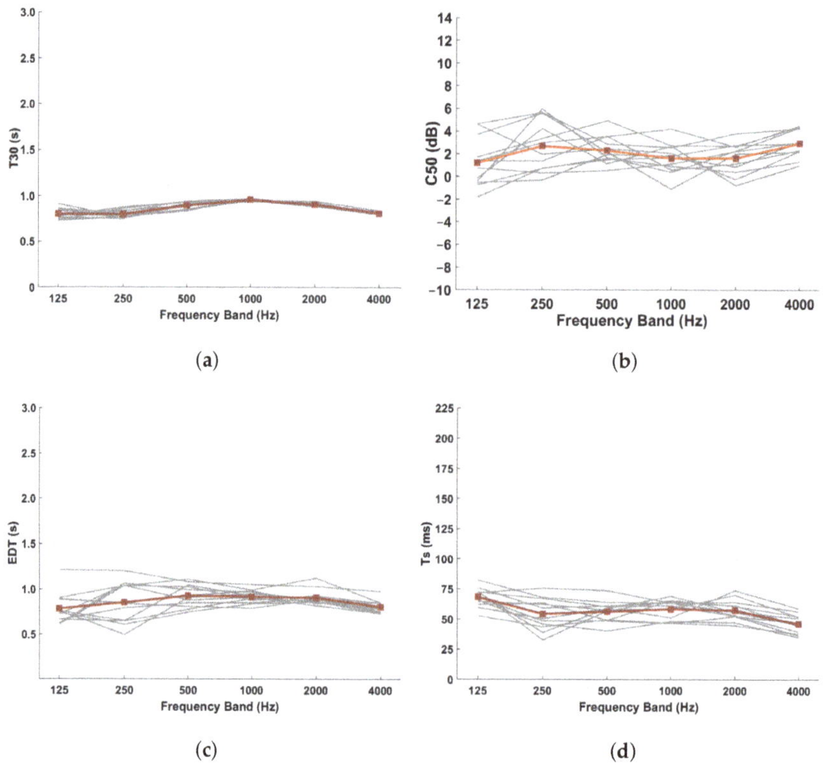

(**a**)　　　　　　　　　(**b**)

(**c**)　　　　　　　　　(**d**)

Figure 20. Results of the twelve individual measured positions in York Guildhall Council Chamber and the average values across frequency bands for (**a**) T30, (**b**) C50, (**c**) EDT and (**d**) Ts.

6. Results and Discussion

For the evaluation of the designed model, we have followed two approaches, and the results are analysed and discussed in this section. First, the results of the two models (the latest version in CATT-Acoustic and the one in ODEON) are presented, and their comparison provides us with confidence in the decisions made for this simulation. Secondly, the results of the acoustic parameters of the five existing spaces, as discussed above, are compared with the final version of the ODEON model, and conclusions are drawn in relation to the validity of the final model.

All the presented results have been analysed using the Aurora tools; both those from the in situ measurements and also those from the impulse responses generated by the two software applications. By using a common calculation tool for all, we aimed to avoid any further uncertainties regarding the calculation methods that the different software may use for the acoustic parameters.

It is also worth explaining that for this study, we used the average results of all the measured positions for all spaces. While the recent practice in the literature discourages this approach, as important information on local reflections and material characteristics can be overlooked when aiming to calibrate acoustic models, in this current study, any comparison of individual locations would have been meaningless, as the compared spaces and in situ positions differ.

6.1. Acoustics of the Simulated Models

The two commercial software applications, CATT-Acoustic and ODEON, have been used for several studies and a significant amount of literature is available discussing their effectiveness in simulating enclosed spaces. While they are both based on geometrical acoustic methods, they have some important differences in their algorithms, which especially affects the scattering and diffusion prediction of the models. Thus, a comparison of the model from the two types of software can help to confirm the validity of the decisions taken for this studied space. For this final version of the model, the material *wooden pews (A/S) with seat cushions (without persons)* (WPC) from [36] was used for the seating areas. Figure 21 presents the results of both models for T30 and C50. The values are very close to each other, and for some frequency bands, the results are within 1 JND value for each of the parameters (these limits are indicated with *X* marks for each point). This gives us significant confidence in the decisions taken for each of the two software applications and their individual settings. The error bars present the standard deviation of the results across the eleven measured positions. It is noticeable that the standard deviation is wider for the ODEON model and especially for the results observed for C50 (which is a position-dependent parameter) [50]. The difference in the standard deviation could be caused by the different methods used to simulate scattering in these two applications, as well as the definition of the scattering coefficients for the individual materials (as discussed in Section 4.2).

6.2. Acoustics of the Measured Spaces

The average results of the five chosen spaces (as discussed in Section 5) are compared with the final version of the c. 1820–1834 House of Commons acoustic model as simulated in ODEON. Results of T30, EDT, C50 and Ts for all the spaces are presented in Figure 22, with error bars indicating the standard deviation of the values across source/receiver locations for the ODEON results. Figure 23 also provides very interesting information by comparing the spaces based on the T30 average values and their volume.

It is observed that the Convocation House (CH) and Divinity School (DS) have the highest values of T30. Considering their materials and shape, this was not a surprise; however, they should not be used as a direct comparison or for a calibration of the House of Commons model c. 1820–1834.

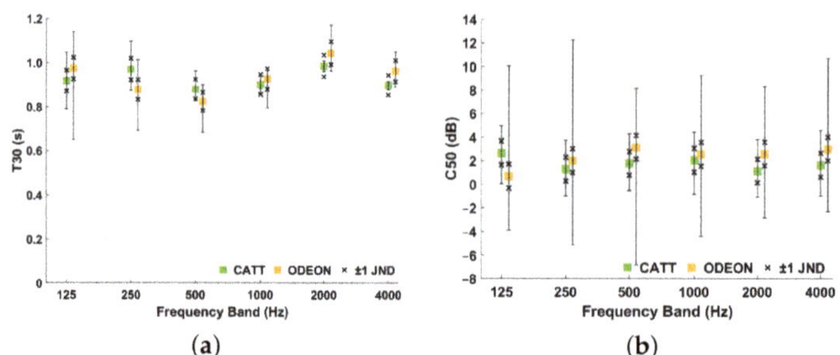

Figure 21. Comparison of the average values across eleven simulated positions in both CATT-Acoustic and ODEON for (**a**) T30, (**b**) C50. The *X* marks indicate the ±1 JND values for each parameter. Error bars at each point indicate the deviation of the results obtained from all eleven simulated positions.

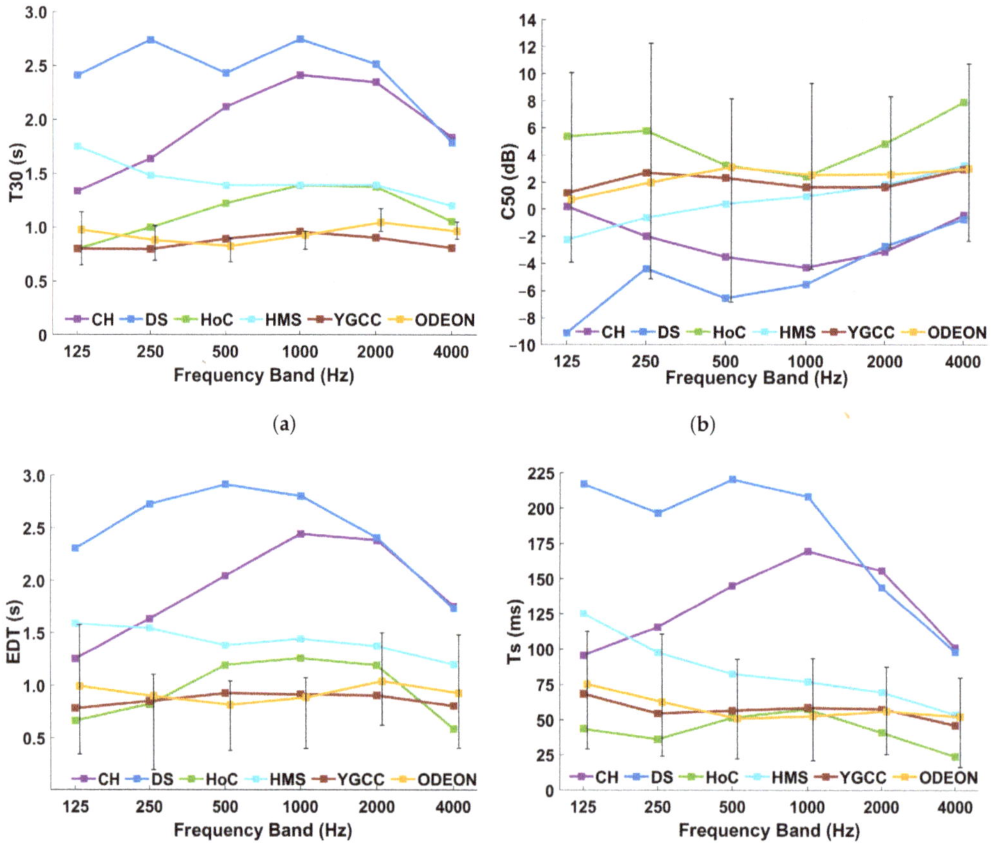

Figure 22. Average values of the five in situ measurements compared with the final model in ODEON: (**a**) T30, (**b**) C50, (**c**) EDT and (**d**) Ts. Error bars indicate the deviation of the values across the different measured positions of the ODEON model.

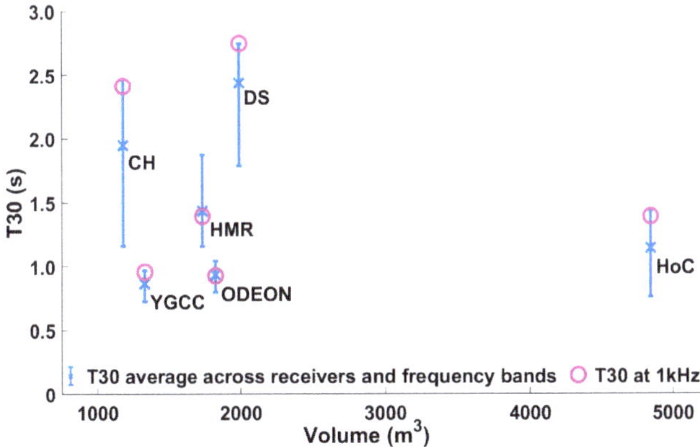

Figure 23. Average T30 results across all frequency bands of the five measured spaces and the ODEON model versus their volume. Error bars represent the deviation of the individual positions for each site, while the pink circle represents the averaged T30 values at 1 kHz.

The modern House of Commons chamber (HoC), while significantly larger in volume than the other spaces (as shown in Figure 23), has a T30 value of 1.14 s (an average across measured positions and across frequency bands). Together with the results obtained from the rest of the acoustic parameters considered here, this indicated that it is a suitable space for the clear perception of speech. This is comparable with the much smaller spaces of the York Guildhall Council Chamber (YGCC) with T30 = 0.86 s, and Holywell Music Room (HMR) with T30 = 1.44 s, as well as the final ODEON c. 1820–1834 House of Commons model with T30 = 0.94 s. This low T30 value for the modern House of Commons was expected, considering the leather materials covering the seating area and the additional use of carpet on the floor. It is also noticeable that the T30 values at 1 kHz for the Convocation House, Divinity School, the modern House of Commons and York Guildhall Council Chamber are at the peak values of the error bars of the average T30 values, indicating the concave curve of the results across frequency bands with maximum values obtained at 1 kHz.

The results of the Holywell Music Room are also as might be expected, considering that the space was designed and used for musical performances. All acoustic parameter values are reasonably higher than the model of the historic House of Commons, and this is useful information and helps to provide verification for the design of the model.

As we had hypothesised from the description of the measured spaces (Section 5), the York Guildhall Council Chamber is the space that not only looks similar to the shape and materials of the designed model of the c. 1820–1834 House of Commons chamber but also has very similar acoustic characteristics to the results achieved for the simulated model. The values of C50 and Ts are also very similar for the two spaces. Ts has a fairly linear shape across all frequency bands, and this is also reflected in the behaviour of the C50 curve. This is something that is observed in both spaces: York Guildhall Council Chamber and the c. 1820–1834 House of Commons model.

The shoe box shape of the modern House of Commons, York Guildhall Council Chamber and the model of the c. 1820–1834 House of Commons, while common, could prove problematic in terms of the distribution and magnitude of early reflections. In all these cases, however, the use of various architectural and design features, such as ornaments, galleries, and coffered wooden panels, help the scattering of the sound across the room.

7. Auralizations

For the evaluation of the simulated model, auralizations are provided as Supplementary Materials as detailed in the section 'Data Availability Statement', enabling the results of the acoustic model for both software (CATT-Acoustics and ODEON), and for all the chosen source-receiver positions to be auditioned. The auralization results of the other presented spaces are also available for comparison, and they are made available via the OpenAir website [51] for posterity. The W channel of the Ambisonic impulse responses (generated from either the commercial software or measured in situ) have been convolved with an anechoic stimulus. The anechoic sample is an excerpt of Henry Beaufoy's speech to the House of Commons in 1792 on the subject of the slave trade, performed by John Cooper (co-author) in the anechoic chamber at the Audiolab, University of York. The perceived differences and similarities of the recorded/simulated spaces as heard in these audio files help to further verify the results of the acoustic parameters presented in this paper.

8. Conclusions

This study has allowed us to present an acoustic reconstruction of the historic House of Commons chamber in the period c. 1820–1834, shortly before its destruction by fire. The acoustic properties have been studied using two commercial software applications based on a 3D model of the reconstructed space. Several implemented versions of the model have allowed us to investigate in depth the most plausible scenarios for the design and choice of construction materials. As in situ measurements of the actual space (either from the past or in a more modern condition) were unavailable, evidence has been gathered from various categories of literature, manuscripts and artistic sources. Information from relevant studies has also been taken into account for the software settings, the chosen materials and the various design decisions.

In situ measurements of spaces with similar architectural characteristics or that have been used similarly, e.g., for parliamentary debates in the 17th century, have offered very helpful data for the comparison and verification of the design of this model.

Considering the above comparisons, we have reassurance that the version of the model presented in this paper provides a plausible representation of the acoustics of the simulated space. It was considered appropriate in this case study to calibrate this model by experimenting mainly on the materials used for the seating area, as confirmed from the results obtained from the in situ measurements of the five spaces. While the assumptions used in defining the absorption and scattering coefficients for the chosen materials can be considered as the main sources of uncertainty for this project, the comparison of the final results with these five spaces has offered confidence towards the final validity of the model.

Future work will take this validated model and use it as a means to investigate the impact of the acoustics of this space on the spatial variation of speech intelligibility. This is with a view of determining the extent by which Members of the UK Parliament, more than two centuries ago, would have been able to hear and contribute effectively to debates based on the location of their seat/position within the Historic House of Commons chamber at this time.

Author Contributions: Conceptualisation, A.F., D.M. and J.P.D.C.; methodology, A.F. and D.M.; software, A.F.; validation, A.F. and D.M.; formal analysis, A.F.; investigation, A.F., D.M. and J.P.D.C.; resources, A.F.; data curation, A.F.; writing—original draft preparation, A.F.; writing—review and editing, A.F., D.M. and J.P.D.C.; visualisation, A.F. and D.M.; supervision, D.M. and J.P.D.C.; project administration, A.F. and D.M.; funding acquisition, D.M., and J.P.D.C. All authors have read and agreed to the published version of the manuscript.

Funding: This project is part of the EU JPI-CH PHE (20-JPIC-0002-FS) (the Past Has Ears) project supported by the UK Arts and Humanities Research Council grant number AH/V001094/1.

Data Availability Statement: Auralization results from all measured positions and spaces are included here https://doi.org/10.5281/zenodo.7560885 (accessed on 20 February 2023).

Acknowledgments: The authors would like to thank Catriona Cooper for the original acoustic model of the House of Commons in 1789 in CATT-Acoustic and relevant literature she had found by the time. Thanks to Frank Stevens, Joe Rees-Jones, Catriona Cooper and OpenAir researchers, who carried out the measurements in situ and provided us with the impulse responses for the purpose of this study. The authors are very grateful to Paul Seaward, Mark Collins, Elizabeth Hallam Smith and Murray Tremellen for their informative discussions while also sharing precious evidence and knowledge with the authors.

Conflicts of Interest: The authors declare no conflict of interest.

References

1. Wyatt, B. Esq. 17 August 1831. In *Report from the Select Committee on House of Commons Buildings Together with the Minutes of Evidence Taken before Them*; House of Commons: London, UK, 1831; p. 6.
2. Smith, E.H. Ventilating the Commons, Heating the Lords, 1701–1834. *Parliam. Hist.* **2019**, *38*, 74–102. [CrossRef]
3. TNA WORK. 29/23, 29/24, 29/25. The National Archives. Available online: https://images.nationalarchives.gov.uk/assetbank-nationalarchives/action/viewAsset?id=35582 (accessed on 23 October 2022).
4. Cooper, C. The Sound of Debate in Georgian England: Auralising the House of Commons. *Parliam. Hist.* **2019**, *38*, 60–73. [CrossRef]
5. Hawksmoor, N. *Longitudinal Section of Proposed Alterations to the Chamber of the House of Commons*; Collection of All Souls College, University of Oxford: Oxford, UK, 1692; AS IV 91.
6. Tillemans, P. *The House of Commons in Session, Oil on Canvas, 137.2 × 123.2 cm*; Parliamentary Art Collection; Palace of Westminster: London, UK, 1709–1714; WOA 2737.
7. Hickel, K.A. *Pitt Addressing the House of Commons, 1793*; National Portrait Gallery: London, UK, 1793–1795; NPG 745.
8. Wallis, W. *Interior of the House of Commons (from 'The Beauties of England and Wales'); Engraving, 142 × 203 mm*; Department of Prints & Drawings, BH/FF/10/London Topography, British Museum: London, UK, 1815; No. 1880, 0911.1241.
9. Scott, J. *View of the Interior of the House of Commons during the Sessions of 1821–1823*; Parliamentary Art Collection: London, UK, 1836; WOA 357.
10. Gillray, J. *Anticipation, or the Approaching Fate of the French Commercial Treaty*; National Portrait Gallery: London, UK, 1787; NPG D12356.
11. Gillray, J. *Parliamentary-Reform,-or-Opposition-Rats, Leaving the House They Had Undermined*; British Museum: London, UK, 1797; No. 1868, 0808.6634.
12. Gillray, J. *The Giant-Factotum Amusing Himself*; British Museum: London, UK, 1797; No. 1868, 0808.6587.
13. Gillray, J. *Stealing Off;-or-Prudent Secession*; British Museum: London, UK, 1798; No. 1851, 0901.944.
14. Virtual St Stephen's Project. Available online: https://www.virtualststephens.org.uk/ (accessed on 14 November 2022).
15. *CATT-Acoustic. User's Manual v9.1*; CATT: Gothenburg, Sweden, 2016.
16. Hallam Smith, E. (English Historian and Librarian of the House of Lords Library from 2006–2016, UK). Personal communication, 2022.
17. Collins, M. (Estates Historian & Archivist of the Estates Historic Archive at the Houses of Parliament, UK). Personal communication, 2022.
18. Seaward, P. (British Historian and Director of the History of Parliament Trust, UK). Personal communication, 2022.
19. Tremellen, M. (Architectural Historian and Ph.D. Candidate, University of York, UK). Personal communication, 2022.
20. Hayter, G. *1833, The House of Commons; Oil on Canvas, 3460 mm × 5420 mm. Overall*; © National Portrait Gallery: London, UK, 1858; Primary Collection NPG 54.
21. Bradley, D.T.; Wang, L.M. Effect of model detail level on room acoustic computer simulations. *J. Acoust. Soc. Am.* **2002**, *111*, 2389. [CrossRef]
22. Duran, S.; Chambers, M.; Kanellopoulos, I. An Archaeoacoustics Analysis of Cistercian Architecture: The Case of the Beaulieu Abbey. *Acoustics* **2021**, *3*, 252–269. Available online: https://www.mdpi.com/2624-599X/3/2/18 (accessed on 20 February 2023). [CrossRef]
23. ODEON. *User Manual, Version 13.0*; ODEON A/S: Lyngby, Denmark, 2016.
24. Naylor, G.M.; Rindel, J.H. Predicting Room Acoustical Behaviour with the ODEON Computer Model. In Proceedings of the 124th ASA Meeting, New Orleans, LA, USA, 31 October–4 November 1992.
25. Alvarez-Morales, L.; Zamarreño, T.; Girón, S.; Galindo, M. A methodology for the study of the acoustic environment of Catholic cathedrals: Application to the Cathedral of Malaga. *Build. Environ.* **2014**, *72*, 102–115. [CrossRef]
26. Alayón, J.; Girón, S.; Romero-Odero, J.A.; Nieves, F.J. Virtual Sound Field of the Roman Theatre of Malaca. *Acoustics* **2021**, *3*, 632–641. Available online: https://www.mdpi.com/2624-599X/3/1/8 (accessed on 20 February 2023). [CrossRef]
27. Weinzierl, S.; Sanvito, P.; Schultz, F.; Büttner, C. The Acoustics of Renaissance Theatres in Italy. *Acta. Acust.* **2015**, *101*, 78–96. [CrossRef]

28. Tronchin, L.; Merli, F.; Dolci, M. Virtual acoustic reconstruction of the Miners' Theatre in Idrija (Slovenia). *App. Acoust.* **2021**, *172*, 107595. Available online: https://www.sciencedirect.com/science/article/pii/S0003682X2030699X (accessed on 20 February 2023). [CrossRef]
29. Iannace, G.; Maffei, L.; Aletta, F. Computer Simulation of the Effect of the Audience on the Acoustics of the Roman Theatre of Beneventum (Italy). In Proceedings of the Acoustics of Ancient Theatres, Patras, Greece, 18–21 September 2011.
30. Iannace, G.; Maffei, L.; Trematerra, P. The Acoustic Evolution of the Large Theatre of Pompeii. In Proceedings of the Acoustics of Ancient Theatres, Patras, Greece, 18–21 September 2011.
31. Gade, A.; Lynge, C.; Lisa, M.; Rindel, J.H. Matching simulations with measured acoustic data from Roman Theatres using the ODEON programme. In Proceedings of the Forum Acusticum 2005, Budapest, Hungary, 29 August–2 September 2005.
32. Beranek, L.L. Analysis of Sabine and Eyring equations and their application to concert hall audience and chair absorption. *J. Acoust. Soc. Am.* **2006**, *120*, 1399–1410. [CrossRef] [PubMed]
33. Martellotta, F.; D'Alba, M.; Crociata, S.D. Laboratory measurement of sound absorption of occupied pews and standing audiences. *App. Acoust.* **2011**, *72*, 341–349. Available online: https://www.sciencedirect.com/science/article/abs/pii/S0003682X10002902 (accessed on 20 February 2023). [CrossRef]
34. Martellotta, F.; Crociata, S.D.; D'Alba, M. On site validation of sound absorption measurements of occupied pews. *App. Acoust.* **2011**, *72*, 923–933. Available online: https://www.sciencedirect.com/science/article/pii/S0003682X11001642 (accessed on 20 February 2023). [CrossRef]
35. Martellotta, F.; Cirillo, E. Experimental studies of sound absorption by church pews. *App. Acoust.* **2009**, *70*, 441–449. Available online: https://www.sciencedirect.com/science/article/abs/pii/S0003682X08001345 (accessed on 20 February 2023). [CrossRef]
36. Carvalho, A.P.O.; Pino, J.S.O. Sound absorption of church pews. In Proceedings of the Inter. Noise 2012, New York, NY, USA, 19–22 August 2012.
37. Galindo, M.; Zamarreño, T.; Girón, S. Acoustic simulations of Mudejar-Gothic churches. *J. Acoust. Soc. Am.* **2009**, *126*, 1207–1218. [CrossRef] [PubMed]
38. Vorländer, M. *Auralization*; Springer: Berlin/Heidelberg, Germany, 2008.
39. AURORA Plug-ins. Available online: http://www.aurora-plugins.com/ (accessed on 24 March 2022).
40. Christensen, C.L.; Rindel, J.H. A new scattering method that combines roughness and diffraction effects. In Proceedings of the Forum Acusticum 2005, Budapest, Hungary, 29 August–2 September 2005.
41. Adeeb, A.H.; Sü Gül, Z. Investigation of a Tuff Stone Church in Cappadocia via Acoustical Reconstruction. *Acoustics* **2022**, *4*, 419–440. Available online: https://www.mdpi.com/2624-599X/4/2/26 (accessed on 20 February 2023). [CrossRef]
42. Facondini, M.; Ponteggia, D. Acoustics of the Restored Petruzzelli Theater. In Proceedings of the 128th AES Convention, London, UK, 22–25 May 2010.
43. Foteinou, A.; Murphy, D.; Cooper, J. Architectural acoustics and parliamentary debate: Exploring the acoustics of the UK House of Commons Chamber. In Proceedings of the Acoustics of Ancient Theatres (2nd Symposium), Verona, Italy, 6–8 July 2022.
44. Farina, A.; Ayalon, R. Recording Concert Hall Acoustics for Posterity. In Proceedings of the 24th AES International Conference, Banff, AB, Canada, 26–28 June 2003.
45. Sert, F.Y.; Karaman, Ö.Y. An Investigation on the Effects of Architectural Features on Acoustical Environment of Historical Mosques. *Acoustics* **2021**, *3*, 559–580. Available online: https://www.mdpi.com/2624-599X/3/3/36 (accessed on 20 February 2023). [CrossRef]
46. Wikimedia Commons Contributors. "File: Convocation House 2, Bodleian Library, Oxford, UK-Diliff.jpg." Wikipedia Commons, the Free Media Repository. 2021. Available online: https://commons.wikimedia.org/wiki/File:Convocation_House_2,_Bodleian_Library,_Oxford,_UK_-_Diliff.jpg (accessed on 30 April 2022).
47. Wikimedia Commons Contributors. "File: Divinity School Interior 1, Bodleian Library, Oxford, UK-Diliff.jpg." Wikipedia Commons, the Free Media Repository. 2021. Available online: https://commons.wikimedia.org/w/index.php?title=File:Divinity_School_Interior_1,_Bodleian_Library,_Oxford,_UK_-_Diliff.jpg&oldid=529292672 (accessed on 30 April 2022).
48. Wikimedia Commons Contributors. "File: House of Commons Chamber 1.png". Wikipedia Commons, the Free Media Repository. 2012. Available online: https://commons.wikimedia.org/wiki/File:House_of_Commons_Chamber_1.png (accessed on 30 April 2022).
49. Music At Oxford. Available online: https://www.musicatoxford.com/venues/holywell-music-room/ (accessed on 26 October 2022).
50. Foteinou, A. Perception of Objective Parameter Variations in Virtual Acoustic Spaces. Ph.D. Thesis, University of York, York, UK, 2013.
51. Open AIR. Available online: https://www.openair.hosted.york.ac.uk/?page_id=1167 (accessed on 30 April 2022).

Article

The Bacinete Main Shelter: A Prehistoric Theatre?

Lidia Alvarez-Morales [1,2], Neemias Santos da Rosa [1,2], Daniel Benítez-Aragón [1], Laura Fernández Macías [1], María Lazarich [3] and Margarita Díaz-Andreu [1,2,4,*]

1. Departament d'Història i Arqueologia, Universitat de Barcelona, 08001 Barcelona, Spain
2. Institut d'Arqueologia, Universitat de Barcelona (IAUB), 08001 Barcelona, Spain
3. Departamento de Historia, Geografía y Filosofía, Universidad de Cádiz, 11001 Cádiz, Spain
4. Institució Catalana de Recerca i Estudis Avançats (ICREA), 08010 Barcelona, Spain
* Correspondence: m.diaz-andreu@ub.edu

Abstract: In the last few years, archaeoacoustic studies of rock art sites and landscapes have undergone significant growth as a result of renewed interest in the intangible aspects of the archaeological record. This article focuses on the acoustic study carried out in the rock art complex of Bacinete, Cádiz (Spain). After describing the archaeological site and its importance, a representative set of monaural and spatial IRs gathered onsite is thoroughly analysed to explore the hypothesis that the sonic component of the site played an important role in how prehistoric people interacted with it. Additionally, we briefly discuss the challenges of analysing the acoustics of open-air spaces following the recommendations of the ISO 3382-1 guidelines, a standard developed not for open-air spaces, but for room acoustics. The results obtained confirm the favourable acoustic conditions of the Bacinete main shelter for speech transmission. The different subjective acoustic impressions obtained in a somewhat similar shelter located nearby, Bacinete III, are also explained, alluding to a lesser degree of intimacy felt in the latter.

Keywords: archaeoacoustics; heritage acoustics; rock art; acoustic measurements

Citation: Alvarez-Morales, L.; Santos da Rosa, N.; Benítez-Aragón, D.; Fernández Macías, L.; Lazarich, M.; Díaz-Andreu, M. The Bacinete Main Shelter: A Prehistoric Theatre? *Acoustics* **2023**, *5*, 299–319. https://doi.org/10.3390/acoustics5010018

Academic Editors: Papatya Nur Dökmeci Yörükoğlu and Rosario Aniello Romano

Received: 27 December 2022
Revised: 16 February 2023
Accepted: 26 February 2023
Published: 6 March 2023

1. Introduction

Beginning as long as forty thousand years ago, and probably even earlier in some places, people all over the world have marked the land with images, a cultural practice that has continued in some places up to our days. The archaeological study of this cultural manifestation has traditionally focused on analysing the style and chronology of the painted or carved motifs or on examining the places where the art is found. However, beginning timidly in the 1980s, and increasingly in the last two decades, there has been growing interest in the intangible aspects associated with rock art sites. In this respect, rock art studies are no different from other fields of archaeology. New theoretical and methodological approaches have led to the appearance of subdisciplines in archaeology, among them archaeoacoustics, which studies the role of sound in past cultures. This field of study brings together knowledge not only from archaeology but also from fields such as architectural acoustics, anthropology and musicology. Archaeoacoustics aims to investigate how past societies perceived and interpreted different types of sound experiences in the context of their cultural practices [1]. From early studies assessing reverberation and echoes measured using simple methods in Palaeolithic caves [2–5] and at post-Palaeolithic rock art sites [6,7], the field has moved on to use more sophisticated methods.

In the last few years, the technological development of acoustic measurement instruments and software for audio processing has allowed a new generation of researchers to expand the studies on the archaeoacoustics of rock art sites and landscapes. This can be illustrated by recent work undertaken by several teams. One of them is that of Riitta Rainio and colleagues, who have been examining acoustic phenomena recorded at Finnish post-Palaeolithic painted sites with impulsive sound sources and binaural recording techniques [8,9]. In these areas, the researchers have identified the presence of echoes and observed that the cliffs chosen as a medium for the images act as potent sound reflectors [9]. Another team was that formed in the context of the Song of the Caves project. A group of experts, including Rupert Till and Bruno Fazenda, analysed the relationship between the location of the rock art and the acoustic properties of five Palaeolithic caves [10,11]. After an exhaustive statistical study, the authors found a subtle—although sufficiently significant—correlation between the acoustics of the spaces and the location of the art. Moreover, since 2012, a team led by Margarita Díaz-Andreu has been carrying out archaeoacoustic research at post-Palaeolithic sites on several continents [12–16]. Based on audio recordings, the registration of site impulse responses (IRs) and the analysis of a wide set of acoustic parameters, the results obtained by this team suggest there was a frequent relationship between the presence of graphic representations and the acoustic properties of rock art sites in many areas of the world. In this respect, it has been argued that places with particular acoustics, such as the Cueva Pintada area in Baja California and the site of Cuevas de la Araña in Spain, may have been chosen to increase the sensorial impact of the social or ritual activities performed in the shelters. Thus, acoustic effects, such as echoes or reverberation, would have intensified sound perception in them. The importance of sound experiences and acoustics at rock art sites has also been corroborated by this team thanks to the presence of musical instruments in some caves and open-air shelters with paintings [17,18] and the depiction of dances in several decorated shelters [19].

The methods followed by all these groups have evolved over time. This is well illustrated by the work of Díaz-Andreu, Lazarich and others in the area of Cádiz and by comparing this article with another published almost a decade ago [20]. The fieldwork for that investigation was undertaken in 2012. It attempted to answer the question of whether the acoustic properties of the Bacinete rock art complex—one of the most emblematic decorated sites in the south of the Iberian Peninsula—could have influenced the activities related to the production and use of images. More specifically, the aim of the research was to test the hypothesis—defended by some rock art experts—that the main shelter of this complex could have functioned as an "auditorium" [20] (p.11), [21] (p. 510). The acoustic tests performed at the time were based on human-made sounds, such as voices, whistles and hand clapping, and indicated the presence of reverberation. However, would their conclusions stand up to measuring the acoustics with a more accurate methodology? This question led us to carry out a new fieldwork campaign in June 2021.

This article, after briefly describing the archaeological site of the Bacinete rock art complex, offers an explanation of the acoustic measurements used to capture monaural and spatial IRs undertaken in the core rock art area of the complex. Before that, we discuss the challenges of analysing the acoustics of open-air spaces following the requirements and specifications of the ISO 3382-1 guidelines, a standard developed not for open-air spaces, but for room acoustics. The acoustic data gathered onsite is then thoroughly analysed to explore the hypothesis that the sonic component of the Bacinete main shelter and Bacinete III played an important role in how prehistoric people interacted with it. Moreover, the results of this research contribute to a better understanding of the activities that could have been carried out at rock art sites and the sociocultural practices of prehistoric societies.

2. The Bacinete Rock Art: An Archaeological Approach

The Bacinete rock art complex is located in the Niño mountain range, in the southeast of the province of Cádiz, Spain. Rock art was produced in this region over millennia, from the Upper Palaeolithic [22,23] to the Late Middle Ages [24]. The site is close to the so-called Bacinete mountain pass at an altitude of 180 m in the Palmones river basin and not far from the sea. The shelters are in an area in which visibility of the surroundings is limited. The only exception is one shelter located at a distance from the others, Shelter I, which has extensive views across the landscape to Algeciras Bay and the Rock of Gibraltar [21] (pp. 494, 499). The site is easily accessed and one of the remarkable features of the area is the spectacular geological rock formations of aljibe sandstone in which the painted shelters are inserted [21] (p. 492). This type of rock is prone to wind erosion and its effects on the sandstone panels are evident. That, and the influence of water and other factors, has resulted in the fading of the paintings and even the possible disappearance of some of them.

The Bacinete rock art has been known for about a century. It was first published in 1929 by Henri Breuil [25] (63–67, Plates 26, 27, 31) and later by many other researchers [21,26–29] (for a historiographical overview, see [21] (pp. 495–497)). Ten different sites have been identified at the complex in an area of some 250 by 150 square metres. Five sites, however, are concentrated in a smaller area of c. 60 by 45 square metres, which will be designated in this article as the core rock art area of the Bacinete complex (Figure 1). The sites in the core area are the main shelter, about which more is said below, and another four that are designated with Roman numerals as Bacinete III, IV, V and VIII. Rock art experts have identified 169/171 motifs for the main site (see below), three sites with 22, 37 and 17 motifs, respectively, and one site with a single motif [29] (pp. 68, 106–130). Whereas all sites have Schematic paintings, exclusively in the main shelter there are some figures produced in the earlier Laguna de la Janda style, also known as the semi-naturalistic style.

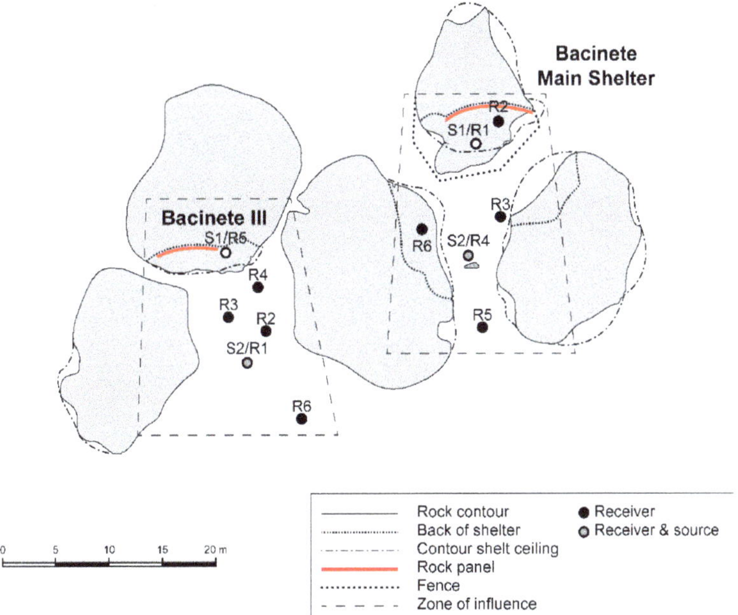

Figure 1. Location of Bacinete main shelter and the Bacinete III shelter in the core rock art area. The approximate sound source (S) and receiver (R) positions set for registering impulse responses are shown. Source: Prepared by the authors based on Lazarich et al. [21] (Figure 5).

2.1. The Main Shelter

In the core rock area of the Bacinete complex, one of its shelters, known as the large or main shelter (Figure 2), stands out for its large number of motifs, styles and production phases, as well as the particular spatial configuration of the rock formations in front of the shelter.

Figure 2. General view of Bacinete main shelter. The red frame indicates the location of the rock art panels. Photographs by N. Santos da Rosa, Artsoundscapes project, 2021.

The number of motifs is severalfold greater than that identified in any of the other shelters, not only in the core area but in the whole complex [29] (p. 68). There is a slight disagreement with respect to the total number of figures, 171 [30] (p. 503) or 169 [29] (p. 68). This minor divergence is probably due to the difficulties associated with working out the nature of certain motifs, given the poor preservation of some of them.

The second reason the main shelter stands out is that the motifs painted in it belong to two distinct styles or rock art traditions: an earlier Laguna de la Janda style, also called semi-naturalist style, and the Schematic style (Figure 3). The Laguna de la Janda style owes its name to the location of what was one of the largest wetlands on the southern Iberian Peninsula until it was drained in the late 1960s [30] (pp. 189–191), [31] (p. 389). In terms of chronology, Solís argues that the earlier style may have originated with the last hunter-gatherer populations in the area and continued after the adoption of agriculture and husbandry [29] (p. 34), the latter being the appropriate date for Lazarich et al. [20] (p. 6), [21]. Regarding the Schematic style, superimpositions make it clear that it is later in time, and both teams date it to the Neolithic and Chalcolithic periods [29] (pp. 45, 150). In addition to a large number of figures and the two distinctive styles, the main shelter is different from all others in the complex because its paintings were produced in at least five different periods. The earliest motifs represent people and animals (principally cervids) that have been classified as Laguna de la Janda style. More motifs were painted later, albeit with a higher degree of schematism. The trend towards schematism is even more marked in the third period, when other motif types, such as dots and different animal species, including canids, were added. More signs possibly symbolising anthropomorphs, as well as dots, were inserted in a fourth phase, and this was followed by a final moment in which the anthropomorphs, zoomorphs (all very schematic) and signs were painted on the surface of the shelter's ceiling and not on the rear wall [29] (pp. 145–146). Spatially, the motifs of the main shelter are located in an area of several metres in which two sectors have been identified. The first comprises the entire rear wall of the shelter, where 13 panels have been distinguished. Sector 2, with only one panel, is on the ceiling and corresponds, as mentioned above, to the most recent painting phase [29] (pp. 106–130).

Figure 3. (**a**) State of conservation of some of the motifs painted in the main shelter of Bacinete with scales added (photograph by M. Díaz-Andreu, Artsoundscapes Project, 2021); and (**b**) record of some of the scenes in the main shelter by Breuil [25] (Plate XXVII).

The final characteristic that distinguishes the main shelter from all the others in the Bacinete complex, with the partial exception of Bacinete III (see below), is its spatial configuration. As explained in the introduction, the site has previously been described as an "auditorium" or a "natural amphitheatre", where the main shelter is considered to be the "stage". Lazarich's team even mentions the visual resemblance of the rock shelter where the panel is to an "acoustic shell". The main shelter is a south-facing concave surface 9.6 m wide by 2.9 m high and 3.10 m deep in which the floor slopes slightly towards the exterior [21] (Figure 2). In front of the painted shelter there is an enclosed flat area surrounded by large rock outcrops (some of them hosting other shelters that are also decorated, Figure 1). This area loosely resembles the "stalls" for "spectators "characteristic of an Italian-style opera house. The area has room for a medium-sized group of people to watch what is taking place at the main shelter. If we calculate about four people per square metre [32] (p. 345), we can estimate a maximum "audience" of some 40 to 50 people. Furthermore, two possible elevated theatre "boxes" (*palcos*) in the rocky outcrops in front of the main shelter are mentioned [21] (p. 510).

Based on all the above, Lazarich et al. and Solís argue that the site could have been used as a gathering place for symbolic or ritual practices [21] (p. 511), [29] (pp. 151, 154), although Bacinete itself cannot be considered as an aggregation site. The archaeological evidence does not indicate that this is a settlement zone, despite the fact that in the area around the rock art complex some flint and sandstone tools (fragments of blades and denticulated blades or sickle teeth) have been found [21] (pp. 512–513), [29] (p. 44). Despite being more restricted in their movements across the landscape in comparison to the previous hunter-gatherer societies, communities with a sedentary way of life also moved around, sometimes as a whole, as happens today in the Catholic "romerías", annual religious pilgrimages to chapels and shrines connected with a particular saint. Events in which the entire populations of villages walk to sanctuaries located in mountainous areas have been known since prehistory on the Iberian Peninsula [33,34]. Moreover, mountain passes are considered special places by many traditional societies (see, for example, [15]).

2.2. Bacinete III

Although there are differences, such as the types and much lower numbers of motifs—only 22, an anthropomorph, dots and a series of unidentified figures—and no evidence of distinct styles and phases [21] (pp. 499–500), [29], the raised platform of the floor of Bacinete III (Figure 4) and the flat area in front of it indicate some interesting similarities with our initial study object, the main shelter. For this reason, it was decided to also acoustically test it. The Bacinete III shelter is 8.3 m long, 3.8 m high and 2.8 m deep. The crowd that would have been able to fit in front of the shelter would also have been similar in size, perhaps a little larger, than at the main shelter. Intriguingly, no author has raised the possibility of it being used as a performance area.

Figure 4. General views of Bacinete III. The location of the rock art panels is framed in red. Photograph by N. Santos da Rosa, Artsoundscapes Project, 2021.

3. Materials and Methods

3.1. Assessing the Acoustics of Open-Air Spaces with ISO 3382-1

Despite referring to indoor performance spaces, numerous studies have successfully adopted ISO 3382-1 [35] as a reference when investigating the acoustic field in open-air spaces, including those from prehistory to our days, even though their sound field differs considerably from the ideal diffuse field considered in the standard [36–44]. In this respect, ancient theatres represent one of the most studied groups of open-air historical sites (see, for example, [36–41], and other references below). Moreover, we can also mention recent investigations conducted in other open spaces that were not intentionally designed as performance areas but where the sound field takes on special relevance, such as the Renaissance Palace of Charles V inside the Alhambra (Granada, Spain) [42] and several historical courtyards in Europe [43,44].

Recently, Astolfi et al. [37] discussed the applicability of ISO 3382-1 for performing acoustical measurements in ancient open-air theatres by using the theatre of Tyndaris (Sicily, Italy) as a case study. They reported excellent reproducibility in terms of the acoustic parameters when using two different measurement techniques (a dodecahedron source and firecrackers) included in ISO 3382-1. However, they expressed their concern regarding the regression analysis of the decay curves proposed in the standard for the estimation of the reverberation parameters. They argued that it is not appropriate for the analysis of the "cliff-decay curve linked to a few strong reflections" commonly registered in open-air sites. Previous studies had already drawn attention to this issue. Farnetani et al. [45] pointed out that the reverberation time (RT) behaviour in an open-air theatre differs from the classical reverberation theory. Moreover, Paini et al. [46] claimed that reverberation time (T_{30}) is not representative of "stepped decays" because, a priori, it does not give enough information about the real sound decay; it misses the initial step caused by the lack of early reflections and underestimates the late part result of the contribution of the isolated late reflections that could be easily heard. Meanwhile, in terms of subjective perception, van Dorp Schuitman and de Vries [47] demonstrated that the Early Decay Time (EDT) accounts poorly for the reverberance of rooms that have uneven absorption distribution. A few years later, Mo and Wang [48] also claimed that conventional reverberation parameters (RT and EDT) are questionable for judging the perceived reverberance in unroofed spaces that have uneven absorption distribution. They recommended that the spatial characteristics of the reflections arriving at each listener position be considered for describing the temporal attribute of reverberance. Despite the above-mentioned limitations, Chourmouziadou et al. [49] and Thomas et al. [50] found the reverberation time (T_{30}) to be an interesting parameter for comparing the sound field in different ancient theatres and urban squares (provided the potential inaccuracy of the obtained values is previously assessed in depth) since T_{30} is strongly dependent on the site geometry and only marginally depends on the listeners' locations.

Despite all the concerns regarding the applicability of ISO 3382-1 and the suitability of the acoustic parameters when dealing with open-air spaces, most authors continue to use ISO 3382-1 as a reference for measuring the acoustics of open-air sites. Despite its limitations, it is regarded as a guarantee of scientific rigour, repeatability and reliability when capturing the impulse responses and studying the acoustic behaviour of sites. Therefore, the methodology to be applied in this article was established by following, as far as practicable, the recommendations included in ISO 3382-1 [35], although we introduced some modifications due to the specific characteristics of rock art sites. In this respect, we followed the recommendations made a few years ago by Till [10]. This scholar stated that sound archaeology needs an alternative approach to the measurement protocol in order to account for the typical complex/irregular shapes of archaeological sites and to make detailed identification of potential acoustic effects possible. Till also claimed that the considerable uncertainties inherent to prehistoric sites force those testing the acoustics to take a flexible approach, both when defining the measurement protocol and in interpreting the results.

3.2. Acoustic Measurements in the Open-Air Core Rock Art Area of the Bacinete Complex

The experimental measurements in the core rock art area of the Bacinete complex were aimed at testing the hypothesis of the suitability of the Bacinete main shelter as a "natural auditorium" [20,21] (pp. 509–511), [29] (p. 152). As explained above, we also decided to test the adjacent site of Bacinete III [21] (pp. 499–500), [29] given the broad similarities between the two sites in terms of spatial arrangement: both sites have a flat, spacious area in front of the painted shelter in which a medium-sized crowd could have gathered and listened. The acoustic tests were undertaken on 10 and 16 June 2021. The purpose of the measurements was to gather a representative set of impulse responses (IR) from the selected sites to be used, in a first instance, to assess their acoustic properties in order to objectively relate the sound component of each site and its rock art. A second objective of the measurements, to gather data for auralisation purposes, will be discussed in a future publication.

An omnidirectional sound source (an IAG DD4 mini dodecahedral loudspeaker together with an IAG AP4GB power amplifier) was used to emit the excitation signal, a 12-s long exponential sine-sweep covering from 50 Hz to 20 kHz. Two positions of the sound source (S) were considered at each rock art site. These were selected according not only to the size and morphology of each of the sites but also to the location of the rock art. Six receiver points (R) were also distributed throughout the "zones of influence" of each source position. For the selection of the receiver locations, the non-diffuse condition of the sites was considered, trying to set enough test points to register the acoustics of the different "zones" created by the particular spatial configurations of the rocks. Inevitably, the number of S-R combinations included in the study was limited due to the time allowed for accessing the sites, which are on private land. Information about the S-R combinations analysed is provided in Figure 1, Tables 1 and 2. Furthermore, the spatial limitations caused by the height of the rocks, as well as the presence of vegetation and other particular elements, such as the protective iron fence installed around the main shelter, needed to be considered. It is also important to reiterate that there is no archaeological evidence regarding the use of the sites, meaning that details of the positioning of the sound source, the audience arrangement or the number of attendees in prehistory are not available. All these peculiarities meant that the equipment positioning had to be flexible, not only in the selection of the S-R locations but also in their height. For the height of the transducers, we worked with the hypothesis of a "standing audience". Only from that position were both the rock art panel and the sound source on the "stage" visible from all parts of the "audience area". Therefore, as far as practicable, the sound source was set at a height of 1.50 m and the receivers at a height of 1.40–1.50 m, which are assumed to be close enough to the average height of the ears of a standing speaker/listener [51,52]. Measurements were taken in unoccupied conditions.

At each selected receiver position, monaural IRs were registered with an omnidirectional microphone (micW n201) together with a high-quality audio device (Zoom F4), by using EASERA 1.2 software tool [53]. Spatial IRs were also gathered using a multichannel microphone (Zylia ZM) with its centre of coordinates always oriented towards the sound source. In this case, the excitation signal was played back using a Zoom H2n handy recorder and recorded with the audio software Plogue Bidule 0.9762. The captured spatial signals were post-processed in Matlab R2022a (The Matlab scripts used for the postprocessing of the IRs are available in the University of Barcelona repository at https://doi.org/10.34810/data649, accessed on 15 September 2022) in order to obtain the first and third-order ambisonic IRs, which were later analysed in several software tools (EASERA, ARTA [54], and IRSpatial [55]).

In the analysis phase, all the IR waveforms were carefully examined, focusing on their reflection pattern to better understand the effect on the acoustics of the site of any strong early reflections potentially coming from the rock formation. In this regard, the third order ambisonic IRs are used to estimate the direction of arrival (DOA) of the most relevant reflections, using the software tool IRSpatial [55]. The standard choices to assess the reverberation conditions of the space (T_{20} and EDT), sound clarity (D, C_{80}, T_s), and perceived loudness (G) were calculated. An in situ calibration method was used to calculate G [16,56]. In addition to these, the early lateral energy fraction (J_{LF}) was studied as a representation of the spatial behaviour of sound at those sites. The spectral behaviour of the selected acoustic parameters was analysed at each S-R combination individually, paying particular attention to the mid values in each of the zones of influence associated with each specific source position. This was performed in order to explore the spatial variability of the acoustics of the space. Spatially averaged values were then used to assess the behaviour of the space as a whole, providing an easy way to establish a comparison between the different rock art sites in the complex.

It is worth mentioning that environmental conditions, including temperature, wind velocity and background noise levels, were monitored during each measurement session. As far as possible, measurements were taken when wind speeds were negligible (less than 0.5 m/s), avoiding wind gusts. The temperature was approximately 34 °C and 31.5 °C, respectively, in the main shelter and Bacinete III. Air temperature variation at each shelter was less than 1.3 °C from the beginning to the end of the session, which is within the limits of the recommended maximum deviation of 2 °C [57]. Background noise at the complex was low enough not to cause any disturbance during the measurement sessions ($L_{Aeq} \approx 30$ dBA).

4. Results and Discussion

4.1. An Overview of the Impulse Response Structures and Reverberation Conditions in the Bacinete Core Rock Art Area

Considering that these shelters are open-air spaces where no diffuse sound field is expected, the analysis first focuses on the IR structures and reflection patterns. In general, the IRs captured at the Bacinete core rock art area show only a few early reflections after the arrival of the direct sound due to the rock formations, and almost negligible scattered sound energy caused by the landscape 100 ms after excitation. No late reflection that could cause an echo in any of the impulse responses registered is identified, and neither can any echo be discerned for speech or music according to the Dietsch and Kraak echo criterion [58]. Furthermore, as sound scattering depends on geometry, the irregular shape of the rock blocks avoids flutter echoes due to parallel walls, despite their proximity.

Even offering high impulse-to-noise ratio values (INR > 45 dB), this IR structure, frequent in open-air places, leads to a non-linear early decay curve (EDC), in which the influence of the direct sound creates an initial "cliff-type" decay in the integrated response [59], followed by about 100 ms of stepped decay produce by the infrequent strong early reflections (see some examples in Figure 5).

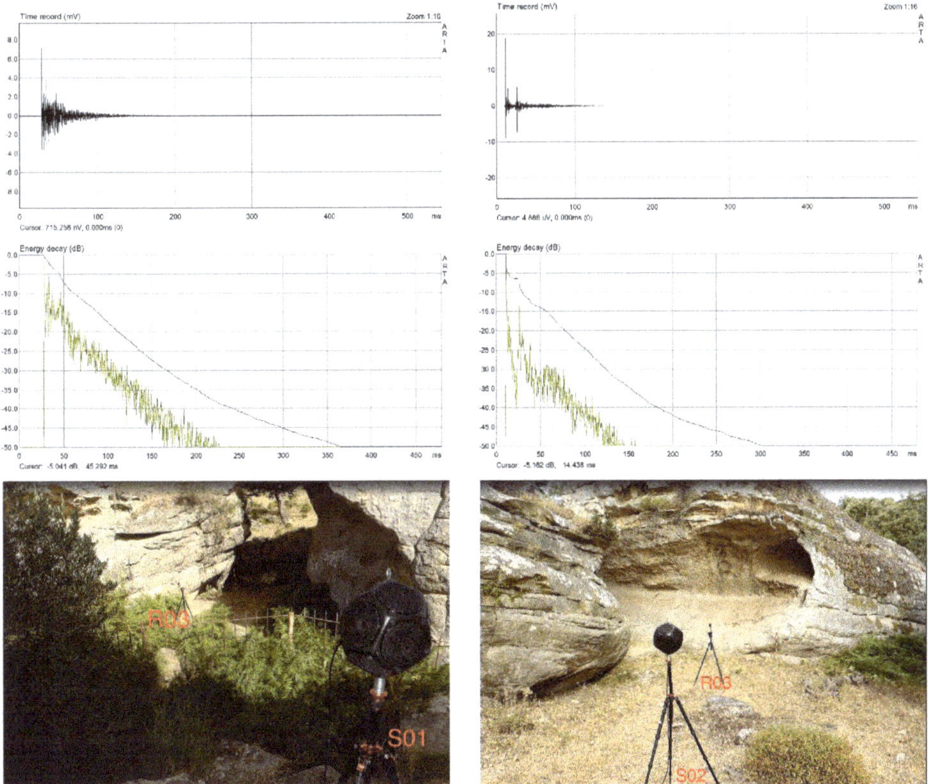

Figure 5. Impulse response (IR) and its early decay curve (EDC) captured at Bacinete main shelter with the source placed in front of the paintings (S1) and the receiver in the "audience" area (R3) (left column); and at Bacinete III, with the source in the middle of the audience area (S2) and the receiver next to the paintings (R3) (right column).

As previously mentioned, this type of EDC leads to an inaccurate estimate of the reverberation parameters. In view of this limitation, and following a recent study by Galindo et al. [60], the degree of non-linearity (ζ) has been studied so that it is considered when interpreting the results of the reverberation parameters obtained at the Bacinete rock art complex. The non-linearity parameter, ζ, as defined in Annex B of the ISO 3382–2 [61], quantifies the degree of EDC deviation (figures given per thousand) with respect to perfect linearity represented by a value of the square of the correlation coefficient (r^2) equal to 1. The standard establishes that ζ values greater than 10‰ indicate that the EDC is far from being a straight line and the value of the reverberation time estimated from this curve may be doubtful. Although in most of the IRs measured at the Bacinete main shelter and Bacinete III, the aforementioned linearity criteria are not strictly matched, meaning that this non-diffuse field affects the reverberation time calculations, ξ values are not extremely far from this criterion ($\xi \leq 10‰$), ranging between 7 and 17‰, except in a few S-R combinations (see Tables 1 and 2). Those S-R combinations that produce higher ξ values will be discussed in detail later. Therefore, although the reverberation parameter results should be interpreted with caution, the reverberation experienced at the sites is assessed through the analysis of the reverberation time, in order to present an initial insight into the Bacinete rock art complex acoustics and to establish a comparison between the different shelters.

Table 1. Mid values measured and averaged according to ISO 3382-1 [35] in the Bacinete main shelter.

ID Test	Reverberation Ratios			Energy Ratios			Loudness Param.	Spatial Param.	S-R Dist. [3]
	T_{20m} (s)	EDT_m (s)	ξ [1] (‰)	D_m (-)	C_{80m} (dB)	T_{Sm} (ms)	G_m [2] (dB)	J_{LFm} (-)	- (m)
BAm_S1_R5	0.37	0.32	16.928	0.89	15.40	21.95	−1.80	0.12	19.3
BAm_S1_R4	0.34	0.32	7.984	0.89	15.25	22.57	3.10	0.19	14.0
BAm_S1_R3	0.32	0.24	10.970	0.93	16.35	18.30	6.10	0.21	10.4
BAm_S1_R2	0.29	0.16	13.951	0.97	20.30	18.88	15.85	0.21	2.9
BAm_S1_R6	0.32	0.40	13.951	0.73	11.55	14.06	1.65	-	12.7
BAm_S2_R1	0.36	0.30	12.958	0.89	14.90	24.70	2.50	0.19	14.5
BAm_S2_R3	0.29	0.33	16.928	0.93	18.20	22.57	10.20	0.21	5.2
BAm_S2_R5	0.30	0.19	23.856	0.96	20.80	18.30	12.10	0.21	4.0
BAm_S2_R6	0.37	0.20	12.958	0.95	17.25	18.88	11.15	-	3.65

[1] Linearity factor calculated as defined in Annex B of the ISO 3382-2: 2008 [61]. [2] Reference signal information: in situ calibration method [16,56]. [3] Approximate values measured onsite.

Table 2. Mid values measured and averaged according to ISO 3382-1 [35] in the Bacinete main shelter.

ID Test	Reverberation Ratios			Energy Ratios			Loudness Param.	Spatial Param.	S-R Dist. [3]
	T_{20m} (s)	EDT_m (s)	ξ [1] (‰)	D_m (-)	C_{80m} (dB)	T_{Sm} (ms)	G_m [2] (dB)	J_{LFm} (-)	- (m)
BA3_S1_R1	0.27	0.11	13.951	0.98	23.5	10.4	9.95	0.13	10.0
BA3_S1_R2	0.24	0.31	12.958	0.91	18.8	20.3	8.85	0.13	7.7
BA3_S1_R3	0.28	0.14	6.988	0.97	21.8	11.5	13.70	0.23	4.8
BA3_S1_R4	0.27	0.30	21.879	0.92	19.7	16.0	12.00	0.16	3.9
BA3_S2_R2	0.31	0.20	15.936	0.95	20.1	10.9	9.55	0.13	4.1
BA3_S2_R3	0.30	0.28	11.964	0.95	18.6	14.6	7.85	0.23	5.3
BA3_S2_R4	0.30	0.29	11.964	0.92	16.7	16.7	5.55	0.16	7.2
BA3_S2_R5	0.29	0.12	13.951	0.97	21.2	12.4	9.50	0.34	10.0
BA3_S2_R6	0.34	0.19	37.639	0.95	20.2	10.3	5.05	0.19	6.5

[1] Linearity factor calculated as defined in Annex B of the ISO 3382-2: 2008 [61]. [2] Reference signal information: in situ calibration method [16,56]. [3] Approximate values measured onsite.

In closed spaces, reverberation time values above 1 s are common, and results in low-frequency bands are often significantly higher than at high frequencies, due to the absorption properties of their finishing materials and the air volume inside them. However, at Bacinete, we are dealing with an open space; thus, different results are expected. Figure 6 shows the spatially averaged values of the reverberation time (T_{20}) obtained in each shelter for each source location in each frequency band. As can be seen, the T_{20} values obtained in all the shelters are very low (well below half a second), with fairly flat behaviour throughout the studied frequency bands. These values, ranging from 0.25 to 0.35 s at mid frequencies (500 Hz and 1 kHz), are even below the common values found in small conference rooms (typically around 0.5 and 0.6 s in rooms with volumes of 50 to 1000 m^3 [62]). Despite this, they lie within the range expected for open conditions, being similar to those found at other open archaeological sites such as Stonehenge in its old configuration [63] (Table 3, Table 4 and Table 5) and other rock art shelters on the Iberian Peninsula [16]. A priori, this lack of reverberation, with T_{20m} of about 0.3 s, makes these spaces suitable for speech transmission [64]. Additionally, the observed spectral behaviour, with no significant enhancement either at low or high frequencies, suggests that the site cannot be considered to add any warmth or brightness to sound [65]. The reverberation time also shows very little dependency on the source–receiver combination (see error bars in Figure 6).

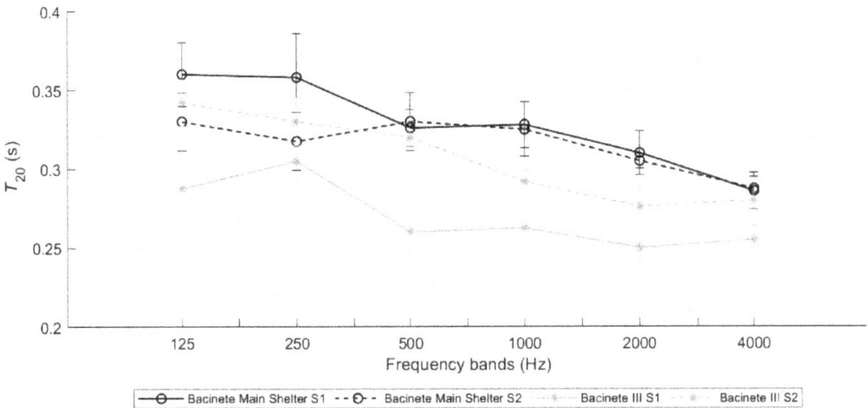

Figure 6. Reverberation time (T_{20}) obtained in each shelter of the Bacinete rock art complex: Bacinete main shelter; Bacinete III. Spatially averaged values for each source location are represented at each frequency band.

Returning to the visual inspection of the IRs, and focusing on their first part, the initial time-delay gap (ITDG) is analysed. The ITDG is defined as the time between the direct sound and the first significant reflection at a certain receiver position and had long been considered to be related to the acoustical intimacy [66] or, in other words, the listener's impression of the size of a room [65]. Furthermore, Mi et al. [67] determined through listening tests not only that the ITDG has a large impact on perceptual reverberation, but also that, for places with a reverberation of around 30 ms as in this case, small variations in ITDG (of about 18 ms) significantly influence the resultant perceived reverberation. Despite the low number of reflections present at Bacinete, determining which should be considered for the calculation of ITDG is not straightforward (see Figure 5 as an example), and for this reason, the exact ITDG values associated with each IR are not included here. Nevertheless, what is clear is that ITDG values are considerably shorter in all the S-R combinations (remaining well below 15 ms), presumably because the rock formations causing the first reflections are relatively close to the receiver positions. Beranek [68] reported that concert halls with IDTG values up to 20 ms were judged to have intimate acoustics, and the shorter the ITDG, the more intimate the experience. He also observed that lateral reflections are crucial for intimacy. It is worth mentioning that Lokki and Pätynen [69] recently discussed the concept of intimacy, claiming that it cannot be judged just through the ITDG, as sometimes considered [70], since this metric ignores other important factors, such as the perceived loudness or the direction of arrival of first reflections. In open spaces with a lack of a reverberant tail, Thomas et al. [50] also found a robust correlation between ITDG and other parameters measuring the fraction of early energy, such as C_{80}. All this suggests that the absence of reverberation and low ITDG values caused by the nearby rock formations may contribute to building a subjective feeling of intimacy in the shelters, despite the fact that they are open spaces.

At this point, and considering that these sites have a non-diffuse acoustic field, the variation in the results in the S-R positions studied is of particular interest for discussing proposed hypotheses for the site acoustics. Hence, individual results at each S-R combination are analysed from now on instead working with spatially averaged values, as recommended in the standard [35] and previous sound archaeology studies [10].

4.2. The Acoustics of the Bacinete Main Shelter

The analysis of the IRs gathered at the Bacinete main shelter was undertaken with the aim of carrying out a more rigorous study of the hypotheses raised by Díaz-Andreu and Lazarich [20,21] (pp. 509–511), and later mentioned by Solís [29] (p. 152), on the use of the main shelter as an auditorium. In their publications, Lazarich et al. and Solís present a thorough spatial study of the site, focusing on the spatial distribution of the rock outcrops. In the core area, both studies identify the space in front of the Bacinete main shelter as a natural theatre or auditorium, establishing in this way a non-explicit formal parallelism between the spatial distribution around the shelter and the traditional arrangement that has predominated in Western theatres and opera houses since the 17th century: the Italian style [71,72]. One of the main architectural characteristics of the so-called Italian style is that the stage or performance area is located in front of the spectators, on a higher level. This is the location of the rock art panels of the Bacinete main shelter in comparison to the area in front (the area identified as A in Figure 1). Lazarich et al. stated that "the concavity of the shelter acts as an acoustic shell amplifying and redirecting sounds from the "stage" area to the "spectators" area" [21] (p. 528). Just in front of the main shelter, the authors identified the audience area, a "lobby" or "stalls" area delimited by two rock outcrops (identified as B and C in Figure 1). The so-called stalls area is a relatively flat surface with a slight slope, and there are two "boxes" on the northern faces of these two rocky elements that could be considered the "upper gallery". Following this parallelism, it can even be considered that there is a space for the "pit" (or orchestra area) between the main shelter (stage) and the "audience", which today is covered by very tall ferns.

It must be clarified that, in this context, the theatre is understood as a space destined for holding scenic events, not necessarily recreational, but ritual, or any other social event that might involve sound. However, in contrast to what Solis appears to imply [29] (p. 151), Bacinete does not present the typical archaeological features of an aggregation site, an aspect discussed in detail in an article recently published by our team [16]. In this respect, although it was probably considered a special place for ritual activities, the absence of a significant number of archaeological finds with stylistic variability suggests that the site was not occupied consistently and repeatedly over a comprehensive period of time.

So that we could explore the initial hypothesis regarding the shelter being used as a theatre, the sound source was placed in front of the paintings, approximately in the centre of the "stage". In order to analyse the acoustics in each of the above-mentioned zones, four listener positions were distributed around the audience area: R3, R4 and R5 in the "stalls"; and R6 in one of the "boxes". Additionally, a receiver location was set on the "stage": R2 (Figure 1). The acoustic parameter results are summarised in Table 1.

As expected, when the sound source and the listener are both in the shelter (S1-R2), the listener perceives a reverberation level even lower than those listeners situated in the "audience" area (EDT_m = 0.16 s), which is due to the short distance between them (about 3 m). Furthermore, the significant influence of direct sound and the strong early reflections produced by the rock shelter itself cause extremely high values of sound clarity (D_m = 0.97 and C_{80m} = 20.3 dB). The rock shelter delimited the space at this site, similarly to the stage wall (*scaena frons*) in an ancient open-air theatre [39,73,74]. Thus, the sound reflecting on the shelter caused a significantly prompt first-order reflection towards the listeners in the "stalls", supporting the direct sound and, consequently, helping to improve the source loudness and sound clarity. The reflections caused by the shelter are of great intensity since it is a hard surface that barely absorbs sound. Nevertheless, the scattering effect of the shelter was limited due to the roughness of the rock formation. Thus, stating that the shelter acted like a shell able to focus the sound evenly on the audience was perhaps an overstatement. This is because the acoustic efficacy of a shell depends to a great extent on its shape and what is made of [75,76].

As the results show (see Table 1), Lazarich et al. [21] and Solis [29] were correct when they identified the "stalls" as a privileged area from an acoustic point of view. This, however, as the analysis of the impulse responses demonstrates, is not exclusively due to the reflections provided by the "acoustic shell" formed by the shelter, as they propose. To investigate the DOA of the main reflections, the spatial IRs captured with the Zylia microphone were analysed with the IRSpatial software tool. The direct sound from the source on the stage reaches the receivers in the stalls area without interference, since there are no physical obstacles that could distort sound propagation towards the receivers, and also due to the absence of a reverberant field. In the first 40 ms (approximately) after the arrival of the direct sound, there is a group of important reflections coming from the front, presumably from the rear panel of the main shelter. Immediately after this, there is a group of intense reflections coming from the rock outcrops designated B and C in Figure 1. These are more lateral and come from the upper plane, due to the shape of the rocks. Figure 7 shows an example of the analysis performed with the IRSpatial software tool.

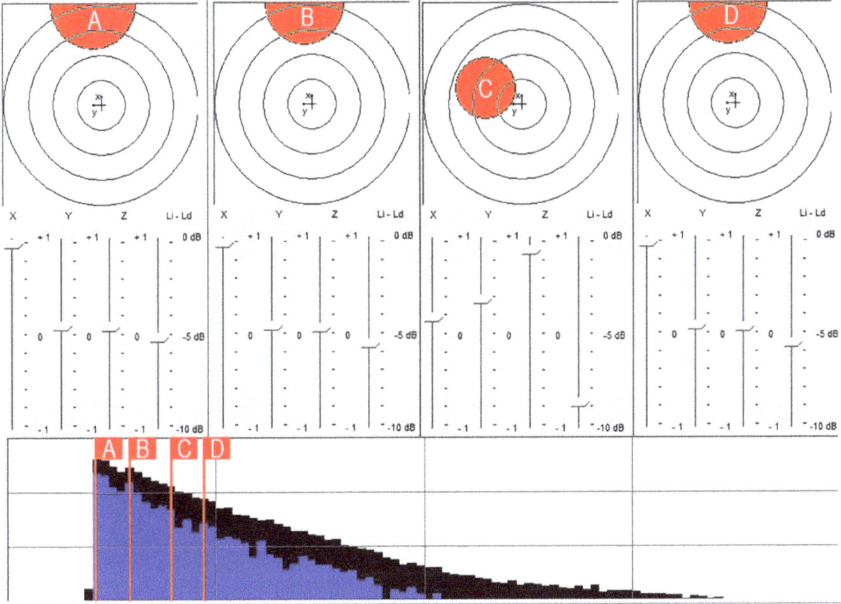

Figure 7. Example of the DOA analysis performed with the IRSpatial software tool. The 1st-order ambisonics impulse response corresponds to the S1-R4 combination in the Bacinete main shelter. Red dots show the DOA and the intensity of each group of reflections (A, B, C or D), the blue bars represent the sound intensity and the black bars the energy density of the signal.

Therefore, potential listeners in the "stalls" would have received significant early reflections from the lateral blocks, which would have reinforced the direct sound providing greater clarity and definition ($D_m > 0.89$, $C_{80m} > 15$ dB and $T_{Sm} < 25$ ms, see Table 1). These results indicate very favourable conditions for speech transmission. Music would also have been heard very clearly in the "stalls". However, as commonly found in other open-air sites, e.g., Roman and Greek theatres, C_{80m} values are possibly too high, exceeding the upper range limit criteria set for roofed concert halls and auditoria taking into account different types of music [77,78]. Additionally, as a considerable number of these early reflections arrive late rally at the reception points R3 and R4, values of $J_{LFm} \approx 0.20$ are obtained. These values are higher than expected at an open-air site, considering that the J_{LF} is related to the spatial impression [79] and such averaged values are similar to those measured in popular concert halls [78,80]. Centre time is considered not only as a cue of clarity of sound as usual;

rather, it is seen as an indicator of the presence of late reflections and echoes due to its sensitivity to late energy. This is because no prior division of the impulse response between early and late sound is considered [46]. In view of the low T_{Sm} values obtained (between 13 and 37 ms), it seems there are no reflections that are able to shift the impulse response centre of gravity to higher values.

Lazarich et al. also referred to a natural amplification in terms of the level subjectively experienced in this area when sound is emitted from the "stage" [21] (p. 510). This subjective feeling is corroborated by the G_m values obtained in R3, R4 and R5, which are 3 to 6 dB higher than the expected level in the free field using the same system configuration (see Figure 8). The just noticeable difference (JND) for G is set as 1 dB [10], and therefore it can be said that there is a significant natural amplification of sound caused by the shelter at those positions. Actually, as has been pointed out in the literature, spatial perception has been proven to increase not only with the density of early lateral reflections but also with the early sound level [81]. Therefore, these high G values would have reinforced the spatial impression created by the lateral early reflections denoted by considerable J_{LFm} values. In combination with the absence of reverberation and the low ITDG values obtained, these greater J_{LFm} values are possibly responsible for the subjective feeling of intimacy reported by the authors.

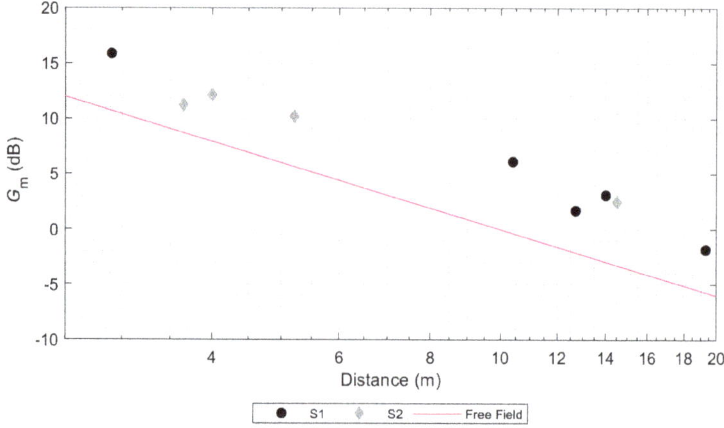

Figure 8. Sound strength, G_m (dB), spectrally averaged values calculated from the measured IR in the Bacinete main shelter. The theoretical free field propagation is represented by the red line.

Lazarich et al. [21] (p. 510) also mention that the "boxes" of the "upper gallery" could be considered privileged positions in terms of visibility and listening conditions. Looking at the results obtained for S1-R6, however, nothing objectively supports the idea that listeners in those positions would have experienced better acoustics than those in the "stalls" due to the acoustic features of the site. Even the sound clarity parameter values (D_m and C_{80m}) are slightly reduced. Nevertheless, an improvement in the visual cues in those upper positions might have led to a subjective feeling of improvement in acoustic perception [82].

Trying to put aside the presumption that the space was used as a theatre since it can be considered an idea influenced by the human ability to establish a parallel between the natural arrangement of the rocky blocks and the architectural elements of a theatre, one further hypothesis on the use of the site was explored. In this case, the sound source was located in the middle of the area previously designated as the "stalls" (S2 in Figure 1), considering the "stage" part of the area for the listeners. The results in Table 1 show that the acoustic experience of the potential listeners is very similar in comparison to the previous hypothesis in terms of reverberation and clarity of sound. The main difference is that the S-R distance is now much shorter, except in the "stage" area, so the sound level registered at the receiver locations is higher. Furthermore, the reflection pattern changes considerably.

Through the analysis of the spatial IRs, it can be seen that, with the source located in S2, frontal and lateral reflections are produced by the upper part of the rock formations B and C, and the role of the painted shelter is almost negligible since it only produces sporadic reflections of very low intensity. In any case, the areas where the listeners would have been located continue to receive a considerable number of early lateral reflections, as indicated by the J_{LFm} values (Table 1). Therefore, although this configuration reduces the number of attendees with visibility of the sound source, it could also have been favourable for speech transmission since the listeners received the message with a higher sound level than when the source is located on the stage.

4.3. The Acoustics of Bacinete III

A similar analysis to the one described in Section 4.2 was performed at Bacinete III. The measurement campaign was designed to answer two main questions: Could the area in front of Bacinete III have also been used as a place for a medium-sized group of people (slightly larger in numbers than at the main shelter) performing rituals and/or social events? Were the acoustic conditions of this site suitable or advantageous for such a use in comparison to those found at the Bacinete main shelter? Two sound source positions and six receiver locations were set for the acoustic measurements: S1 was placed next to the paintings on the rock formation, in a position raised about 2.70 m from the floor, and S2 was placed approximately in the middle of the rock formations. Following the same inspiration in Italian-style theatres as in the Bacinete main shelter, S1 would have been on the "stage" and S2 in the middle of the audience area. The receiver designated R5 was also on the stage, and the receivers from R1 to R4 were distributed throughout the audience area, withR6 being located in the farthest position from the paintings, right in front of a small boulder.

The results obtained for the different acoustic parameters are summarised in Table 2. As expected, the acoustics experience at the site is again defined by the lack of reverberation ($T_{20\mathrm{m}}$ of about 0.30 s) and the presence of strong early reflections that produce considerable clarity of sound ($D_{\mathrm{m}} > 0.91$, $C_{80\mathrm{m}} > 16$ dB and $T_{\mathrm{Sm}} < 20$ ms). However, as denoted by the J_{LFm} values, at Bacinete III fewer early lateral reflections than at the Bacinete main shelter reach the listener positions (except those set next to the rock formations) since the distance between the rock formations at the sides is greater in this case. Hence, lower J_{LF} values, meaning lesser feelings of a spatial impression than at the Bacinete main shelter, would have been experienced. Furthermore, the shape of the rocks leaves, in this case, the audience area completely uncovered, which leads to a prevalence of reflections on the horizontal plane. This means that at Bacinete III, reflections from above those of the Bacinete main shelter that might have supported such subjective feelings of "being inside a place" were not received either. Consequently, even though at Bacinete III there is more space for congregation, and reverberance and clarity conditions are comparable to those experienced at the Bacinete main shelter, the audience area is more open than that considered in the other shelter. As a result, listeners would have perceived a lesser feeling of spatial impression in Bacinete III than in the Bacinete main shelter. In terms of loudness, in both configurations considered at Bacinete III, a high enough sound level would have reached the listeners. Only when the source or the receiver was placed in the rock shelter next to the paintings (meaning the source was placed at S1 at all the receiver points; and at R5 when the source was placed at S2) did the intense early reflections produced by the shelter result in a perceptible natural amplification of sound (G_{m} values well above the free field curve, see Table 2 and Figure 9).

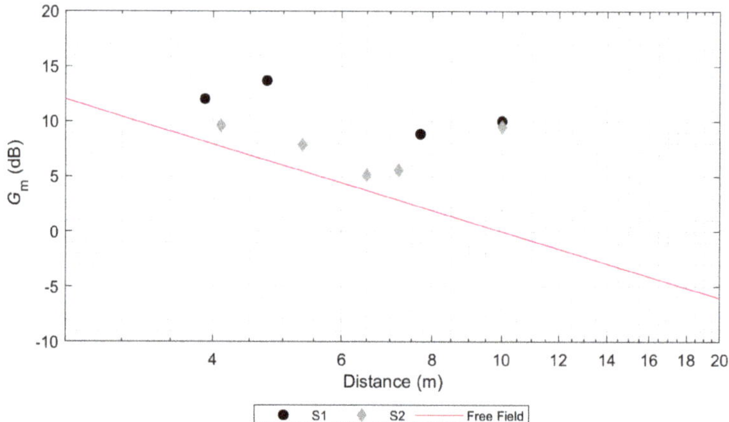

Figure 9. Sound strength, G_m (dB), spectrally averaged values calculated from the measured IR at Bacinete III. The theoretical free field propagation is represented by the red line.

Furthermore, examining the DOA of the main reflections in IRSpatial in the same way as at the Bacinete main shelter, it is possible to observe that when the source is located next to the paintings (S1 position), most of the sound energy that reaches the potential listeners comes from the front, except for some lateral reflections. Conversely, when the source is located in S2, potential listeners receive the direct sound from the front, followed by about 30 ms dominated by strong lateral reflections produced by the lateral blocks. After these reflections, a group of lower-intensity reflections coming from the rear (i.e., the painted shelter) is discerned. When the reflections coming from the rear are intense and delayed enough in comparison to the direct sound, they affect intelligibility in the receiver positions closest to the painted shelter. As a final comment, it is worth mentioning that this lack of lateral reflections could be the reason why the averaged reverberation time estimated with S1 in Bacinete III is even lower than that found in the other configurations studied, both in Bacinete III and in the Bacinete main shelter (see Figure 6).

5. Conclusions

This article has presented the results of a thorough acoustic measurement campaign carried out in the rock art complex of Bacinete, Cádiz province (Spain). The discussion has focused on what we have designated as the core rock art area in the complex, which contains five painted shelters. The so-called main shelter stands out over all the others for several reasons, the first being the presence of two different artistic styles, with the oldest —the Laguna de la Janda style—being unique in the whole complex. The site also contrasts with all the others because it has the largest number of motifs and they were painted in at least five distinct phases. The Bacinete main shelter is also remarkable because it is more elevated than the area in front of it, loosely reminding us of the layout of an Italian-style theatre, which has led rock art researchers to propose that it acted as an auditorium [21,83]. An initial analysis of its acoustics appeared to indicate that there was a certain amount of reverberation that may have helped create the right atmosphere for any events that may have taken place there [20]. However, the simplicity of the method used at that time called for a revision of the tests to identify the acoustic properties of the site making use of more sophisticated field methods. In addition to the main shelter, the study was expanded to include the nearby site of Bacinete III, also located in the core rock art area of the complex. The latter is more modest in terms of the number of motifs, style (only Schematic) and has no distinct phases. Nevertheless, both sites have in common panels located in shelters with a raised base and an area in front that could have been used by a potential audience.

In the acoustic measurement campaign, a representative set of monaural and spatial IRs was gathered onsite and analysed according to the ISO 3382-1 standard, adapting the methodology to the special requirements of open-air rock art sites. The use of a high-order ambisonic microphone allowed for the acoustic analysis of the 3D sound maps obtained for each source–receiver combination in the software tool IRspatial. Such maps helped us investigate the role of the painted shelter in the acoustics of each site by identifying the direction of arrival of the main early reflections present in the measured IRs. The results obtained suggest that the Bacinete main shelter offered favourable conditions for speech transmission, and that music would also have been heard very clearly when the sound source is located either next to the paintings or in the middle of the listeners' area. Moreover, when the sound is emitted next to the paintings, the perceived loudness reinforces the spatial impression created by the significant lateral early reflections produced by the rock formations at the sides. Considering the absence of reverberation and the low ITDG values obtained, these reflections are possibly responsible for the subjective feeling of intimacy experienced onsite, as previously reported by Lazarich et al. [21]. Similar acoustic conditions in terms of reverberation, sound clarity and perceived loudness were found at Bacinete III when the source was located next to the paintings. Given these results, we would argue that the reason archaeologists have never considered this area as a possible "auditorium" is related to the differential aural impression obtained in each of them. In the main shelter, the combination of acoustic features and the unique visual impact of being surrounded by those large rock formations could also have exclusively enhanced this site with a subjective feeling of intimacy in front of the shelter, the area with the potential for being used as an "audience area" [69,84]. In contrast, at Bacinete III the spatial impression or subjective sensation of spaciousness would have been less significant, as the space was more open. It must be borne in mind that the lack of archaeological evidence regarding the uses of the sites, makes it difficult to accurately assess the acoustics environment experienced by prehistoric people in real conditions, since details such as the position and nature of the sound source and the number and positioning of attendees, may have implied changes in sound propagation.

Was the Bacinete main shelter ever used for social or ritual activities? Intangible practices leave no apparent evidence, although archaeology has developed ways to deduce them. We would like to argue that a careful excavation of the areas in front of the shelters, including a series of micromorphological analyses, may assist in identifying with a higher degree of certainty the use of the space in that area. For now, we can only say that the universal social understanding of the senses [85], and in particular of hearing [86], makes it highly likely that the past communities who created and used the Bacinete rock art complex were sensitive to its acoustic properties. This means that, if it were somehow special, it is likely that its sonic component played an important part in how people interacted with it.

Author Contributions: Conceptualisation, M.D.-A. and M.L.; Methodology, L.A.-M. and M.D.-A.; Software, D.B.-A. and L.A.-M.; Validation, all; Formal Analysis, L.A.-M.; Investigation, L.A.-M., M.D.-A., N.S.d.R. and L.F.M.; Resources, M.D.-A.; Data Curation, M.D.-A., L.A.-M. and D.B.-A.; Writing—Original Draft Preparation, L.A.-M., M.D.-A. and N.S.d.R.; Writing—Review & Editing, all; Visualisation, L.A.-M. and N.S.d.R.; Supervision, M.D.-A.; Project Administration, M.D.-A.; Funding Acquisition, M.D.-A. All authors have read and agreed to the published version of the manuscript.

Funding: Research for this article was funded by the Artsoundscapes Advanced ERC project (Grant Agreement No. 787842, Principal Investigator: ICREA Research Professor Margarita Díaz-Andreu) funded by the European Research Council (ERC) under the European Union's Horizon 2020 research and innovation programme. As the PI, Díaz-Andreu signs this article as the last author.

Data Availability Statement: Version 01 of the data set "ERC Artsoundscapes project—Impulse responses measured at the Bacinete rock art complex (Bacinete main shelter and Bacinete III rock art sites)" can be downloaded from https://doi.org/10.34810/data591 (accessed on 27 December 2022). Measurements are available under Attribution-NonCommercial-ShareAlike 4.0 International (CC BY-NC-SA 4.0) https://creativecommons.org/licenses/by-nc-sa/4.0/ (accessed on 19 December 2022).

Acknowledgments: The fieldwork was undertaken with permission from the Cultural Goods Service, of the Consejería de Cultura y Patrimonio Histórico (Department of Culture and Historical Heritage) of the Junta de Andalucía (Autonomous Government of Andalusia) (202199900646587—05/04/2021). We are grateful to the owner of the land on which Bacinete is located for permission to access the area, and to Los Barrios archaeologist, María Aguilera, who kindly accompanied us in the fieldwork. The authors would like to give special thanks to our colleagues from the PAIDI HUM-812 research group of the University of Cádiz, Antonio Ramos-Gil, Antonio Ruiz-Trujillo, Mercedes Versaci and Alba Salceda Pino for their valuable help during the different phases of the fieldwork, and Angelo Farina for his precious help with some of the technical aspects.

Conflicts of Interest: The authors declare no conflict of interest.

References

1. Scarre, C.; Lawson, G. (Eds.) *Archaeoacoustics*; McDonald Institute for Archaeological Research: Cambridge, MA, USA, 2006.
2. Reznikoff, I. Sur la dimension sonore des grottes à peintures du Paléolitique. *Comptes Rendus L'académie Sci.* **1987**, *304*, 53–156, 307–310.
3. Reznikoff, I.; Dauvois, M. La dimension sonore des grottes ornées. *Bull. Société Préhistorique Française* **1988**, *85*, 238–246. [CrossRef]
4. Dauvois, M. Evidence of sound-making and the acoustic character of the decorated caves of the western Palaeolithic world. *Int. Newsl. Rock Art (INORA)* **1996**, *13*, 23–25.
5. Waller, S.J. Sound Reflection as an Explanation for the Content and Context of Rock Art. *Rock Art Res.* **1993**, *10*, 91–101.
6. Reznikoff, I. On the sound dimension of prehistoric painted caves and rocks. In *Musical Signification*; Taratsi, E., Ed.; Mouton de Gruyter: Berlin, Germany, 1995; pp. 541–557.
7. Waller, S.J. Spatial correlation of Acoustics and Rock Art in Horseshoe Canyon. *Am. Indian Rock Art* **2000**, *24*, 85–94.
8. Rainio, R.; Lahelma, A.; Äikäs, T.; Lassfolk, K.; Okkonen, J. Acoustic Measurements at the Rock Painting of Várikallio, Northern Finland. In *Archaeoacoustics: The Archaeology of Sound. Publication of Proceedings from the 2014 Conference in Malta*; Eneix, L.C., Ed.; The OTS Foundation: Myakka City, FL, USA, 2014; pp. 141–152.
9. Rainio, R.; Lahelma, A.; Äikäs, T.; Lassfolk, K.; Okkonen, J. Acoustic measurements and Digital Image Processing suggest a link between sound rituals and sacred sites in northern Finland. *J. Archaeol. Method Theory* **2018**, *25*, 453–474. [CrossRef]
10. Till, R. Sound archaeology: Terminology, Palaeolithic cave art and the soundscape. *World Archaeol.* **2014**, *46*, 292–304. [CrossRef]
11. Fazenda, B.; Scarre, C.; Till, R.; Jiménez Pasalodos, R.; Rojo Guerra, M.; Tejedor, C.; Ontañon Peredo, R.; Watson, A.; Wyatt, S.; García Benito, C.; et al. Cave acoustics in prehistory: Exploring the association of Palaeolithic visual motifs and acoustic response. *J. Acoust. Soc. Am.* **2017**, *142*, 1332–1349. [CrossRef] [PubMed]
12. Díaz-Andreu, M.; García Benito, C. Acoustics and Levantine Rock Art: Auditory Perceptions in La Valltorta Gorge (Spain). *J. Archaeol. Sci.* **2012**, *39*, 3591–3599. [CrossRef]
13. Mattioli, T.; Farina, A.; Armelloni, E.; Hameau, P.; Díaz-Andreu, M. Echoing landscapes: Echolocation and the placement of rock art in the Central Mediterranean. *J. Archaeol. Sci.* **2017**, *83*, 12–25. [CrossRef]
14. Díaz-Andreu, M.; Gutiérrez Martínez, M.d.l.L.; Mattioli, T.; Picas, M.; Villalobos, C.; Zubieta, L.F. The soundscapes of Baja California Sur: Preliminary results from the Cañón de Santa Teresa rock art landscape. *Quat. Int.* **2021**, *572*, 166–177. [CrossRef]
15. Díaz-Andreu, M.; Jiménez Pasalodos, R.; Rozwadowski, A.; Alvarez Morales, L.; Miklashevich, E.; Santos da Rosa, N. The soundscapes of the Lower Chuya River area, Russian Altai. Ethnographic sources, indigenous ontologies and the archaeoacoustics of rock art sites. *J. Archaeol. Method Theory* **2022**, *633*, 40–999. [CrossRef]
16. Santos da Rosa, N.; Alvarez Morales, L.; Martorell Briz, X.; Fernández Macías, L.; Díaz-Andreu García, M. The acoustics of aggregation sites: Listening to the rock art landscape of Cuevas de la Araña (Spain). *J. Field Archaeol.* **2023**, *48*, 130–143. [CrossRef]
17. Díaz-Andreu, M.; Mattioli, T. Rock Art, music and acoustics: A global overview. In *The Oxford Handbook of the Archaeology and Anthropology of Rock Art*; David, B., McNiven, I.J., Eds.; Oxford University Press: Oxford, UK, 2019; pp. 503–528.
18. Jiménez Pasalodos, R.; Alarcón Jiménez, A.M.; Santos da Rosa, N.; Díaz-Andreu, M. Los sonidos de la prehistoria: Reflexiones en torno a las evidencias de prácticas musicales del paleolítico y el neolítico en Eurasia. *Vínculos Hist.* **2021**, *10*, 17–37. [CrossRef]
19. Santos da Rosa, N.; Fernández Macías, L.; Mattioli, T.; Díaz-Andreu, M. Dance scenes in Levantine Rock Art (Spain): A critical review. *Oxf. J. Archaeol.* **2021**, *40*, 342–366. [CrossRef]
20. Díaz-Andreu, M.; García Benito, C.; Lazarich, M. The sound of rock art: The acoustics of the rock art of southern Andalusia (Spain). *Oxf. J. Archaeol.* **2014**, *33*, 1–18. [CrossRef]
21. Lazarich, M.; Ramos-Gil, A.; Ruiz Trujillo, A.; Gómar, A.M.; Torres, F.; Narváez Cabeza de Vaca, M. Bacinete: Un escenario de arte rupestre al aire libre. In *1915–2015. 100 Anys. Real Academia Valenciana*; Aparicio Pérez, J., Ed.; Sección de Estudios Arqueológicos V. Serie Arqueológica. Varia XII; Diputación Provincial de Valencia: Valencia, Spain, 2015; pp. 487–534.
22. Fernández-Sánchez, D.S.; Collado Giraldo, H.; Vijande Vila, E.; Domínguez-Bella, S.; Luque Rojas, A.; Cantillo Duarte, J.J.; Mira, H.A.; Escalona, S.; Ramos-Muñoz, J. A contribution to the debate about prehistoric rock art in southern Europe: New Palaeolithic motifs in Cueva de las Palomas IV, Facinas (Tarifa, Cádiz, Spain). *J. Archaeol. Sci. Rep.* **2021**, *38*, 103086. [CrossRef]
23. Mira Perales, H.A. Arte paleolitico y postpaleolitico en el extremo sur de la Peninsula Iberica, la comarca del Campo de Gibraltar, Cádiz (España). *Cuad. Arte Prehistórico* **2021**, *11*, 97–123.

24. Gomar Barea, A.M. La escena naval del abrigo de Laja Alta (Jimena de la frontera, Cádiz). Una nueva propuesta cronocultural. *Zephyrus* **2021**, *88*, 209–234. [CrossRef]
25. Breuil, H.; Burkitt, M.C. *Rock Paintings of Southern Andalusia. A Description of a Neolithic and Copper Age Art Group*; Clarendon Press: Oxford, UK, 1929.
26. Topper, U.; Topper, U. *Arte Rupestre en la Provincia de Cádiz*; Diputación de Cádiz: Cádiz, Spain, 1988.
27. Mas Cornella, M.; Jordá, J.; Cambria, J.; Mas, J.; Lobarte, A. La conservación del arte rupestre en las sierras del Campo de Gibraltar. Un primer diagnóstico. *Espac. Tiempo Forma. Prehist. Arqueol.* **1994**, *7*, 93–128.
28. Mas Cornella, M. (Ed.) *Las Manifestaciones Rupestres Prehistóricas de la Zona Gaditana*; Consejería de Cultura de la Junta de Andalucía: Sevilla, Spain, 2000.
29. Solís Delgado, M. *El Conjunto Rupestre de Bacinete (Los Barrios, Cádiz). Pinturas Prehistóricas Para la Reunión*; Instituto de Estudios Campogibraltarenos: Algeciras, Spain, 2020.
30. Lazarich, M.; Gomar, A.M.; Ruiz, A.; Torres, F.; Ramos, A.; Cruz, M.J. Las manifestaciones postpaleolíticas del entorno de la Laguna de la Janda (Cádiz). Nuevas perspectivas de estudio. In *Ponencias del Seminarios de Arte Prehistórico de 2011*; Aparicio Pérez, J., Ed.; Varia X; Diputación Provincial de Valencia: Valencia, Spain, 2012; pp. 181–207.
31. Lazarich, M.; Ramos-Gil, A.; González-Pérez, J.L. Prehistoric Bird Watching in Southern Iberia? The Rock Art of Tajo de las Figuras Reconsidered. *Environ. Archaeol.* **2019**, *24*, 387–399. [CrossRef]
32. Kopij, K.; Pilch, A. The Acoustics of Contiones, or How Many Romans Could Have Heard Speakers. *Open Archaeol.* **2019**, *5*, 340–349. [CrossRef]
33. Machause, S.; Sanchis, A. La ofrenda de animales como práctica ritual en época ibérica: La Cueva del Sapo (Chiva, Valencia). In *Preses Petites i Grups Humans en el Passat: II Jornades d'Arqueozoologia del Museu de Prehistòria de València*; Sanshis, A., Pascual Benito, J.L., Eds.; Diputació de València: València, Spain, 2015; pp. 261–286.
34. Machause, S.; Skeates, R. Caves, Senses, and Ritual Flows in the Iberian Iron Age: The Territory of Edeta. *Open Archaeol.* **2022**, *8*, 1–29. [CrossRef]
35. *ISO 3382-1*; Acoustics-Measurement of Room Acoustic Parameters—Part 1: Performance Spaces. International Organization for Standardization: Geneva, Switzerland, 2009.
36. Tronchin, L.; Merli, F.; Bevilacqua, A.; Dolci, M.; Berardi, U. Measurements of Acoustical Parameters in the Roman Theatre of Verona. *Can. Acoust.* **2021**, *49*, 7–14.
37. Astolfi, A.; Bo, E.; Aletta, F.; Shtrepi, L. Measurements of Acoustical Parameters in the Ancient Open-Air Theatre of Tyndaris (Sicily, Italy). *Appl. Sci.* **2020**, *10*, 5680. [CrossRef]
38. Bo, E.; Shtrepi, L.; Pelegrin-Garcia, D.; Barbato, G.; Aletta, F.; Astolfi, A. The Accuracy of Predicted Acoustical Parameters in Ancient Open-Air Theatres: A Case Study in Syracusae. *Appl. Sci.* **2018**, *8*, 1393. [CrossRef]
39. Girón, S.; Alvarez-Corbacho, A.; Zamarreño, T. Review Paper. Exploring the Acoustics of Ancient Open-Air Theatres. *Arch. Acoust.* **2020**, *45*, 181–208. [CrossRef]
40. Rindel, J.H. Roman Theatres and Revival of Their Acoustics in the ERATO Project. *Acta Acust. United Acust.* **2013**, *99*, 21–29. [CrossRef]
41. Sukaj, S.; Ciaburro, G.; Iannace, G.; Lombardi, I.; Trematerra, A. The Acoustics of the Benevento Roman Theatre. *Buildings* **2021**, *11*, 212. [CrossRef]
42. Almagro-Pastor, J.A.; García-Quesada, R.; Vida-Manzano, J.; Martínez-Irureta, F.J.; Ramos-Ridao, Á.F. The Acoustics of the Palace of Charles V as a Cultural Heritage Concert Hall. *Acoustics* **2022**, *4*, 800–820. [CrossRef]
43. Iannace, G. The use of historical courtyards for musical performances. *Build. Acoust.* **2016**, *23*, 207–222. [CrossRef]
44. Wall, J. Recovering Lost Acoustic Spaces: St. Paul's Cathedral and Paul's Churchyard in 1622. *Digit. Stud. Champ Numérique* **2014**, *4*. [CrossRef]
45. Farnetani, A.; Prodi, N.; Pompoli, R. On the acoustic of ancient Greek and Roman Theatres. *J. Acoust. Soc. Am.* **2008**, *124*, 157–167. [CrossRef] [PubMed]
46. Paini, D.; Gade, A.C.; Rindel, J.H. Is Reverberation Time Adequate for Testing the Acoustical Quality of Unroofed Auditoriums? In Proceedings of the 6th International Conference on Auditorium Acoustics, Dublin, Ireland, 20–22 May 2011; Institute of Acoustics: Copenhagen, Denmark, 2011; Volume 28, pp. 66–73.
47. van Dorp Schuitman, J.; de Vries, D. An Artificial Listener for Assessing Content-Specific Objective Parameters Related to Room Acoustical Quality. *Build. Acoust.* **2011**, *18*, 145–157. [CrossRef]
48. Mo, F.; Wang, J. The Conventional RT is Not Applicable for Testing the Acoustical Quality of Unroofed Theatres. *J. Acoust. Soc. Am.* **2012**, *131*, 3492. [CrossRef]
49. Chourmouziadou, K.; Kang, J. Acoustic evolution of ancient Greek and Roman theatres. *Appl. Acoust.* **2008**, *69*, 514–529. [CrossRef]
50. Thomas, P.; van Renterghem, T.; Botteldooren, D. Using room acoustical parameters for evaluating the quality of urban squares for open-air rock concerts. *Appl. Acoust.* **2011**, *72*, 210–220. [CrossRef]
51. *ISO 12913-1*; Acoustics—Soundscape—Part 1: Definition and Conceptual Framework. International Organization for Standardization: Geneva, Switzerland, 2014.
52. Cox, S.L.; Ruff, C.B.; Maier, R.M.; Mathieson, I. Genetic contributions to variation in human stature in prehistoric Europe. *Proc. Natl. Acad. Sci. USA* **2019**, *116*, 21484–21492. [CrossRef] [PubMed]

53. EASERA. Electronic and Acoustic System Evaluation and Response Analysis (EASERA) AFMG Technologies GmbH, Germany. Available online: https://www.afmg.eu/en/afmg-easera (accessed on 15 September 2022).

54. Mateljan, I. *ARTA Program for Impulse Rresponse Measurement and Real Time Analysis of Spectrum and Frequency Response User Manual*; Electroacoustics Laboratory, Faculty of Electrical Engineering, R. Boskovica: Bijeljina, Bosnia and Herzegovina, 2011.

55. Farina, A.; Tronchin, L. 3D Sound Characterisation in Theatres Employing Microphone Arrays. *Acta Acust. United Acust.* **2013**, *99*, 118–125. [CrossRef]

56. Katz, B. In-situ calibration of the sound strength parameter G. *J. Acoust. Soc. Am.* **2015**, *138*, EL167–EL173. [CrossRef]

57. Guski, M. Influences of External Error Sources on Measurements of Room Acoustic Parameters. Ph.D. Thesis, RWTH Aachen University, Aachen, Germany, 2015.

58. Dietsch, L.; Kraak, W. Ein objektives kriterium zur erfassung von echostörungen bei musik-und sprachdarbietungen. *Acta Acust. United Acust.* **1986**, *60*, 205–216.

59. Barron, M.B. Interpretation of Early Decay Time in concert auditoria. *Acustica* **1995**, *81*, 320–331.

60. Galindo, M.; Girón, S.; Cebrián, R. Acoustics of performance buildings in Hispania: The Roman theatre and amphitheatre of Segobriga, Spain. *Appl. Acoust.* **2020**, *166*, 107373. [CrossRef]

61. *ISO 3382-2*; Acoustics-Measurement of Room Acoustic Parameters—Part 2: Reverberation Time in Ordinary Rooms. International Organization for Standardization: Geneva, Switzerland, 2008.

62. Long, M. *Architectural Acoustics*; Elsevier: Amsterdam, The Netherlands, 2006.

63. Till, R. Sound Archaeology: A Study of the Acoustics of Three World Heritage Sites, Spanish Prehistoric Painted Caves, Stonehenge, and Paphos Theatre. *Acoustics* **2020**, *1*, 661–669. [CrossRef]

64. Ando, Y.; Okura, M.; Yuasa, K. On the preferred reverberation in auditoriums. *Acustica* **1982**, *50*, 134–141.

65. Beranek, L. *Concert Halls and Opera Houses: Music, Acoustics, and Architecture*, 2nd ed.; Springer: New York, NY, USA, 2004.

66. Ando, Y.; Gottlob, D. Effects of early multiple reflections on subjective preference judgments of music sound fields. *J. Acoust. Soc. Am.* **1979**, *65*, 524–527. [CrossRef]

67. Mi, H.; Kearney, G.; Daffern, H. Impact Thresholds of Parameters of Binaural Room Impulse Responses (BRIRs) on Perceptual Reverberation. *Appl. Sci.* **2022**, *12*, 2823. [CrossRef]

68. Beranek, L. Concert hall acoustics. *J. Acoust. Soc. Am.* **1992**, *92*, 1–39. [CrossRef]

69. Lokki, T.; Pätynen, J. Auditory Spatial Impression in Concert Halls. In *The Technology of Binaural Understanding*; Blauert, J., Braasch, J., Eds.; Modern Acoustics and Signal Processing; Springer: Cham, Switzerland, 2020; pp. 173–202.

70. Hyde, J.R. Discussion of the relation between initial time delay gap (ITDG) and acoustical intimacy: Leo Beranek's final thoughts on the subject, documented by Jerald R. Hyde. *Acoustics* **2019**, *1*, 561–569. [CrossRef]

71. Mullin, D.C. *The Development of the Playhouse: A Survey of Theatre Architecture from the Renaissance to the Present*; University of California Press: Berkeley, CA, USA; Los Angeles, CA, USA, 1970.

72. Prodi, N.; Pompoli, R.; Martellotta, F.; Sato, S.-i. Acoustics of Italian Historical Opera Houses. *J. Acoust. Soc. Am.* **2015**, *138*, 769–781. [CrossRef] [PubMed]

73. Berardi, U.; Iannace, G. The acoustic of Roman theatres in Southern Italy and some reflections for their modern uses. *Appl. Acoust.* **2020**, *170*, 107530. [CrossRef]

74. Haddad, N.A.; Fakhoury, L.F. Conservation and preservation of the cultural heritage of ancient theatres and odea in the eastern Mediterranean. *Stud. Conserv.* **2010**, *55*, 18–23. [CrossRef]

75. Bouvet, G.A.; Shtrepi, L.; Bo, E.; Méndez Echenagucia, T.; Astolfi, A. Computational design: Acoustic shells for ancient theatres. In *Forum Acusticum. Lyon 2020*; European Acoustics Association: Madrid, Spain, 2020; pp. 1581–1585.

76. Palma, M.; Sarotto, M.; Méndez Echenagucia, T.; Sassone, M.; Astolfi, A. Sound-Strength Driven Parametric Design of an Acoustic Shell in a Free Field Environment. *Build. Acoust.* **2014**, *21*, 31–41. [CrossRef]

77. Adelman-Larsen, N.W. *Rock & Pop Venues*; Acoustic and Architectural Design; Springer: Cham, Switzerland, 2014.

78. Barron, M.B. *Auditorium Acoustics and Architectural Design*; Spon Press: London, UK, 2009.

79. Barron, M.B.; Marshall, A.H. Spatial impression due to early lateral reflections in concert halls: The derivation of a physical measure. *J. Sound Vib.* **1981**, *77*, 211–232. [CrossRef]

80. Bradley, J.S. Comparison of concert hall measurements of spatial impression. *J. Acoust. Soc. Am.* **1994**, *96*, 3525–3535. [CrossRef]

81. Marshall, A.H.; Barron, M.B. Spatial responsiveness in concert halls and the origins of spatial impression. *Appl. Acoust.* **2001**, *62*, 91–108. [CrossRef]

82. Hyde, J.R. *Acoustical Intimacy in Concert Halls: Does Visual Input Affect the Aural Experience? (Multisensory Integration and the Concert Experience)*; Paul S. Veneklasen Research Foundation: Santa Monica, CA, USA, 2003.

83. Solís Delgado, M. Procesos de abreviación en los diseños de arte rupestre postpaleolítico del estrecho de Gibraltar. El ejemplo de la Sierra del Niño. *Almoraima* **2020**, *52*, 153–167.

84. Kuusinen, A.; Lokki, T. Auditory distance perception in concert halls and the origins of acoustic intimacy. In Proceedings of the Institute of Acoustics, 9th International Conference on Auditorium Acoustics, Paris, France, 29–31 October 2015; Volume 37, pp. 151–158.

85.   Ackerman, D. *A Natural History of the Senses*; Vintage Books: London, UK, 1990.
86.   Schafer, R.M. *Our Sonic Environment and the Soundscape. The Tuning of the World*; Destiny Books: Rochester, NY, USA, 1977.

 *acoustics*

Case Report

The Acoustics of the Palace of Charles V as a Cultural Heritage Concert Hall

Jose A. Almagro-Pastor [1,*], Rafael García-Quesada [2], Jerónimo Vida-Manzano [3], Francisco J. Martínez-Irureta [4] and Ángel F. Ramos-Ridao [5]

[1]   Servicio de Acústica, Cámara Anecoica Acústica (CIC-UGR), University of Granada, C/ Periodista Fernando Gómez de la Cruz, 61, 18014 Granada, Spain
[2]   E.T.S. Arquitectura, University of Granada, Campo del Príncipe, 11, 18009 Granada, Spain
[3]   Facultad de Ciencias, University of Granada, Av. de Fuente Nueva, s/n, 18001 Granada, Spain
[4]   Irumar Arquitectura, C/ Juan de Herrera, 53, 29009 Málaga, Spain
[5]   Edificio Politécnico, University of Granada, Calle Dr. Severo Ochoa, s/n, 18001 Granada, Spain
*   Correspondence: jaalmagro@ugr.es

Abstract: This paper analyses the acoustic behaviour of the Palace of Charles V from a room acoustics perspective but also ponders the uniqueness of the space and its ability to engage and enhance the audience experience. The Palace of Charles V is a relevant part of the historical heritage of Granada. It has an architectural but also an acoustic uniqueness that deserves research. A measurement campaign was made to calculate parameters such as T30, IACC, C_{80} or Gm, and to explain the behaviour of the Palace. The BQI is quite high, but the late part of the impulse response (t > 80 ms) has strong unwanted reflections causing low clarity (C_{80}) and listener envelopment (LEV). Nevertheless, the Palace is a successful concert venue with good feedback from musicians and the audience.

Keywords: room acoustics; open-air auditorium; heritage acoustics

Citation: Almagro-Pastor, J.A.; García-Quesada, R.; Vida-Manzano, J.; Martínez-Irureta, F.J.; Ramos-Ridao, Á.F. The Acoustics of the Palace of Charles V as a Cultural Heritage Concert Hall. *Acoustics* **2022**, *4*, 800–820. https://doi.org/10.3390/acoustics4030048

Academic Editors: Margarita Díaz-Andreu and Lidia Alvarez Morales

Received: 14 July 2022
Revised: 19 August 2022
Accepted: 31 August 2022
Published: 15 September 2022

Publisher's Note: MDPI stays neutral with regard to jurisdictional claims in published maps and institutional affiliations.

1. Introduction

The Palace of Charles V (from now on referred to as the Palace, in capital letters), inside the Alhambra fortification, has been used as a concert hall for a long time with quite a degree of success. Previous acoustic research in the Alhambra covers soundscape [1] and concert noise [2] but not room acoustics. Analysing the Palace as a concert hall is not straightforward for several reasons: it is open-air, part of the cultural heritage of the city and it was not designed for speech or music transmission.

Open-air venues lack reflections from a ceiling, so most of the energy is reflected from walls that usually have low absorption. Scattering depends on the geometry. Some researched cases are Greek or Roman theatres [3,4], or public squares [5]. The shape of the Palace and its porticoed gallery has some similitudes to the use of arcades in squares [6]. Using room acoustic parameters in urban squares is useful, according to Thomas et al. [7]. The listener position has a strong influence on the space wideness assessment, and C_{50} and T_{30} are important in urban spaces according to the research of Calleri et al. [8]. Paini et al. [6] conclude that the addition of arcades to a public square increases T_{30}, while decreasing C_{80}. Previous research has discussed the applicability of the ISO 3382-1 [9] standard to unroofed spaces [10] and the relevant objective parameters to describe them [11].

The use of heritage buildings as concert venues is a common practice. Being in a historical place can improve the concert experience of attendants from an emotional point of view [12]. Brezina [13] divides the studies of historical places into two: the measurement of acoustic parameters and the storage of acoustics as audio heritage. The safeguard of the acoustic behaviour was pioneered by M. Gerzon [14] and continued by others such as Farina or Katz [15–17]. This work proved to be very important when the Gran Teatro La Fenice in Venice burned in 1996, but its sonic behaviour was saved because several acoustic measurements had been performed prior by Tronchin and Farina [18]. Furthermore, a fire destroyed the Notre Dame Cathedral in Paris and works by Katz et al. stored the original acoustics [19]. Recording of Ambisonic RIR allows the reconstruction of the sound field and the estimation of the direction-of-arrival (DOA) of the reflections [20]. This also allows the calculation of the impulse responses of virtual microphones, such as dipoles for lateral fraction or binaural microphones by Menzer and Faller [21], enabling the calculation of IACC or auralisation. Other reasons to keep Ambisonic RIRs are documentation and safeguarding of the historical heritage, visualisation of spatial information such as the research by Martellotta [22] or Alary and Valimaki [23], or using different available 3D reproduction techniques to recreate concerts, as shown by Tronchin and Farina [16].

There is not a lot of research regarding the room acoustics of heritage places that are not designed for music or speech transmission. Previous work by Iannace [24] showed that historical courtyards can be used for concerts without acoustic issues and good feedback from the performers. Heritage places have different sizes and shapes, and can even be open-air or squares. Most of them were not thought to be used for music or even speech transmission. Those singularities may suggest that their acoustics and their suitability for different kinds of music should be studied in each heritage place. Suitability, in this case, should be interpreted as 'eignung', used in sound quality. Blauert [25] groups sound-quality aspects into several degrees of abstraction. The same author [26] explores the idea of the composers and performers using the acoustic properties of a room to relay messages to the audience. Musical programmers should also take decisions based on the venues they have available. Farina [27] and Pätynen-Lokki [28] moved forward to improve the understanding of the relationship between measurements and preferred acoustics.

Lots of research has been carried out for historical concert hall acoustic measurements and there are some guidelines such as those by Pompoli and Prodi [29], but not too much concerning cultural heritage places not built as concert halls, excluding churches and different religious buildings as some guidelines for churches [30], cathedrals [31] and mosques [32] exists. The main musical use of the Palace is for orchestral music but it is also used for opera, jazz, flamenco or even rock [33].

This paper aims to review the physical descriptors that may explain the different and high aural quality of the Palace. The claim of good acoustics in the place is something explained in every guided visit, but it had never been scientifically researched. The descriptors used in this paper are included in the ISO 3382-1 [9] and the IEC 60268-16 [34] standards. The focus will be on measuring the objective parameters and discussing the results but the chance of safeguarding the acoustics must not be wasted. Ambisonic room impulse responses (RIR) were computed, used to estimate the direction of arrival of several reflections and to calculate the IACC. The Palace and its use as a concert hall will be discussed as part of this introduction. The material and Methods section will explain the tests made during the measurement campaign. The results and Discussion section will explain the outcomes with attention to the singularities of the space. Finally, some conclusions will be set forth.

1.1. The Palace of Charles V and the Alhambra of Granada

On both sides of the river Darro rise two hills that have seen several cultures throughout the history of the city. The Albaycín hill, where the city started, and the Sabica hill. The fortification of the Alhambra is on the Sabica hill. Inside, the Palace of Charles V is located (see Figure 1), an example of the best Italian Renaissance in Spain. Names such as Enrique de Egas, Diego de Siloé and Pedro Machuca have imprinted the history of the construction of the Palace in the style of the best Italian Renaissance. More information about the building and its historical circumstances can be found in the work by Rosenthal [35] and Brothers [36].

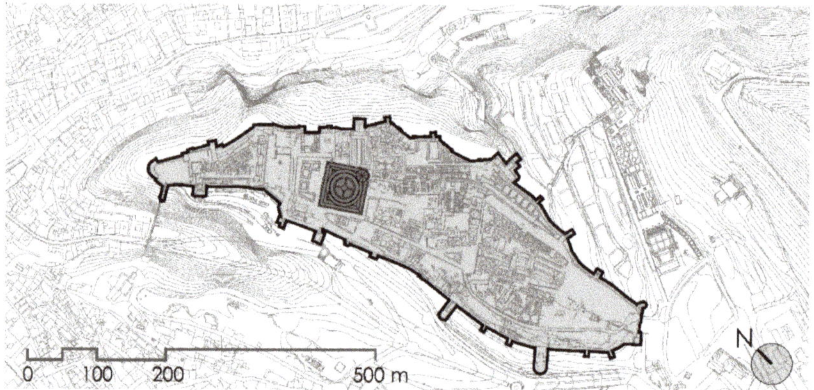

Figure 1. Location plan of the Palace inside the Alhambra.

In 1637, the construction process was finally abandoned due to the decline in the Spanish Empire and political factors. We owe the appearance of the Palace that we admire today to Leopoldo Torres Balbás and later to Francisco Prieto Moreno, who finally carried out a master plan for the Palace restoration, including the covering (the roof), all between 1923 and 1958. The first known musical event held in the Palace of Charles V was in 1883 as part of the city's Corpus Christi festival. It was also used to hold international flamenco competitions, such as the one held in 1922 by García Lorca, Manuel de Falla, Andrés Segovia and other intellectuals of that time. More recently, every year since 1952, the International Festival of Music and Dance of Granada has been using it as a concert hall. In this important musical event, the Palace of Charles V always occupies a central position. According to the local press [37], Daniel Barenboim said "The sound of the Palace of Charles V is much better than that of many enclosed halls. The shape of its walls makes it a wonderful acoustic shell."

The space used as a concert hall is circular and open-air with a diameter of 30 m (see Figure 2). On its perimeter, there is a 5 m wide porticoed gallery covered with a toroidal vault with basket-handle arches whose height from the keystone to the floor is 5.80 m. On the upper floor, another porticoed gallery crowns the building, this time covered by recent wooden porticoes and wooden coffered ceilings from 1958. The entire solid cylinder enclosing the interior of the arcaded galleries is around 17,907 m^3 (see Table 1), with built-in stone with bas-reliefs, half-columns and pediments framing various openings to the interior of the palace.

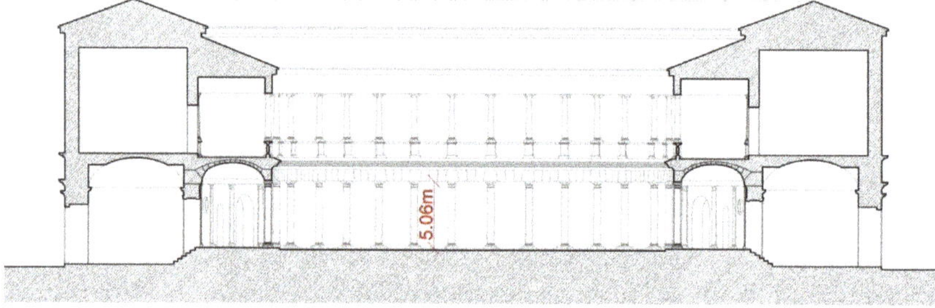

Figure 2. Elevation section of the Palace. Reproduced with permission from the authors.

Table 1. Volumes and area of the Palace.

	Surfaces [m^2]				Volume [m^3]
	Seating Area	Scenario	Residual Spaces	Total Area	Total Volume
Main floor (uncovered)	430	257	-	687	9391
Main floor	303	-	433	736	4136
Upper floor	303	-	313	616	4380
Total	1036	257	746	2038	17,907

The cylinder enclosing the interior of the courtyard is a sequence of voids between stone columns and other elements enclosing the galleries. In the interior cylinder of the courtyard, the closing element located halfway up the building, between heights of 5 and 8.12 m, stands out, with a thickness of more than 3 m; this is the group of friezes, triglyphs and metopes that cover the development of the bell-shaped arches of the toroidal vault, together with the height of the parapet. Most of the concerts are orchestral music but some others can include public address systems. The sound engineers working at the Palace deal with the strong delayed reflections and reverberation using high-directivity line arrays. The PA projects are not straightforward as they must cover the central area but also the second floor. This is problematic for rock but especially flamenco concerts.

1.2. Audience

The layout for the concerts has slight variations every year. The maximum audience is 1200 people, but restrictions due to COVID-19 had a big impact on the audience size in terms of distance among members of the audience. Being an open-air venue helped to safely keep enough seats and did not affect the layout. A big proportion of the audience is located in the patio; there are side stalls in the lower gallery and an audience arch in the upper floor gallery. All of the audience is seated on plastic chairs. On the upper floor, the chairs are on grandstands to enable the visibility of the stage (see Figure 3).

Figure 3. The Palace during the measurement campaign.

1.3. Stage

The stage is wide and covers an important part of the open-air central patio (Figure 4). The particular geometry of the stage makes the stage acoustics of the Palace quite singular. Some performers can be quite near, while others can be more than 20 m away from a given musician in the orchestra.

Figure 4. The stage during the measurement campaign.

The long distance from closer walls can exacerbate this problem as the direct sound path will predominate over the first reflections among near performers. A quick estimation would predict an attenuation of more than 25 dB, comparing a musician 20 m away to another one at a 1 m distance (under free-field conditions). The high reverberation is expected to reduce this issue by enhancing the strength (G) between distant positions.

Gade [38–40] recommends measuring ST (see Section 2.2.4) with chairs on stage. In addition, Dammerud [41] deepens into the different results obtained in real-condition experiments (with musicians). Sadly, this set of measurements was made without musicians or chairs on stage, as was explained previously. Uncertainty of stage measurements can be higher than expected for other descriptors according to Giovannini and Astolfi [42].

2. Materials and Methods

A measurement campaign was carried out on the 2nd and 6th of July 2021. The measurements followed the recommendations of the ISO 3382-1 standard [9]. It took place between 10 PM and 1 AM without the presence of the public. The late hours allowed for low noise levels as the Palace was closed for visits. Wind speeds were negligible (less than 0.5 m/s) and the temperature was lower than during hot summer afternoons. All the tests were taken without musicians or an audience. Only the chairs in the audience and the platform of the stage were mounted. The Palace does not have those elements unless a performance is programmed. Giving reliable and enlightening results has therefore been a concern. The recommendations of Pompoli-Prodi [29] and Astolfi et al. [10] were interpreted and followed as far as possible and some of the descriptors were not averaged to provide more information. Measuring with an audience was not possible. Performing the measurements in the Palace required permission from Patronato de la Alhambra and Festival de Granada, but performing them with an audience and orchestra would have required extra permissions from the visiting orchestras and would have been disturbing for the audience.

2.1. Measurement Setup and Methodology

Three source positions were selected on the stage. They were kept for both the audience and the stage measurements. Furthermore, sixteen microphone positions on the stage and ten in the audience (see Figure 5) were selected following the guidelines of the ISO 3382-1 standard [9]. All the source–receiver combinations were measured.

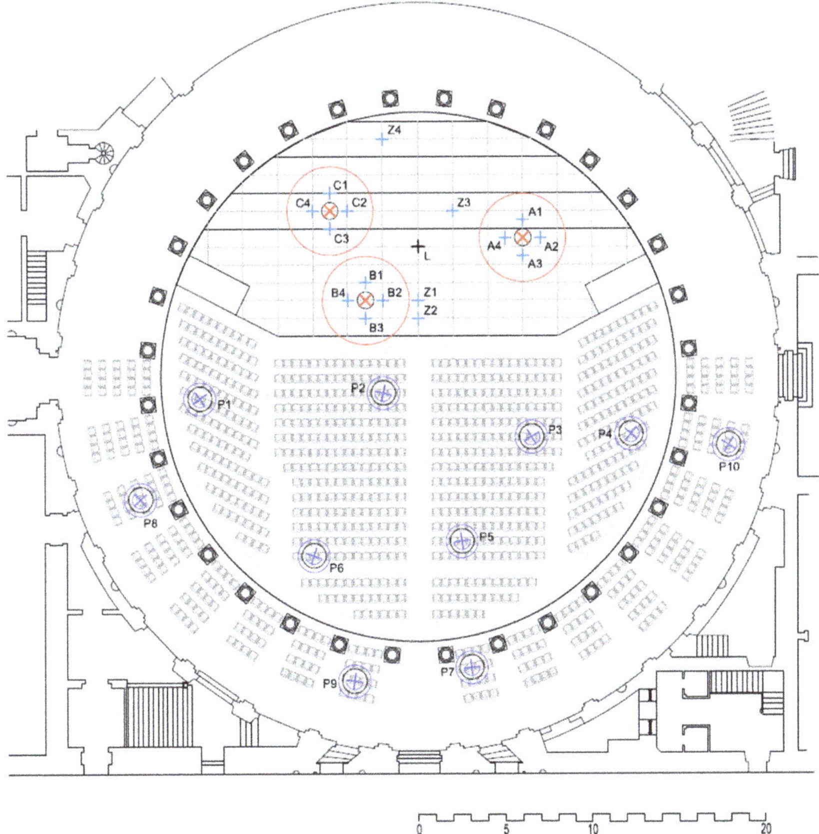

Figure 5. Layout of the concerts and the measurement campaign with source and receiver positions. (Red crosses indicate source positions; blue crosses indicate microphone positions).

Measurements in selected audience seats took place during the second night. A Lookline DL-203 dodecahedral source was used. Calibration of the source in an anechoic chamber enabled the calculation of G (strength) and other energetic parameters. A Genelec 8040 studio monitor was used for half of the seats to improve the frequency and phase response of the computed RIR in addition to the omnidirectional source. The sweeps were recorded using one of the omnidirectional microphones together with a Rode NT-SF1 first-order Ambisonic array.

The measurements of the stage were carried out on the first night using the dodecahedral source. The signals were recorded using four 378B02 PCB half-inch microphones. The air conditions are stated in Table 2.

Table 2. Air conditions during the measurement campaign.

	Audience		Stage	
	Beginning	**End**	**Beginning**	**End**
Temperature, °C	28.1	24.9	30.5	24.7
Humidity, %	33	49	29	31
Barometric Pressure, mB	929.0	928.8	931.4	929.7

Exponential sine sweeps were used because of their segregation of harmonic distortion, documented by Farina [43], and their higher impulse-to-noise ratio under usual test circumstances. Recorded signals were deconvolved by post-processing using Aurora plugins [44] to obtain the RIR. Extracting room acoustic indices from the RIR was easy and convenient.

2.2. Acoustic Indices

Unless otherwise specified, all indices have been calculated according to ISO 3382-1 [9].

2.2.1. Level Parameters

The impulse-to-noise ratio (INR) describes the quality of the RIR measured, as it gives the range of the usable decay. Therefore, the minimum values of each measurement are more interesting to know than the average values.

Strength (G) is the logarithmic ratio of the squared pressure of the measured impulse response to that of the response measured in a free field at a distance of 10 m. A graphical plot of Gm as a function of the source–receiver distance can be useful, as it varies with distance. Moreover, a comparison with the theoretical free field and the summation of it with the room constant (R) helps to visualise the contribution of the reverberation to the acoustic level.

All of the strength values displayed are the average values of the octave bands of 500 and 1000 Hz (noted as Gm).

2.2.2. Reverberation and Energy Ratios

The T20 and T30 reverberation times (RTs) and the EDT (early decay time) have been calculated together with their standard deviations (σ(T20) and σ(T30)). T20 and T30 are normalised reverberation times, calculated from linear regression of different sections of the Schröder curve, while EDT is calculated using the first 10 dB drop when the signal is not yet diffuse.

Definition (D_{50}) is the 50 ms-to-total arriving sound energy ratio. D_{50} relates to the perceived definition or speech intelligibility. Clarity (C_{80}) is the logarithmic early-to-late arriving sound energy ratio, being 80 ms, the time limit between "early" and "late". C_{80} relates to the perceived musical clarity. According to Adelman-Larsen [45], the reverberation time of the 63 Hz octave is important for rock music.

2.2.3. Speech Transmission

This paper covers the use of the Palace for music but the voice is usually part of the music. The speech transmission index (STI) is based on modulation transfer functions as defined by the IEC 60268-16 standard [34]. The values were calculated for female and male voices using the impulse responses without any corrections due to background noise, meaning the SNR is assumed to be infinite.

2.2.4. Stage Parameters

Stage conditions are important for the performers to hear themselves and each other. Early support (ST_{Early}) describes the ensemble conditions while late support (ST_{Late}) describes the reverberance. The results are averaged from 200 to 2000 Hz.

2.2.5. Spatial Impression Parameters

Inter-aural cross-correlation coefficients (IACC), correlate well with the subjective quality of "spatial impression" in a concert hall. IACCE stands for the early (<80 ms) IACC, while L stands for late (>80 ms). The results averaged in the octave bands of 500, 1000 and 2000 Hz are noted as "3".

The binaural quality index (BQI), calculated as (1-IACCE3) and introduced by Beranek [46], has the highest correlation of all the physical measures with the subjective judgments of acoustic quality in opera houses by the conductors, and it is related to the apparent width of the sound source (AWS) sensation. Listener envelopment (LEV) was calculated as (1–IACCL3). Binaural RIRs were processed from Ambisonic RIR using Ambi Head HD [47] with a Neumann KU 100 SOFA (Spatially Oriented Format for Acoustics) and then calculated using ARTA software [48].

3. Results and Discussion

The results included in this section were calculated from the RIR obtained from the omnidirectional and first-order Ambisonic microphones.

3.1. Level Parameters

Table 3 shows the INR results of the measurements in the audience and stage areas, with both the average and minimum of each band. Some stage measurements showed low INR values. Therefore, T20 will be used to calculate the stage reverberation time instead of T30. Measurements on the stage were recorded with lower input levels to avoid clipping at the positions at a 1 m distance. That explains the lower INR values.

Table 3. Average and minimum INR values.

	Audience		Stage	
f, Hz	**Average, dB**	**Minimum, dB**	**Average, dB**	**Minimum, dB**
125	49	42	50	43
250	53	48	54	47
500	57	53	53	45
1000	58	54	49	42
2000	62	58	49	43
4000	64	60	44	36

Strength (G) was averaged in the 500 and 1000 Hz octaves (Gm). Higher values were expected near the source. The spatial average of Gm does not provide meaningful information. Table 4 displays the 30 individual values for the selected positions in the audience.

Table 4. Gm values of each source–receiver pair in the audience.

Gm (dB)	Rec. 1	Rec. 2	Rec. 3	Rec. 4	Rec. 5	Rec. 6	Rec. 7	Rec. 8	Rec. 9	Rec. 10
Source A	−0.7	−1.3	0.3	1.3	0.4	−0.1	−1.7	−1.5	−1.7	−1.6
Source B	0.4	5.0	−1.3	−1.1	0.3	−0.5	−1.4	−0.7	−2.0	−1.4
Source C	0.4	−0.3	−0.6	−0.2	0.8	1.0	−0.1	−0.2	−1.1	−1.2

Under free-field conditions, the pressure level from an omnidirectional source only depends on its sound power and distance. Reverberant noise must be added to this free-field level. The easiest way is to consider perfect diffuse conditions, represented by a room constant (R), see (1).

$$L_p = L_w + 10 \cdot lg\left(\frac{1}{4\pi r^2} + \frac{4}{R}\right) \tag{1}$$

Figure 6 shows the measured strength versus the distance to the source of the measurements in the audience area.

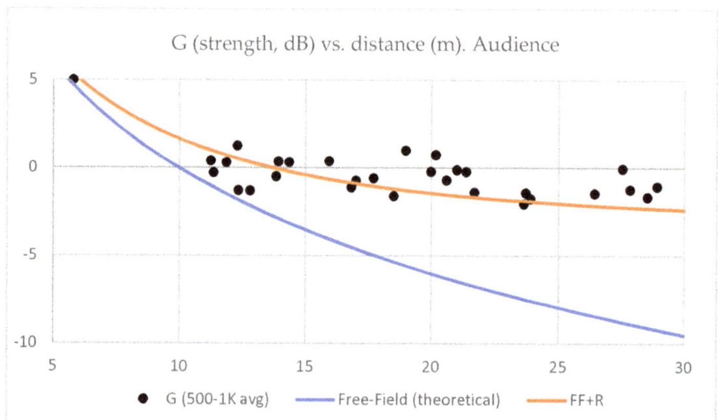

Figure 6. Gm in the audience versus source–receiver distance (m). Free-field values for comparison.

Figure 6 also shows the curves of the free-field behaviour and the free field plus the averaged reverberant noise for comparison (called FF + R). The contribution of the reverberation was calculated by subtracting the free-field contribution of each measurement and averaging the level excess. Positions near to walls, such as those around 20 m and 28 m, are over the curve, while the positions far from reflective areas, around 12 m, are below the curve. This Equation (1), known as the "classical theory", only accounts for free field and diffuse field. A revised theory for concert spaces was formulated by Barron and Lee [49], including the contribution of the early reflections. This topic will be explained in Section 15.6.

Figure 7 shows the same calculated values for the stage positions. Similar behaviour was observed on stage, but the highest distance was 12 m. The 12 measurements at a 1 m distance showed deviations from the expected value of 19.9 dB under free-field conditions. The average was 19.8 dB and the standard deviation was 0.90. Using 1 m distance measurements on stage to calibrate G instead of an average in an anechoic chamber is common practice. Averaging some of them can help to minimise systematic errors.

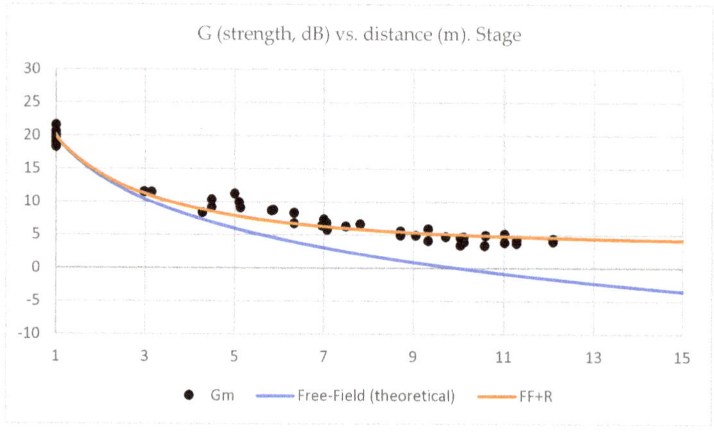

Figure 7. Gm on stage versus source–receiver distance (m). Free-field values for comparison.

3.2. Reverberation and Energy Ratios

Reverberation times averaged from audience points under unoccupied conditions (RT^U): (T20 and T30) are quite similar. T30 and C (which is T30/T20-1) are shown in Table 5. The reverberation time of each measurement is little dependent on the location, as previously reported by Thomas et al. [7]. Table 6 shows the values for the stage measurements.

Table 5. Reverberation times and energy ratios (audience).

f, (Hz)	T_{30}, (s)	σ (T_{30})	EDT, (s)	σ (EDT)	C, %	D_{50}	C_{80}, (dB)
63	2.78	0.309	2.50	0.69	−0.04%	0.34	−0.68
125	2.43	0.101	2.20	0.43	0.04%	0.33	−0.63
250	2.40	0.083	2.10	0.29	−0.04%	0.25	−1.71
500	2.25	0.055	2.08	0.31	1.12%	0.28	−1.26
1000	2.24	0.038	2.08	0.24	0.09%	0.35	−0.1
2000	2.09	0.031	1.93	0.22	0.58%	0.41	0.77
4000	1.81	0.04	1.61	0.24	1.40%	0.46	1.98

Table 6. Reverberation times and energy ratios (stage).

f, (Hz)	T_{20}, (s)	σ (T_{20})	EDT, (s)	σ (EDT)	D_{50}	C_{80}, (dB)
63	2.47	0.416	1.62	1.094	0.76	7.61
125	1.83	0.627	1.60	0.952	0.7	7.83
250	2.06	0.138	1.67	1.015	0.65	6.41
500	1.91	0.152	1.55	0.924	0.69	7.47
1000	1.95	0.14	1.50	0.902	0.74	7.98
2000	1.81	0.081	1.45	0.882	0.74	7.71
4000	1.47	0.176	1.13	0.692	0.78	9.27

The energy–time curves are quite diffuse; more so than expected in an open-air place. In fact, the echo criterion by Dietsch and Kraak [50] EC (1,14) of every measurement was 0.70 averaged, with a maximum of 0.89. For 10% of the audience to perceiving an echo with the music, a value of 1.5 would be needed.

ITDG measured at the central points of the audience was around 60 ms, if we take the first reflection arriving from the parapet, and up to 90 ms with the more energetic reflection from the wall. This can be expected from the geometry and is too much to be a great hall, according to Beranek's research [46]. Figure 8 shows the energy–time curve of the source in B and the microphone in P2, the most central positions of those measured. The top right of the figure shows a closer look at the first 200 ms. Early energy (t < 80 ms) is in dark grey, late energy (t > 80 ms) in light grey, and there are several detection thresholds for single reflections with different signals (clicks in green, pink noise in blue, speech in orange from the work of Olive and Toole [51], and classical music from Barron [52] in red). Thresholds are included as a reference, as reflections are as important as they are audible. The most energetic reflections arrive in the "late" fraction of the impulse response contributing to a high T30, but a particularly low $C_{80,}$ and every other parameter that segregates early-to-late energy. The main effect on subjective judgements is the lack of intimacy. The downward slope of thresholds means that the later a reflection arrives, the more audible it will be.

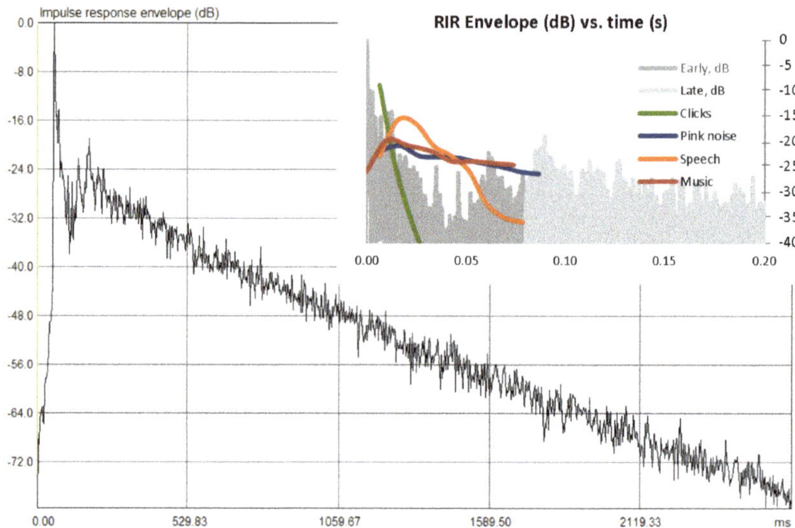

Figure 8. Impulse response envelope. The source in B and microphone in P2.

Figure 9 shows the values of C_{80} versus the source-to-receiver distance. Clarity tends to decrease with distance, with the exception of the farthest positions and the measurement with the source in the C position and the microphone at P6, which is red-coloured (see Figure 5 for the positioning layout)

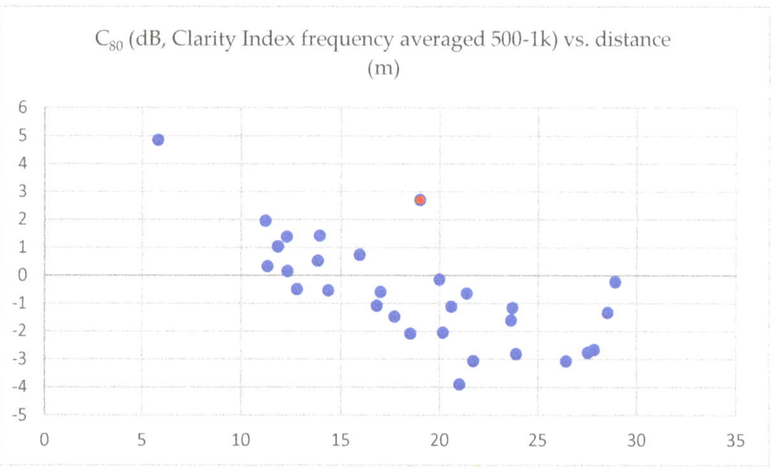

Figure 9. Clarity index (C_{80}) vs. source–receiver distance (m).

A quick estimation of the reverberation time with an audience (RT^O) can be calculated using Sabine's equation [53]. The results will have high levels of uncertainty, as the field is not diffuse and the audience areas are not flat surfaces. Iannace et al. used the audience absorption coefficients of Table 7 to estimate the absorption of the audience in the Roman theatre of Beneventum [54].

Table 7. Audience absorption coefficients (mid frequencies).

Frequency, Hz	500	1000
Minimum values	0.56	0.69
Maximum values	0.88	0.98

The estimation of Tm with Sabine's equation led to a result of 1.28 to 1.49 s. That result does not match the sensation during the concerts. Using data from Table 8, another way to estimate the RT^O is by calculating the equivalent absorption per occupant and using the average from the 25 halls that have data for RT^O, RT^U, number of seats (S) and volume (V). That average was 0.14 m^2 (metric Sabins) and the resulting RT^O was 1.98 s. This result is expected to be higher than the real value as the Palace has chairs, and concert hall seats have some absorption when unoccupied. Table 8 shows the reference values for concert halls with volumes between 15,000 and 20,000 m^3. There are 29 cases from the 100 presented by Beranek [55] in that volume range. It is noteworthy that 28 of the 29 concert venues have a volume per occupant much lower than the Palace's.

Table 8. Technical details of the 29 halls between 15,000 and 20,000 m^3 from Beranek [55].

Concert Hall	V [m^3]	Seats	RT^O [s]	RT^U [s]	RT^U/RT^O	V/S [m^3]
Palace of Charles V	17,907	1208	-	2.25	-	14.80
Benedict Music Tent, Aspen, Colorado, USA	19,830	2050	-	3.40	-	9.67
Kleinhans Music Hall, Buffalo, USA	18,240	2839	1.60	1.94	1.21	6.24
Severance Hall, Cleveland, USA	16,290	2101	1.65	2.10	1.27	7.75
Orchestra Hall, Minneapolis, USA	18,970	2450	1.90	2.35	1.23	7.74
Academy of Music, Philadelphia, USA	15,700	2921	1.20	1.40	1.17	5.38
Abravanel Symphony Hall, Salt Lake City, USA	19,500	2812	1.80	2.03	1.13	6.93
Beranoya Hall, Seattle, USA	19,263	2500	1.80	2.23	1.24	7.70
Festspielhaus, Salzburg, Austria	15,500	2158	1.50	1.96	1.31	7.18
Grosser Musikvereinssaal, Vienna, Austria	15,000	1680	2.56	3.60	1.41	8.93
Konzerthaus, Vienna, Austria	16,600	1865	1.96	2.30	1.17	8.90
Sala Sao Paulo, Sao Paulo, Brazil	20,000	1620	2.00	-	-	12.40
Barbican Concert Hall, London, England	17,000	1924	1.40	1.70	1.21	8.84
Sibelius, Talo Lahti, Finland	15,500	1250	2.30	-	-	12.40
Salle Pleyel, Paris, France	15,500	2386	1.55	2.12	1.37	6.50
Shauspielhaus, Berlin, Germany	15,000	1677	2.00	2.30	1.15	9.53
Beethovenhalle, Bonn, Germany	15,728	1407	1.65	1.80	1.09	11.18
Liederhalle, Stuttgart, Germany	16,000	2000	1.60	2.03	1.27	8.00
Megaron, Athens, Greece	19,100	1992	1.90	2.30	1.21	9.73
Concert Hall, Kyoto, Japan	20,000	1840	2.00	2.20	1.10	10.90
Symphony Hall, Osaka, Japan	17,800	1702	1.80	2.20	1.22	10.45
Bunka Kaikan, Tokyo, Japan	17,300	2327	1.50	1.89	1.26	7.42
Opera City Concert Hall, Tokyo, Japan	15,300	1636	1.99	2.72	1.37	9.40
Filharmonik Petronas, Kuala Lumpur, Malaysia	17,860	850	2.05	2.30	1.12	21.00
Concertgebow, Amsterdam, Netherlands	18,780	2037	2.05	2.59	1.26	9.20
Usher Hall, Edinburgh, Scotland	15,700	2502	1.80	2.55	1.42	6.27
Auditorio Nacional de Música, Madrid, Spain	20,000	2293	1.85	2.07	1.12	8.72
Palau de la Música, Valencia, Spain	15,400	1790	2.10	3.35	1.60	8,60
CCC Concert Hall, Lucerne, Switzerland	17,823	1892	1.6–2.2	-	-	9.42
Cultural Centre Concert Hall, Taipei, Taiwan	16,700	2074	2.00	2.46	1.23	8.05

The recommended reverberation times for halls of 17,900 m^3, according to Arau [56], range from 1.88–2.20 s for concert music, 1.38–1.86 s for opera and 1.03–1.61 s for speech. Although Adelman-Larsen [45] has a recommendation for rock-pop music, all the halls are smaller and cannot be extrapolated to the size of the Palace.

3.3. Speech Transmission

STI values were found to be quite constant, given the size of the Palace. It makes sense to provide single measurement values of every point instead of averaging. Table 9 also displays the measurement distance and Figure 10 shows that the STI decreases with distance until around 20 m, and rises a little when the distance is higher than 28 m because of the contribution of strong early reflections.

Table 9. STI values for source/receiver combinations.

	Source A			Source B			Source C		
	Female	Male	d, m	Female	Male	d, m	Female	Male	d, m
P1	0.50	0.50	20.59	0.59	0.58	11.23	0.56	0.55	13.93
P2	0.53	0.52	12.80	0.73	0.71	5.79	0.54	0.53	11.34
P3	0.55	0.54	11.85	0.55	0.54	12.33	0.48	0.48	17.71
P4	0.56	0.55	12.29	0.50	0.49	16.82	0.53	0.52	21.37
P5	0.51	0.51	15.94	0.52	0.51	14.36	0.53	0.53	20.17
P6	0.52	0.52	21.00	0.54	0.53	13.84	0.59	0.59	18.98
P7	0.45	0.45	23.89	0.46	0.46	21.70	0.52	0.51	27.52
P8	0.46	0.45	26.42	0.49	0.49	17.01	0.51	0.51	19.98
P9	0.50	0.49	28.52	0.46	0.45	23.63	0.51	0.50	28.90
P10	0.45	0.45	18.51	0.48	0.47	23.70	0.47	0.46	27.83

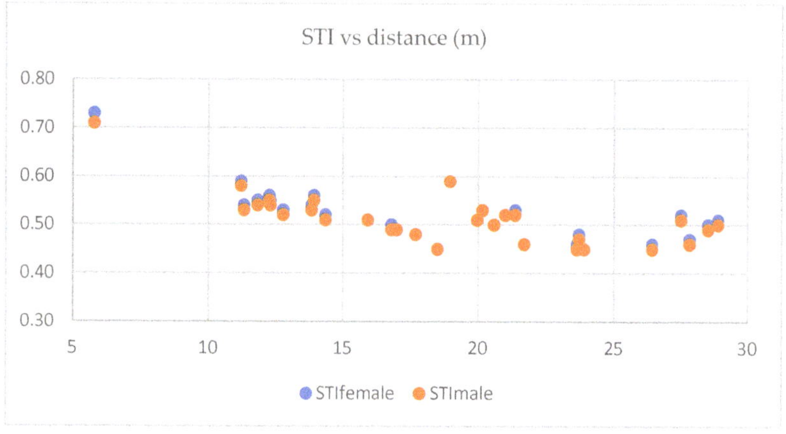

Figure 10. Speech transmission index vs. source–receiver distance (m).

3.4. Stage Parameters

As an open-air auditorium, a lack of overhead reflections is expected. Those reflections are important for the musicians to hear each other. Reflections coming from the sides can be easily affected by the presence of the performers. Early and total support has been averaged in frequency (250–2000 Hz octave bands) but single values of every source–mic combination are displayed in Tables 10 and 11. The 1 m distance positions are marked in grey.

Table 10. Early support ST_{Early} (250–2000 Hz frequency averaged).

ST_{Early}	Microphones in A				Microphones in B			
	mic 1	mic 2	mic 3	mic 4	mic 1	mic 2	mic 3	mic 4
Source A	−17.3	−17.1	−18.3	−16.9	−3.0	−3.1	−2.2	−1.9
Source B	−2.9	−1.6	−4.2	−2.9	−18.5	−18.1	−18.7	−17.0
Source C	−1.4	0.3	−0.5	−0.9	−6.2	−6.3	−4.9	−7.0
	Microphones in C				Microphones in Z			
	mic 1	mic 2	mic 3	mic 4	mic 1	mic 2	mic 3	mic 4
Source A	−0.7	0.1	0.4	1.0	−2.6	−3.4	−4.0	−1.4
Source B	−3.6	−7.8	−8.0	−5.5	−11.6	−12.6	−3.7	−0.5
Source C	−15.9	−17.3	−17.9	−17.2	−4.1	−4.2	−1.9	−7.0

Table 11. Total support ST_{Total} (frequency averaged).

AVG	Microphones in A				Microphones in B			
	mic 1	mic 2	mic 3	mic 4	mic 1	mic 2	mic 3	mic 4
Source A	−15.0	−14.8	−15.4	−14.8	0.6	0.4	1.5	1.5
Source B	1.2	2.2	0.1	1.0	−14.8	−14.8	−14.8	−14.1
Source C	2.3	3.6	3.1	2.3	−3.2	−2.2	−1.3	−3.4
	Microphones in C				Microphones in Z			
	mic 1	mic 2	mic 3	mic 4	mic 1	mic 2	mic 3	mic 4
Source A	3.0	3.2	3.9	4.3	0.6	−0.2	−1.5	1.9
Source B	−0.1	−4.1	−4.1	−1.9	−7.3	−8.5	−0.1	2.8
Source C	−13.7	−13.7	−15.0	−14.5	−0.2	−0.7	0.8	−4.3

Support is also averaged from the 12 positions in which the source is a 1 m distance from the microphones. The results are ST_{Early}: −17.5 dB, ST_{Late}: −18.1 dB and ST_{Total}: −14.7 dB. Information in octave bands for late support shows little influence on the frequency (see Table 12). These two tables contain a lot of information about the stage at every point. The overall view shows a big stage with a lack of strong reflections. In consequence, the musicians will have a balance of the orchestra, in which the closer instruments will prevail.

Table 12. Late support, ST_{Late}.

f, Hz	ST_{Late}, dB
125	−20.4
250	−18.4
500	−18.6
1000	−18.3
2000	−17.3
4000	−19.5
Avg (250–2000)	−18.1

3.5. Spatial Impression Parameters

The IACC results were averaged from all of the 30 source–receiver combinations in the audience. The BQI average was 0.53 and its standard deviation was 0.11. This is a good to excellent result according to Beranek. The LEV average was 0.72 with a standard deviation of 0.03. The existence of strong reflections after 80 ms can explain this low value. It must be noted that the uncertainties of these two results can be high as they have been calculated with a non-standard method.

3.6. Spatial Distribution of the Reflections

Reverberation times show very little dependency on the source–receiver combination (see Table 5). Previous figures show interesting trends in the behaviour of several parameters with distance (G, STI and C_{80}) that suggest a deeper sight on reflections should be carried out. Barron and Lee [49] observed a deficiency in the reflected sound level at distant seats. These local differences are caused by strong reflections, which in turn are caused by geometry. The circular geometry of the Palace creates the opposite effect: an excess of sound level at distant seats.

The computation of first-order Ambisonic room impulse responses (FOA-RIR) with IRIS [20] enabled the estimation of the direction of arrival (DOA) of the reflections. A few measurements were chosen (Figure 11).

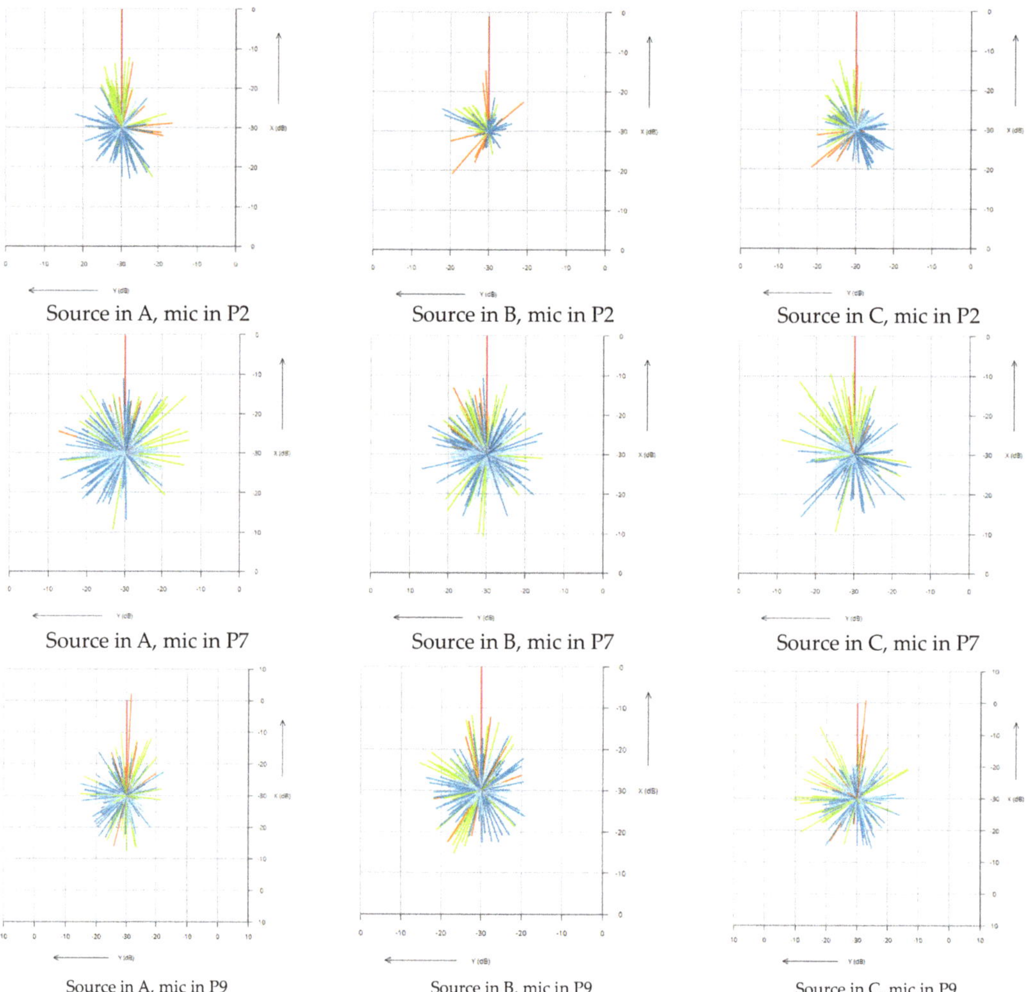

Figure 11. Directions of arrival of reflections. 30 dB range.

P2 is near the centre of the patio, very near to B (see Figure 5). That increased G and C_{80} from B, but not so from A or C, due to the dimensions of the stage ($C_{80,\text{A-P2}} = -0.5$ dB, $C_{80,\text{B-P2}} = -4.8$ dB, $C_{80,\text{C-P2}} = 0.3$ dB). Those close musicians will stand out from the rest of the orchestra. P7 is in the corridor on the main floor, under the toroid vault. C_{80} was low ($C_{80,\text{A-P7}} = -2,8$ dB, $C_{80,\text{B-P7}} = -3.1$ dB, $C_{80,\text{C-P7}} = -2.8$ dB). The blend of the musicians seems more adequate and the DOA of the first reflections is highly dependent on the position of the source on stage. That may improve the spaciousness sensation. P9 on the grandstands of the first floor (see Figure 12). C_{80} was higher ($C_{80,A\text{-}P9} = -1.3$ dB, $C_{80,B\text{-}P9} = -1.6$ dB, $C_{80,C\text{-}P9} = -0.2$ dB) and the blend level of the three sources was quite similar. However, reflections do not tend to come from the sides as in P7. Some returning spectators show a preference for seats in this area.

Figure 12. View from P9 featuring A, B and C positions.

Figure 9 shows a remarkably high C_{80} value for the C-P6 combination that deserves some explanation. The energy–time curve (Figure 13) shows two groups of strong reflections arriving in the first 80 ms (yellow and red lines) causing a C_{80} and STI rise. Figure 14 shows the directions of arrival for the same combination.

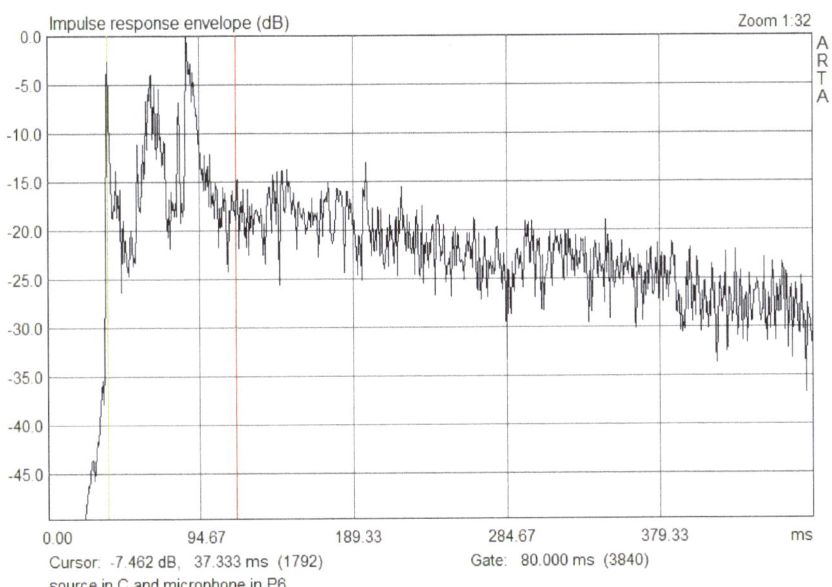

Figure 13. Energy–time curve (source in C, microphone in P6).

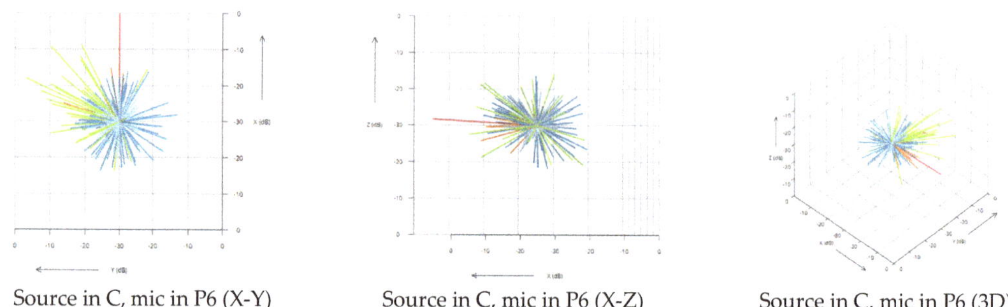

Source in C, mic in P6 (X-Y)   Source in C, mic in P6 (X-Z)   Source in C, mic in P6 (3D)

Figure 14. DOA of reflections. Source in C, microphone in P6. 30 dB range.

There are two groups of strong reflections with delays around 30 and 50 ms. IRIS shows they both come from the left side (Figure 15). The source was positioned in C (red dot and line), while the first group came from the parapet (green lines and dot, 15–50 ms delayed) and the second from the outer ring (blue lines and dot, >50 ms delay). The delays match the length of the geometrical paths.

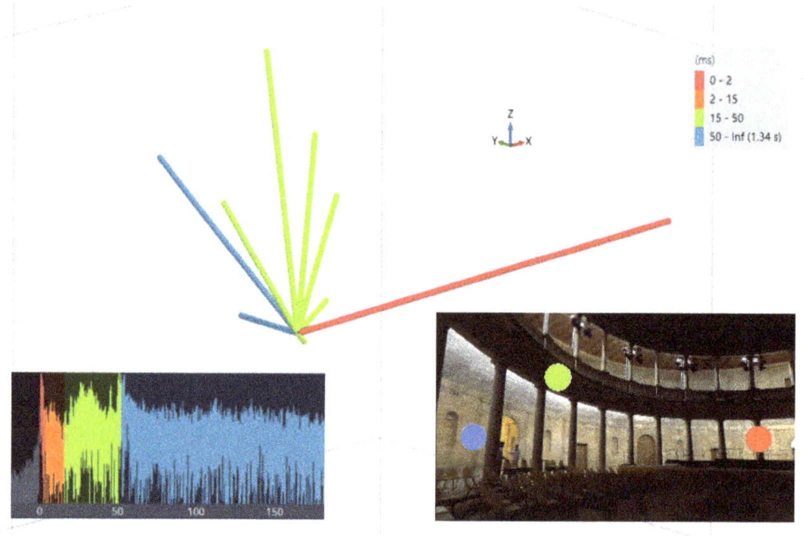

Figure 15. Directions of arrival of reflections. Time regions and 360° photograph included. 10 dB range.

3.7. Overall Attributes Discussion

The auditorium is quite reverberant. Taking into account it is open-air, a lot of the reflections come from the sides. This has been found by Pätynen and Lokki [57] to increase the emotional impact. The Palace is also wide, so the ITDG is very high as well. Taking a look at Figure 8, it is not straightforward to decide which reflection should be considered the first for the calculation of ITDG, especially if we think about the implications of psychoacoustical pre-masking and post-masking [58]. The balance of frequencies in the reverberation time is adequate: the bass ratio (BR) was 1.08 and brightness (Br) was 0.87. Only Br is too low for pop and rock according to Adelman-Larsen [45], but BR and Br are optimum for different kinds of music according to Arau [56]. The differences between T30 and EDT are not significant when averaged (see Table 5), but they are high in the centre of the hall due to the high ITDG. Of note, the standard deviation of EDT was much higher than that for T30. No echoes or strong reflections that could cause artefacts were detected in any of the points measured, other than the first-order high-energy reflections already mentioned.

Previous research by Paini et al. [59] claims reverberation time (T30) and clarity index (C_{80}) not to be accurate for unroofed auditoriums and suggests strength (G) together with auralisations to be satisfactory for finding possible echoes. Moreover, Mo and Wang [60] support this claim. Cabrera and Martens [61] proposed using loudness models to predict reverberance (the subjective perception of reverberation). In the particular case of this palace, T30 was high and diffuse; the dimensions are not big compared to typical squares, and the circular shape avoids flutter echoes due to parallel walls. The only exception is when the source and receiver are close to the centre. Only then, a huge echo is audible and it makes sense to neglect the T30 values.

One of the particularities of the Palace is the high ITDG in every seat. This is caused by the long distance from the side walls. Some first reflections are strong and surpass the 80 ms limit of the early ones, affecting parameters such as clarity (C_{80}) or envelopment (LEV). LEV has a low value and it is paradoxical, given that the room is circular and open-air. This low ITDG is expected to be judged as a lack of intimacy and a defect. The best-liked halls in the world have an ITDG around or below 25 ms in the centre of the main floor, according to Beranek [46]. At over 35 ms, halls are considered lower grade, and over 60 ms, lower results are expected. However, the Palace is an imposing and monumental venue. Furthermore, clarity (C_{80}) in the mid frequencies had a negative value (Table 5), meaning that energy arriving late to the receiver was higher than early energy.

This blend of characteristics makes the orchestra sound big and not intimate. Good acoustics involve definition and intimacy but also reverberation, loudness and spatial impression. This multidimensional nature of the preference approach is not new; Hawkes and Douglas [62], and others such as Beranek [46] have researched in that direction. All these attributes should be blended to some extent, but some of them are opposed to others. Blauert [25] divides the quality of the acoustics into functional adequacy, typicality, listening tradition and aesthetics. The concept of 'eignung', named in the introduction, makes sense here. Good halls sound intimate but "should the palace of the emperor who ruled territories on which the Sun never set sound intimate?" We believe that a place with those visual characteristics should have monumental, not intimate acoustics.

4. Conclusions

This paper tries to answer the question of whether the Palace of Charles V has good acoustics, and if not, why the audience thinks it sounds so good. It is a heritage building where musical performances are held. Concerts include different genres such as classical music, opera, flamenco or rock, with different acoustic needs. Several acoustic parameters were measured without the public or musicians in the audience and stage areas. The most remarkable findings are related to the high energy and delayed reflections due to the circular shape of the inner patio.

The $RTU_{500\text{-}1000}$ was 2.24 s, while the RTO was expected to be between 1.28 and 1.98 s, which is not excessive given the size of the building. The high RT is due to the massive stone building. The absence of a ceiling leads to a predominance of reflections in the horizontal plane. This non-diffuse field affects the calculations of reverberation times under different conditions (occupied or adding treatment).

Clarity is very low due to the delay of the first strong reflections after 80 ms. Intimacy is supposed to be very important for concert halls. This palace is the opposite of intimate. Furthermore, the LEV was very low but the BQI was quite good. The lack of earlier reflections from the sides, compared to a similar shoebox-style concert hall, causes all these issues.

Using standard descriptors has been found useful in open-air spaces and heritage buildings but it is still not clear if the recommendations regarding their value should be applied. The emotional response to historical architecture or suitability of the place and the music must be taken into account. This research includes an extensive set of descriptors and also Ambisonic RIR that can be used for auralisation, documentation and safeguarding. Further research can use these data to assess specific recommendations for heritage buildings. Which acoustic indices are important for music in heritage places deserve further research.

Auralisations with visual content can be an effective tool for checking whether the sound of the Palace can be improved with the use of absorbers or by reconsidering the design of the stage to give earlier reflections. There is no doubt that the acoustic indices would improve, but it would be interesting to know if the audience would find it appropriate as the architecture of the Palace is monumental, not intimate.

Programmers need to understand the acoustics of singular heritage halls such as this. A search for appropriately sized and aesthetically pleasing heritage sites can result in possible high-quality venues. Further research is needed, including semiology and audio-visual interactions, to understand why halls that were not designed as auditoria and are far from perfect in terms of hall acoustic recommendations sound so good.

Author Contributions: J.A.A.-P. conceived of the presented idea. J.A.A.-P., R.G.-Q., J.V.-M. and F.J.M.-I. took part in the field measurements. J.A.A.-P. performed the computations. R.G.-Q., J.V.-M. and Á.F.R.-R. verified the analytical methods. J.A.A.-P. and R.G.-Q. wrote the manuscript with support from the rest of the authors. R.G.-Q., J.V.-M. and Á.F.R.-R. supervised the findings of this work. All authors have read and agreed to the published version of the manuscript.

Funding: This research received no external funding.

Institutional Review Board Statement: Not applicable.

Informed Consent Statement: Not applicable.

Data Availability Statement: The data that support the findings of this study are available from the corresponding author upon reasonable request.

Acknowledgments: The authors would like to thank the institutions and individuals who helped make this article possible: Festival Internacional de Música y Danza de Granada, Patronato de la Alhambra y el Generalife, Centro de Instrumentación Científica de la Universidad de Granada (CIC-UGR), Teófilo Zamarreño, Lidia Álvarez, Alexis Campos, Dario Paini, Niels Adelman-Larsen, Suso Ramallo, Cástor Rodríguez (Sound of Numbers), Paco Pretel and Daniel Ortiz.

References

1. Pérez-Martínez, G.; Torija, A.J.; Ruiz, D.P. Soundscape Assessment of a Monumental Place: A Methodology Based on the Perception of Dominant Sounds. *Landsc. Urban Plan.* **2018**, *169*, 12–21. [CrossRef]
2. Almagro-Pastor, J.A.; García-Quesada, R.; Ramallo, S.; Martínez-Irureta, F.J.; Ramos-Ridao, Á.F. Assessing Environmental Noise Impact of PA Systems with the Swept-Sine Method. A Case Study in the Heritage Site of the Alhambra. *Appl. Acoust.* **2021**, *176*, 107897. [CrossRef]
3. Bevilacqua, A.; Ciaburro, G.; Iannace, G.; Lombardi, I.; Trematerra, A. Acoustic Design of a New Shell to Be Placed in the Roman Amphitheater Located in Santa Maria Capua Vetere. *Appl. Acoust.* **2022**, *187*, 108524. [CrossRef]
4. Galindo, M.; Girón, S.; Cebrián, R. Acoustics of Performance Buildings in Hispania: The Roman Theatre and Amphitheatre of Segobriga, Spain. *Appl. Acoust.* **2020**, *166*, 107373. [CrossRef]
5. Paini, D.; Rindel, J.H.; Gade, A.C.; Turchini, G. The Acoustics of Public Squares/Places: A Comparison between Results from a Computer Simulation Program and Measurements in Situ. *Inter-Noise* **2004**, *2004*, 1–8.
6. Paini, D.; Gade, A.C.; Rindel, J.H. Agorá Acoustics—Effects of Arcades on the Acoustics of Public Squares. In Proceedings of the Forum Acusticum Budapest 4th European Congress of Acoustics, Budapest, Hungary, 2 August–29 September 2005; pp. 1813–1818.
7. Thomas, P.; Van Renterghem, T.; Botteldooren, D. Using Room Acoustical Parameters for Evaluating the Quality of Urban Squares for Open-Air Rock Concerts. *Appl. Acoust.* **2011**, *72*, 210–220. [CrossRef]
8. Calleri, C.; Shtrepi, L.; Armando, A.; Astolfi, A. Evaluation of the Influence of Building Façade Design on the Acoustic Characteristics and Auditory Perception of Urban Spaces. *Build. Acoust.* **2018**, *25*, 77–95. [CrossRef]
9. *ISO 3382-1*; Acoustics—Measurement of Room Acoustic Parameters—Part 1: Performance Spaces. International Standardization Organization: Geneve, Switzerland, 2009; ISBN 9788578110796.

10. Astolfi, A.; Bo, E.; Aletta, F.; Shtrepi, L. Measurements of Acoustical Parameters in the Ancient Open-Air Theatre of Tyndaris (Sicily, Italy). *Appl. Sci.* **2020**, *10*, 5680. [CrossRef]

11. Bo, E.; Shtrepi, L.; Alerta, F.; Puglisi, G.E.; Astolfi, A. Geometrical Acoustic Simulation of Open-Air Ancient Theatres: Investigation on the Appropriate Objective Parameters for Improved Accuracy. *Build. Simul. Conf. Proc.* **2019**, *1*, 18–25. [CrossRef]

12. Smith, L.T.A.-T.T. *Emotional Heritage: Visitor Engagement at Museums and Heritage Sites LK*; Routledge: London, UK, 2021; Available online: https://Tamulibraries.on.Worldcat.Org/Oclc/1154133804 (accessed on 30 August 2022).

13. Brezina, P. Acoustics of Historic Spaces as a Form of Intangible Cultural Heritage. *Antiquity* **2013**, *87*, 574–580. [CrossRef]

14. Gerzon, M.A. Recording Concert Hall Acoustics for Posterity. *AES J. Audio Eng. Soc.* **1975**, *23*, 569.

15. Farina, A.; Ayalon, R. Recording Concert Hall Acoustics for Posterity. In Proceedings of the AES 24th International Conference Multichannel Audio, Banff, AB, Canada, 26–28 June 2003.

16. Farina, A.; Tronchin, L. Measurements and Reproduction of Spatial Sound Characteristics of Auditoria. *Acoust. Sci. Technol.* **2005**, *26*, 193–199. [CrossRef]

17. Katz, B.F.G.; Murphy, D.; Farina, A. The Past Has Ears (PHE): XR Explorations of Acoustic Spaces as Cultural Heritage. In Proceedings of the Augmented Reality, Virtual Reality, and Computer Graphics: 7th International Conference, AVR 2020, Lecce, Italy, 7–10 September 2020; Proceedings, Part II. De Paolis, L.T., Bourdot, P., Eds.; Springer International Publishing: Cham, Switzerland, 2020; pp. 91–98.

18. Tronchin, L.; Farina, A. Acoustics of the Former Teatro "La Fenice" in Venice. *J. Audio Eng. Soc JAES* **1997**, *45*, 1051–1062.

19. Katz, B.F.; Postma, B.N.; Poirier-Quinot, D.; Meyer, J. Experience with a Virtual Reality Auralization of Notre-Dame Cathedral. *J. Acoust. Soc. Am.* **2017**, *141*, 3454. [CrossRef]

20. Marshall Day IRIS. Available online: https://www.iris.co.nz/ (accessed on 30 August 2022).

21. Menzer, F.; Faller, C. Obtaining Binaural Room Impulse Responses from B-Format Impulse Responses. In Proceedings of the 2008 Audio Engineering Society 125th Convention, San Francisco, CA, USA, 2–5 October 2008; Volume 2, pp. 912–919.

22. Martellotta, F. On the Use of Microphone Arrays to Visualize Spatial Sound Field Information. *Appl. Acoust.* **2013**, *74*, 987–1000. [CrossRef]

23. Alary, B.; Valimaki, V. A Method for Capturing and Reproducing Directional Reverberation in Six Degrees of Freedom. In Proceedings of the International Conference on Immersive and 3D Audio: From Architecture to Automotive (I3DA), Bologna, Italy, 8–10 September 2021; pp. 1–8. [CrossRef]

24. Iannace, G. The Use of Historical Courtyards for Musical Performances. *Build. Acoust.* **2016**, *23*, 207–222. [CrossRef]

25. Blauert, J. Conceptual Aspects Regarding the Qualification of Spaces for Aural Performances. *Acta Acust. United Acust.* **2013**, *99*, 1–13. [CrossRef]

26. Blauert, J.; Raake, A. Can Current Room-Acoustics Indices Specify the Quality of Aural Experience in Concert Halls? *Psychomusicol. Music. Mind. Brain* **2015**, *25*, 253–255. [CrossRef]

27. Farina, A. Acoustic Quality of Theatres: Correlations between Experimental Measures and Subjective Evaluations. *Appl. Acoust.* **2001**, *62*, 889–916. [CrossRef]

28. Pätynen, J.; Lokki, T. Evaluation of Concert Hall Auralization with Virtual Symphony Orchestra. *Build. Acoust.* **2011**, *18*, 349–366. [CrossRef]

29. Pompoli, R.; Prodi, N. Guidelines for Acoustical Measurements inside Historical Opera Houses: Procedures and Validation. *J. Sound Vib.* **2000**, *232*, 281–301. [CrossRef]

30. Martellotta, F.; Cirillo, E.; Carbonari, A.; Ricciardi, P. Guidelines for Acoustical Measurements in Churches. *Appl. Acoust.* **2009**, *70*, 378–388. [CrossRef]

31. Álvarez-Morales, L.; Zamarreño, T.; Girón, S.; Galindo, M. A Methodology for the Study of the Acoustic Environment of Catholic Cathedrals: Application to the Cathedral of Malaga. *Build. Environ.* **2014**, *72*, 102–115. [CrossRef]

32. Suárez, R.; Sendra, J.J.; Navarro, J.; León, A.L. The Sound of the Cathedral-Mosque of Córdoba. *J. Cult. Herit.* **2005**, *6*, 307–312. [CrossRef]

33. Adelman-Larsen, N.W. *Rock and Pop Venues—Acoustic and Archictectural Design*; Springer: New York, NY, USA, 2014; ISBN 978-3-642-45235-2.

34. IEC. *60268-16-2003*; Sound System Equipment—Part 16: Objective Rating of Speech Intelligibility by Speech Transmission Index. International Electrotechnical Commission: London, UK, 2003; ISBN 9780580761638.

35. Rosenthal, E.E. *The Palace of Charles V in Granada*; Princeton University Press: Princeton, NJ, USA, 1985.

36. All, U.T.C. The Renaissance Reception of the Alhambra: The Letters of Andrea Navagero and the Palace of Charles V Author (s): Cammy Brothers Published by: Brill Stable URL. *JSTOR* **1994**, *11*, 79–102.

37. Tapia, J.L.; El Oído Del Carlos, V. Available online: https://www.ideal.es/granada/20090705/mas-actualidad/cultura/palacio-carlos-v-granada-acustica-oido-200907050909.html (accessed on 30 August 2022).

38. Líndez, B.; Rodríguez, M. La Bóveda Anular Del Palacio de Carlos V En Granada. Hipótesis Constructiva; The Annular Vault of Carlos V Palace in Granada. Constructive Hypothesis. *Inf. La Construcción* **2015**, *67*, e125. [CrossRef]

39. Gade, A.C. Investigations of Musicians' Room Acoustic Conditions in Concert Halls. Part II: Field Experiments and Synthesis of Results. *Acustica* **1989**, *69*, 249–262.

40. Gade, A.C. Practical Aspects of Room Acoustic Measurements on Orchestra Platforms. In Proceedings of the 14th International Congress on Acoustics, Beijing, China, 3–10 September 1992.

41. Dammerud, J.J. Stage Acoustics for Symphony Orchestras in Concert Halls. Ph.D. Thesis, University of Bath, Bath, UK, September 2009.
42. Giovannini, M.; Astolfi, A. The Acoustical Characterization of Orchestra Platforms and Uncertainty Estimation of the Results. *Appl. Acoust.* **2010**, *71*, 889–901. [CrossRef]
43. Farina, A. *Simultaneous Measurement of Impulse Response and Distortion with a Swept-Sine Technique*; Audio Engineering Society: New York, NY, USA, 2000.
44. Farina, A. Aurora for Audacity. Available online: http://www.angelofarina.it/Public/Aurora-for-Audacity/ (accessed on 15 February 2020).
45. Adelman-Larsen, N.W.; Thompson, E.R.; Gade, A.C. Suitable Reverberation Times for Halls for Rock and Pop Music. *J. Acoust. Soc. Am.* **2010**, *127*, 247–255. [CrossRef]
46. Beranek, L.L. *Concert Halls and Opera Houses: Music, Acoustics, and Architecture*, 2nd ed.; Springer: New York, NY, USA, 2005; Volume 117, ISBN 9781441930385.
47. Noise Makers Ambi Head HD Software. Available online: https://www.noisemakers.fr/ambi-head-hd/ (accessed on 30 August 2022).
48. Mateljan, I. ARTA User Manual. 2007. Available online: https://artalabs.hr/download/ARTA-user-manual.pdf (accessed on 30 August 2022).
49. Barron, M.; Lee, L.-J. Energy Relations in Concert Auditoriums. I. *J. Acoust. Soc. Am.* **1988**, *84*, 618–628. [CrossRef]
50. Dietsch, L.; Kraak, W. Ein Objektives Kriterium Zur Erfassung von Echostörungen Bei Musik- Und Sprachdarbietungen. *Acustica* **1986**, *60*, 205–216.
51. Olive, S.E.; Toole, F.E. The Detection of Reflections in Typical Rooms. *J. Audio Eng. Soc.* **1989**, *37*, 539–553.
52. Barron, M. The Subjective Effects of First Reflections in Concert Halls—The Need for Lateral Reflections. *J. Sound Vib.* **1971**, *15*, 475–494. [CrossRef]
53. Sabine, W.C.; Dwight, C.H. Collected Papers on Acoustics. *Am. J. Phys.* **1966**, *34*, 370. [CrossRef]
54. Iannace, G.; Maffei, L.; Aletta, F. Computer Simulation of the Effect of the Audience on the Acoustics of the Roman Theatre of Beneventum (Italy). In Proceedings of the Acoustics of Ancient Theatres Conference, Patras, Greece, 18–21 September 2011; pp. 1–6. [CrossRef]
55. Beranek, L.L. Analysis of Sabine and Eyring Equations and Their Application to Concert Hall Audience and Chair Absorption. *J. Acoust. Soc. Am.* **2006**, *120*, 1399–1410. [CrossRef]
56. Arau-Puchades, H. *ABC de La Acústica Arquitectónica*; CEAC: Barcelona, Spain, 1999; ISBN 9788432920172.
57. Pätynen, J.; Lokki, T. Concert Halls with Strong and Lateral Sound Increase the Emotional Impact of Orchestra Music. *J. Acoust. Soc. Am.* **2016**, *139*, 1214–1224. [CrossRef]
58. Fastl Hugo, E.Z. *Psycho-Acoustics Facts and Models*; Springer: New York, NY, USA, 2017; Volume 91, ISBN 9783642517655.
59. Paini, D.; Gade, A.C.; Rindel, J.H. Is Reverberation Time Adequate for Testing the Acoustical Quality of Unroofed Auditoriums? In Proceedings of the Institute of Acoustics. 6th Auditorium Acoustics. International Conference, Dublin, Ireland, 20–22 May 2011; Volume 28, pp. 66–73.
60. Mo, F.; Wang, J. The Conventional RT Is Not Applicable for Testing the Acoustical Quality of Unroofed Theatres. *Build. Acoust.* **2013**, *20*, 81–86. [CrossRef]
61. Lee, D.; Cabrera, D.; Martens, W.L. The Effect of Loudness on the Reverberance of Music: Reverberance Prediction Using Loudness Models. *J. Acoust. Soc. Am.* **2012**, *131*, 1194–1205. [CrossRef]
62. Hawkes, R.; Douglas, H. Subjective Acoustic Experience in Concert Auditoria. *Acta Acust. United Acust.* **1971**, *24*, 235–250.

acoustics

MDPI

Article

Measuring the Acoustical Properties of the BBC Maida Vale Recording Studios for Virtual Reality

Gavin Kearney *, Helena Daffern, Patrick Cairns, Anthony Hunt, Ben Lee, Jacob Cooper, Panos Tsagkarakis, Tomasz Rudzki and Daniel Johnston

AudioLab, Department of Electronic Engineering, University of York, York YO10 5DD, UK
* Correspondence: gavin.kearney@york.ac.uk

Abstract: In this paper we present a complete acoustic survey of the British Broadcasting Corporation Maida Vale recording studios. The paper outlines a fast room acoustic measurement framework for capture of spatial impulse response measurements for use in three or six degrees of freedom Virtual Reality rendering. Binaural recordings from a KEMAR dummy head as well as higher order Ambisonic spatial room impulse response measurements taken using a higher order Ambisonic microphone are presented. An acoustic comparison of the studios is discussed, highlighting remarkable similarities across three of the recording spaces despite significant differences in geometry. Finally, a database of the measurements, housing the raw impulse response captures as well as processed spatial room impulse responses is presented.

Keywords: recording studio acoustics; spatial impulse resposes; kemar; binaural; audio for virtual reality; ambisonics

Citation: Kearney, G.; Daffern, H.; Cairns, P.; Hunt, A.; Lee, B.; Cooper, J.; Tsagkarakis, P.; Rudzki, T.; Johnston, D. Measuring the Acoustical Properties of the BBC Maida Vale Recording Studios for Virtual Reality. *Acoustics* **2022**, *4*, 783–799. https://doi.org/10.3390/acoustics4030047

Academic Editors: Margarita Díaz-Andreu and Lidia Alvarez Morales

Received: 30 June 2022
Revised: 13 August 2022
Accepted: 17 August 2022
Published: 14 September 2022

Publisher's Note: MDPI stays neutral with regard to jurisdictional claims in published maps and institutional affiliations.

1. Introduction

There is an increasing motivation to replicate spaces for Virtual Reality (VR) with the rapid advancement of VR technologies and uptake in the use of the hardware in home situations. Especially for musical situations, the acoustic auralisation of the space needs to be as representative of the real space as possible. These can be created using simulated or measured Room Impulse Responses (RIRs), each technique having its limitations: simulated auralisations based on geometric models created in room acoustics simulation software such as ODEON have been shown to be inaccurate at certain room acoustic parameters [1,2], with perceptually noticeable impact on clarity compared to measured spaces [3]; recorded RIRs however, whilst more representative of the real space, are highly time- and resource-consuming [4]. Consequently, for the latter, fast and efficient workflows for capturing detailed room acoustics for use in VR rendering are necessary. Having efficient frameworks and processes to conduct complete acoustic surveys of recording studios is also important for the posterity of these spaces. It has implications for broadcasting companies worldwide, as they are provided with the opportunity to illustrate their unique and often iconic acoustics as well as connect people virtually within them. This paper presents the acoustic measurement and comparison of four studio spaces at the British Broadcasting Corporation (BBC) Maida Vale studios through RIR capture tailored for use in VR applications. The work was conducted as part of a collaboration between the AudioLab at the University of York and the BBC Research and Development Audio Team.

2. BBC Maida Vale Recording Studios

2.1. History

Originally a Roller skating palace built in 1909, the BBC took over the building on London's Delaware Road in 1934 to be home of the BBC Symphony Orchestra and created the now iconic Maida Vale studios. During the Second World War it became the standby centre of the radio news service and as a result was targeted and bombed during the blitz, after which it was extensively restored [5]. The building houses seven sound studios, of which four have been measured for this paper. The studios are labelled throughout the paper as they are named by the BBC: by number following the initials MV to represent Maida Vale.

MV2, shown in Figure 1a, is a large rectangular space described as having 'a lively sound that suits chamber music and piano sessions better than pop music' [6] and, although no longer used frequently for recording, remains home to the BBC Singers. MV3, shown in Figure 1b, is a similar size to MV2 and used for Radio 1 sessions with a live audience. It was home to the BBC Radio Orchestra and was the last place Bing Crosby recorded in 1977, three days before his death. MV4, which was home of the famous John Peel Live Lounge sessions, is shown in Figure 1c, and from 1967 it hosted sessions by artists from the Beatles to David Bowie, Nirvana and Adele. It is a smaller space with an isolated vocal booth and small balcony. This studio continues to be used by BBC Radio 1 alongside MV5, which is shown in Figure 1d, and which has the smallest, wood-paneled live room.

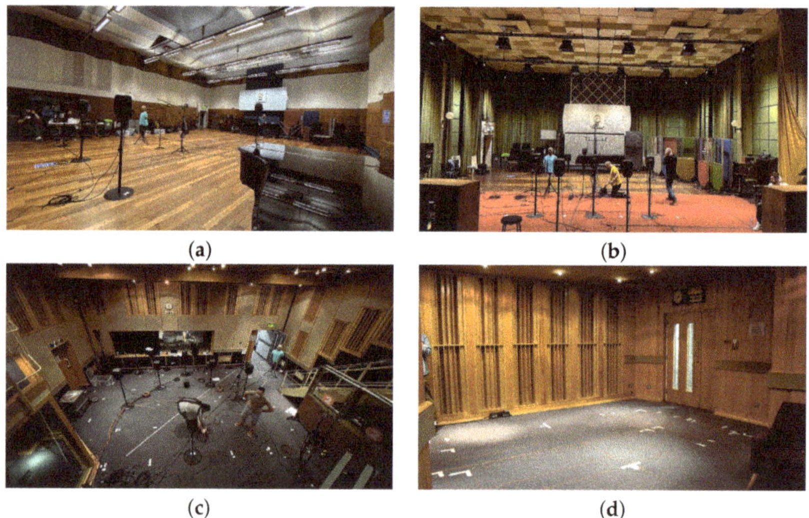

Figure 1. Maida Vale measured live rooms: (**a**) MV2, (**b**) MV3, (**c**) MV4, (**d**) MV5.

The importance of BBC Maida Vale Studios as a heritage space is undisputed, with its claim 'to have recorded more famous artists than any other' British studio [6]. It is therefore timely to conduct the acoustic measurements of studios within this iconic space, both for acoustical heritage, and to also provide opportunities to exploit the latest technologies to allow future musicians to be a part of its musical history in VR.

2.2. Studio Dimensions and Characteristics

MV2 and MV3's live rooms (Figures 1a,b and 5) have very similar dimensions; considering all walls to be flat surfaces, MV2's dimensions would be 13.69 m × 21.96 m and MV3's 13.38 m × 22.69 m. Despite this, the construction of both spaces and the materials used are quite contrasting. MV2 features 'zig-zagged' wooden walls that extend up to 0.5 m into the room and a hard wooden floor. Although there are not any absorbers or diffusers as such in the space, the design of the walls (and the ceiling) causes them to act as geometric diffusers. Given the materials that they are made from, this leads to MV2 being a very reflective space. In contrast to this, MV3 has a much more controlled reverberation time; its walls are mainly constructed using absorptive panelling and preserve the rectangular shape of the room. Large curtains are also hung from walls and add to the absorption in the space. The ceiling features acoustic panelling placed at different heights; this will therefore cause both absorption and diffusion. This studio also features a wooden floor, although a significant portion of this has a carpet placed over the top. This is used in recording sessions in the space; hence, it was not removed for the acoustic measurements. Both studios featured a significant amount of studio equipment, freestanding acoustic baffles and chairs that were present when measurements were taken. In both cases this was all moved to one side of the space.

The live room for MV4 (See Figures 1c and 6) consists of the main downstairs area, a glass vocal booth and a mezzanine level. The downstairs area is based around a rectangle of dimensions 8.50 m × 11.04 m with false walls being used to produce a less regular shape; there is also an additional wall present beneath the mezzanine level that divides that section of the space roughly in half. This wall contains three panel windows. The mezzanine level itself sits 2.69 m above the ground level. Both the ground floor and mezzanine level are carpeted and the walls and ceilings of the space are generally made up of acoustic panelling and wooden diffusers. The same wooden diffusers are used throughout the entire space. It is also worth noting that the space features a stepped ceiling. When the acoustic measurements were taken for this space, two freestanding acoustic baffles were present downstairs and there were three sofas, four chairs and four tables on the mezzanine level.

MV5's live room (Figures 1d and 7) features two 'chambers' (one larger and one smaller) joined along a single line. This room contains no parallel surfaces, however it can be entirely encompassed by a rectangle of dimensions 6.89 m × 6.70 m. The room itself is made up of 12 walls, 11 of which are made from wood and the 12th from acoustic panelling with thin glass panels. The studio's floor is carpeted, and its ceiling made from acoustic panels. A single sofa and piano were present in the space during the recording of the acoustic measurements. The piano was covered with heavy fabric to prevent interference with the measurements.

3. Acoustic Measurements

3.1. Impulse Response Capture

Impulse response capture requires an excitation signal to be played into the acoustic space, recorded back into a digital audio workstation (or equivalent recording device) and then subsequently run through a deconvolution routine to extract the final impulse response. The nature of the source signal is important and many approaches have been discussed in the literature, ranging from Maximum Length Sweeps [7] or Golay Codes [8] to non-signal processing methods of room excitation such as starter pistols, baloon pops or firecrackers [9]. However, for most acoustic measurements, the Exponential Sine Sweep (ESS) method has proven superior, due to the fact that any non-linearities in the reproduction system can be removed from the final impulse response as well as the logarithmic nature of the sweep ensuring more energy is catered for in the lower end of the spectrum than in linear sweeps, which correlates better with the human auditory response [10].

As Maida Vale is a practicing and busy recording environment, only 4 days total were allocated to the measurement of studio acoustics. Thus, in order to capture comprehensive data, a fast measurement protocol was required that facilitated detailed capture of measurement grids within the studios. A desirable feature of the framework was to facilitate varied source directivity of the sources in post-processing, and is discussed further in Section 5.2. This required each measurement point to be undertaken four times, with the loudspeaker oriented in the North, East, South and West directions each time. It was therefore quite apparent that singular impulse response measurements would take far too long, and would not give the spatial resolution required for 6 Degrees of Freedom (DoF) VR capability. MV4, for example, had a total of 45 specific receiver points and 7 source points, times 4 directions, yielding a minimum of 1260 individual measurements. Prior estimates of the reverberation time of the room were approximately 0.3 s, meaning that at least 3 s sweeps would ideally be used [11]. However, the existence of low-level background noise from air conditioning meant that sweep lengths greater than 3 s were required to maximise signal to noise ratio;

hence, 20 s was chosen as a desired sweep length. Considering also a minimum interval time of 20 s between measurements to reconfigure the sources and receivers, this would result in over 64 h of pure measurement time with no breaks or room for errors.

Consequently, an overlapped sweep framework was implemented, which has seen recent success in binaural measurement procedures [12,13]. In this process, multiple sources are excited in each measurement pass, with each source emitting an ESS with a short interval δ_{int} between each consequent activated sweep as shown in Figure 2. When the resultant recording is deconvolved, the impulse responses line up one after the other at a time equal to δ_{int}. For this, δ_{int} must be at least greater than the reverberation time of the room or else the tail from one RIR measurement will run into the direct sound of the consecutive one. Furthermore, if the loudspeakers are overdriven into harmonic distortion, then the harmonic distortion products will manifest as a series of impulse responses before the direct sound of each of the RIRs. Consequently, δ_{int} must therefore be big enough to ensure that any harmonic distortion products from each impulse response do not manifest in the reverberation tail of the preceding impulse response.

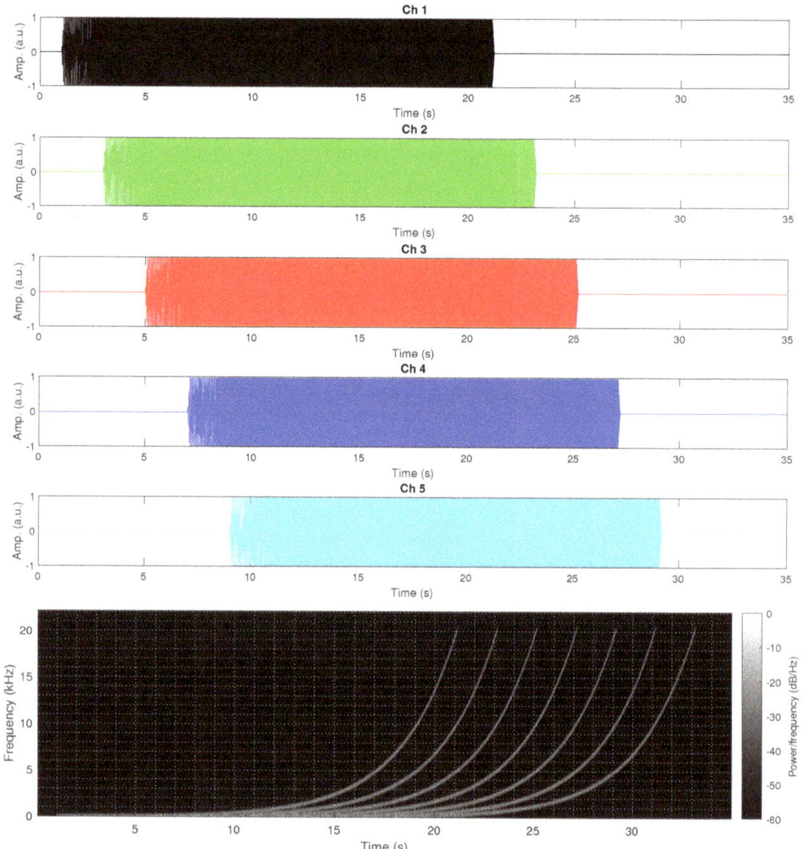

Figure 2. Multiple Source Exponential Sine Sweep excitation signal.

3.2. Methodology and Instrumentation

The capture of each room consisted of two main measurement phases—measurements for virtual-reality representation of the acoustics, as well as ISO-3382-based reference measurements [14].

The primary goal for VR capture was to facilitate virtual studio representation of the spaces from the performer perspective where movement around the space is typically not necessary (3DoF) and from the audience perspective, with ability to move about the space (6DoF). For example, one VR permutation utilising 3DoF is the case of networked music interactions, where a performer could 'dial in' to a virtual version of the studio and occupy one of the pre-defined performer positions, whilst hearing other members of their ensemble with the correct acoustic perception as if they were all together in the real space. Or, with a 6DoF setup, an audience member could move through the VR space through interpolation of the impulse responses across the defined audience area.

ISO-3382 measurements of each space allow for comparison of the acoustics of the studios to assess their similarity and appropriateness for the musical genres recorded there. Typical performer positions were captured that reflected how each studio is commonly used.

A summary of the measurement process for each room is shown in Figure 3. In general, measurement positions are first defined using architectural plans or geometric measurements. The excitation stimulus is then prepared offline ready for playback to the loudspeakers. The source and receiver points are then physically marked up in the space and any environmental noises that can be eliminated are checked. Loudspeakers and microphones are then calibrated and the measurement routine begins. Once all measurement points have been collected, data post-processing can be implemented.

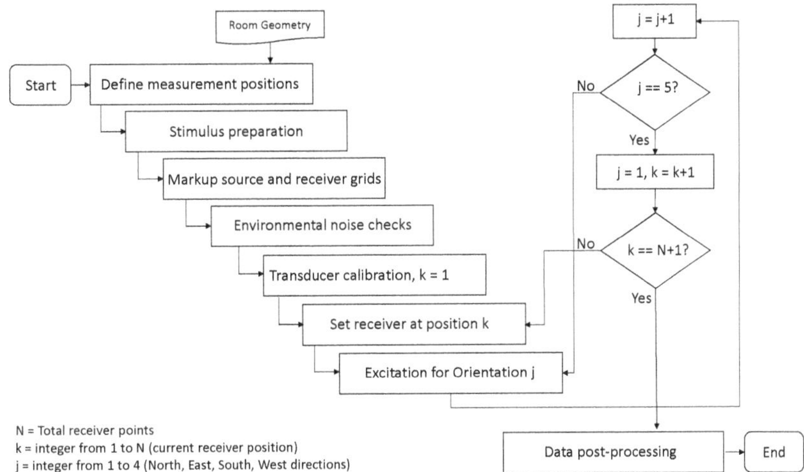

Figure 3. Flowchart of the measurement process.

Table 1 summarises the instrumentation used for each measurement phase. For VR measurements, Genelec 8040A loudspeakers were used to represent typical source positions. These loudspeakers have a frequency response of 45 Hz to 20 kHz. An MH-Acoustics Eigenmike, which is a higher order Ambisonic microphone (up to fourth order) and reference omnidirectional microphone (AKG CK-77, mounted on top of the Eigenmike) were used for measurements that would later facilitate 3DOF and 6DOF rendering in VR. A KEMAR binaural manikin also facilitated reference measurements at the performer positions, so as to later compare true binaural room impulse responses to Ambisonically derived binaural room impulse responses from a binaural-based Ambisonic decode [15], as is regularly used in virtual reality applications [16].

For ISO-3382 measurements, the source was changed to an NTI DS3 dodecahedron loudspeaker. Both the Eigenmike and the KEMAR manikin were used as receivers for these measurements.

All measurements were recorded on a Macbook Pro running the Digital Audio Workstation (DAW) Reaper v6.12 for recording and simultaneous playback of the sweeps. The

KEMAR and AKG microphones were connected to an RME Fireface UFX interface which has digitally controlled preamplifiers. All loudspeakers were also driven from this interface. The Eigenmike was recorded via its own proprietary TCAT interface and clocked to the Fireface via ADAT. The Macbook Pro aggregated the Eigenmike and Fireface interfaces to one virtual soundcard for recording in Reaper.

Table 1. Instrumentation utilised during measurements.

Measurement Phase	Loudspeaker	Receivers	Recording Device
VR	7× Genelec 8040A	Eigenmike, AKG CK-77 , KEMAR	Reaper v6.12, Macbook Pro, Fireface UFX II, TCAT interface
ISO-3382	NTI DS3 Dodecahedron	Eigenmike, KEMAR	

3.3. Practical Measurement

Prior to measurement, the architectural plans to the studios were obtained and measurement points were defined for each studio. Setup on the day required marking out the grids of points, which was implemented using a crosshair laser guide on initial string placement and laser distance measures as shown in Figure 4. Once the grid was defined, sources were set in place. Genelec 8040 loudspeakers were calibrated to each other at 85 dBC pink noise at 1 m. The sweeps were played out from Reaper at −20 dBFS.

Playback levels were then adjusted to ensure maximum signal level without any audible harmonic distortion occuring as a result of the sweeps, in particular from the low-end response. Microphone gain for all microphones was set to ensure no clipping for the closest source reciever points, which, for the Eigenmike, occurs when it is located above the loudspeakers.

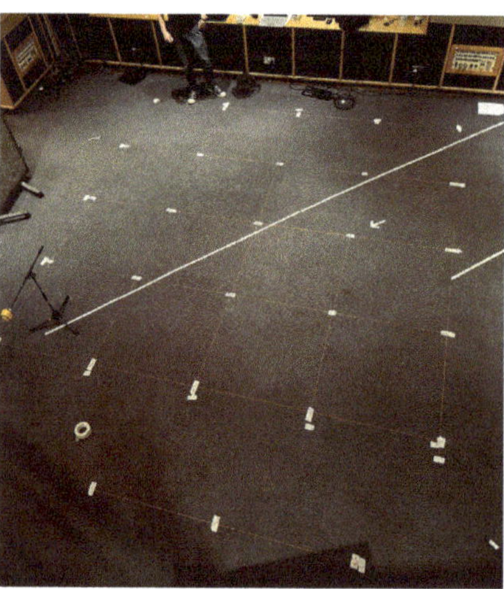

Figure 4. Markup for measurement session in MV4 live room. The image depicts the markup of the main PA grid.

Each measurement set was preceeded by a synthesised vocal idenfifier documenting the upcoming measurement for the benefit of the recordings. After this, the overlap sweeps played out, each with 20 s duration with 2 s intervals. Sweep frequency range was from 20 Hz to 20 kHz. Then, 3 s after the last sweep ends, another voice identifier occurs, indicating the next source–receiver combination. The measurement team then had 20 s to reconfigure the source and receivers to their new orientations before the next new measurement voice identifier. For more complex source–receiver reconfiguring, the recording was paused.

The highest density of measurements was captured in MV4 because it is regularly used for BBC 'Live Lounge' sessions, co-existing during productions as a recording studio and live performance space. Focus on this studio was also particularly important as it was decided to create a full Virtual Reality model of this space for networked music performance experiments [17]. MV2 and MV3 are larger spaces and comparable in size. Thus, the measurement grids were less dense than in MV4 in order to capture the spaces completely within the allocated time. Finally MV5 has the smallest live room, and therefore had the smallest concentration of measurements. Measurements for each room are now discussed in detail:

3.3.1. MV2 and MV3 Measurement Phases

The measurement configuration for MV2 and MV3 is shown in Figure 5. Fully defined Cartesian coordinates for all points in the studios are presented as part of the online database.

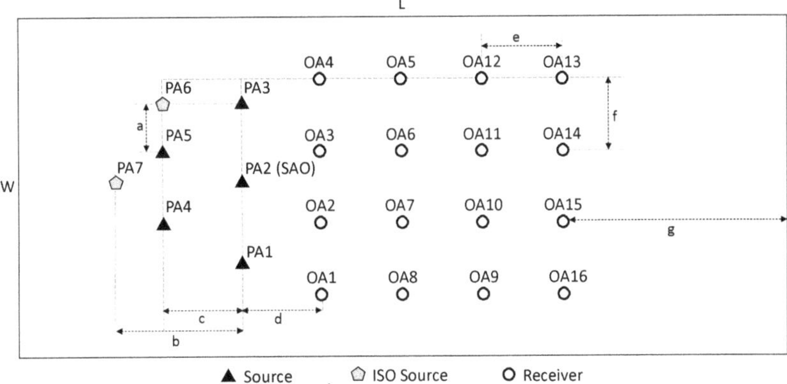

Figure 5. Source and receiver points for Studios MV2 and MV3.

- MV2/MV3 Virtual Reality Capture:
  5 Genelec 8040a sources, representing sources at typical recording positions were set up on a stage area, at points labelled PA1 to PA5 (PA stands for Performance Area), all at 1.5 m height to tweeter. A grid of 16 receiver points, labelled OA1-16 (OA stands for Outside performance Area) were set up, each 3 m apart. The Eigenmike and reference omninidirectional microphone were measured at each of these points at a height of 1.7 m. Both were north facing. No sitting audience positions were captured in the studios due to time constraints.
- MV2/MV3 ISO-3382 Capture:
  Referring to Figure 5, an additional two source points at PA6 and PA7 were set for the ISO measurements. Receiver locations for the these measurements were at OA1, OA6 and OA14. The Eigenmike, reference lav and a KEMAR binaural manikin were measured at these positions. The Eigenmike was at 1.7 m height and KEMAR was set to 1.5 m height to the ear canal, which is in range for the typical heights of UK men (175 cm) and women (161 cm) [18] with consideration of ear canal offset. Both were north-facing.
  For MV2, points OA13 to OA16 were 3.29 m from the back wall (distance g). The grid spacing was 2.5 m (distances c, d, e and f). Line PA1–PA3 was 2 m from line PA4–PA6 (distance c). Dodecahedron point PA7 was 4 m from PA2 (distance b). Dodecahedron point PA6 was 1.25 m from PA5 (distance a). Width (W) is 13.69 m and length (L) is 21.96 m.

For MV3, the line of points OA13 to OA16 were 4.3 m from the back wall (distance g). Grid spacing was 1.5 m between points (distances c, d, e and f).Dodecahedron point PA7 was 3 m from PA2 (distance b). Dodecahedron point PA6 was 0.75 m from PA5 (distance a). Width (W) is 13.38 m and length (L) is 22.69 m.

3.3.2. MV4 Measurement Phase

- MV4 Virtual Reality Capture: This was the largest measurement phase of all studios. Four performer positions were defined, with three sources representing performers at 1.5 m height and a fourth position representing a drum kit. These positions are labelled PA3, PA11, PA15 and PA23 as shown in Figure 6. The drummer position consists of 4 sources, representing a triangular spread of drums and a kick drum at point PA9, PA11, PA19 and PA12 respectively. The triangular configuration covering the drum area allows for IR interpolation to be undertaken to more clearly define drum source positions for any given virtual drum kit.

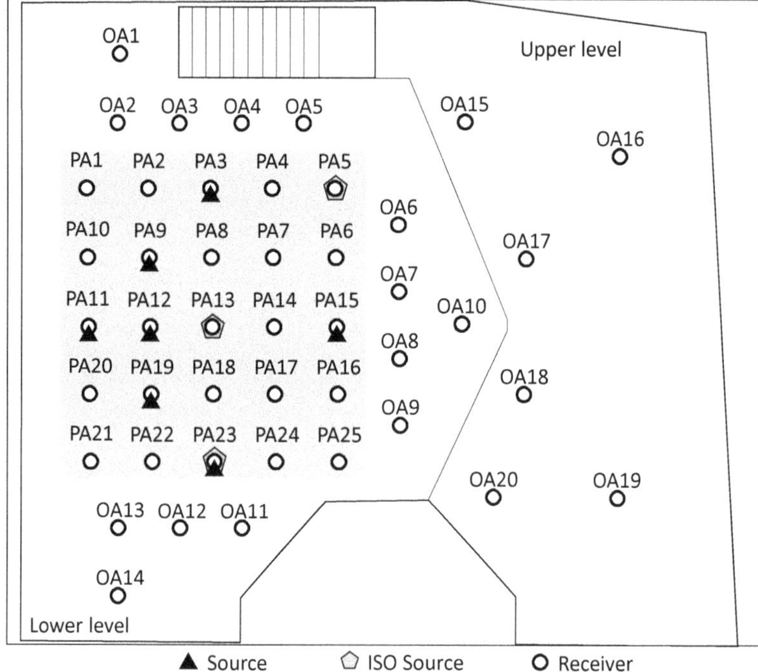

Figure 6. Source and receiver points for Studio MV4.

Impulse responses using the Eigenmike were captured at each performer position, including simultaneous source–receiver points. This was to facilitate foldback of a musician's own acoustic within the virtual space, i.e., when they perform, they should hear back the room response to their own performance at their performance position. Consequently, for these measurements, the Eigenmike was placed 2 cm above the Genelecs. Care was taken to ensure that the signal did not distort. However a noticeable low frequency boost was captured due to the off-axis (on top) position of the Eigenmike relative to the loudspeaker and the proximity of the transducers. This effect was removed in post-processing.

For 6DoF-enabled measurements, the receiver area was extended in between the performers to a grid of 25 measurement points, equally spaced 1 m apart. An additional 20 measurement points were captured outside of the performance area (OA points) to facilitate audience perspective. These include points on the main studio floor as well as on the upper balcony. All receiver points are at a height of 1.5 m, with the exception of PA11 (drummer position), which is at a height of 1.2 m. Full coordinates of each datapoint are available in the online database.

- MV4 ISO-3382 measurements: The NTI dodecahedron loudspeaker was again used for measurements at points PA5, PA13 and PA23. Three receiver positions on the lower level (OA2, OA7 and OA13) and three positions on the upper level (OA15, OA16 and OA20) were utilised. Both the Eigenmike and KEMAR were again used to capture the ISO measurements and were both north-facing, with heights of 1.6 m. The dodecahedron was at a height of 1.5 m.

- MV4 Performer Reference Positions: The KEMAR binaural head utilised in these sessions also includes a voice box (model GRAS 45BC). Measurements were taken by running an equalised sweep through the voice box and capturing the returning room acoustic at the manikin's ears. Equalisation for this sweep was performed in the anechoic chamber at the University of York [19]. This gives an excellent reference for understanding the self-direct-to-reverberant ratio as experienced at each performer position. It also provides a calibration level for performer-direct sound-to-reverberation auralisation in the virtual environment. The four performer positions were measured for this setup. KEMAR faced PA13 in each measurement.

3.3.3. MV5 Measurement Phase

- MV5 VR Capture: MV5 consisted of 12 measurement points altogether, as shown in Figure 7. Two performer positions were defined at PA3A, set 1.5 m into the room with a height of 1.5 m, and PA7B, set 5 m into the room at a height of 1 m. As the studio is often used to record intimate acoustic guitar or piano performances, these points were augmented with further measurements to simulate these instruments. Points PA7A and PA7C were included for simulation of a piano (at a height of 1.2 m), and points 3B and 7C were included to simulate acoustic guitars (at a height of 1 m). The Eigenmike was set to 1.6 m and faced east for all measurements.

- MV5 ISO-3382 Capture: The docecahedron was set to a height of 1.5 m. For ISO-3382, two source–receiver combinations were captured. First with the dodecahedron loudspeaker at PA3A and receivers at PA4, PA5, PA6 and PA7B. Second, with the dodecahedtron at PA6, with receivers PA7B, PA3A, PA4 and PA5. The Eigenmike was set to 1.7 m facing east and KEMAR to 1.5 m facing south.

- MV5 Performer Reference Positions: Similar to MV4, reference measurements were captured at each of the performer positions using 5 Genelec loudspeakers and the KEMAR binaural head with voicebox as the sources. The in-ear microphones of the KEMAR were used as receivers. KEMAR was positioned at PA3A and PA7B for these measurements, facing south then north, respectively, at a height of 1.5 m.

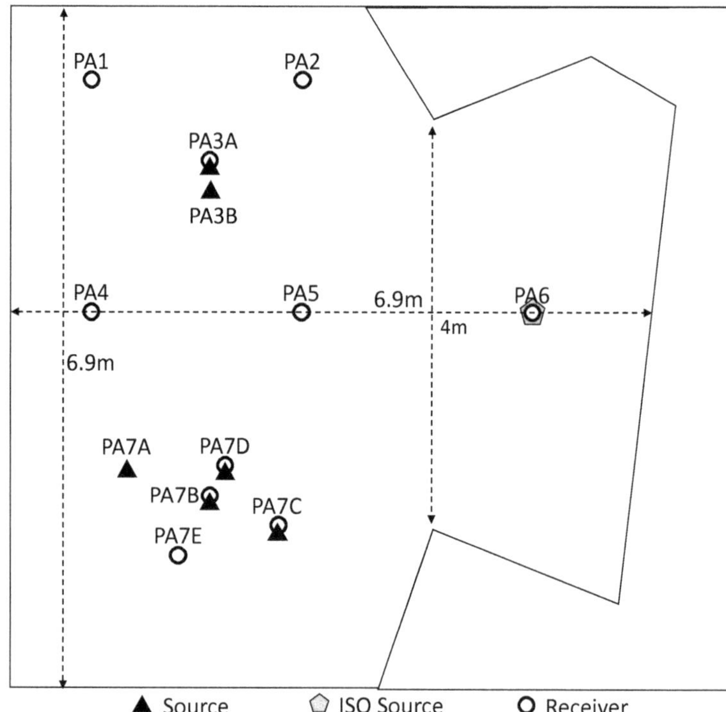

Figure 7. Source and receiver points for Studio MV5.

4. Data Post-Processing

4.1. Data Extraction and Cleanup

Recorded sweeps were exported from Reaper in batches according to the measurement setup. For example, for MV4, the entirety of the VR recording pass for the Eigenmike was exported as a single 32-channel file. Also exported in the same time windows were the other corresponding microphone channels, (KEMAR, reference omni microphone) as well as the original sweep playback stem for the entire recording pass.

Eigenmike recordings were then split into mono files in Matlab and batch-processed along with the other microphones in Izotope RX8 audio editor [20] for cleaning. A major advantage of working with tonal sweeps is that transients are not a desired part of the recorded signal, and will therefore result in 'phantom reversed' sweeps if deconvolved. Occasional low-level transients can occur during measurements due to air conditioning rattle or inadvertant movement from a member of the measurement team. In cleaning such events, we utilised the RX8 'Deconstruct' feature, which seperates the signal into tonal, noise and transient characteristics through spectral decomposition. Further cleaning of the recordings can be achieved through spectral denoising and equalisation if required, although care must be taken to ensure that the low-level tail of the impulse response is not compromised. A frequency-dependent noise profile must first be learned for a silent portion of the recorded signal to set an appropriate frequency-dependent threshold. Noise can then be suppressed through frequency-dependent expansion of the signal. We recommend for this process that once the signal level within a frequency bin has gone below the threshold (ideally 60 dB below the main signal) that the expander release be set to a high value (in this case 350 ms) to ensure the reverberation decay is not unduly gated. Furthermore, care should be taken in the amount of expansion, as -10 dB is usually sufficient to hear a significant improvement in perceived signal-to-noise ratio upon deconvolution. Finally, any signal below (in this case 60 Hz) or above the excitation frequency range of the transducer can also be removed through high-pass filtering.

4.2. Data Parsing and Transcoding

Once the data was cleaned, it was then ready for the deconvolution routine. As previously mentioned, Channel 1 of the sweep playback consisted of a short burst of low-level metadata, which identified the start of the playback. Since the exported original sweeps and recorded sweeps have the same timestamps (i.e., they were both exported simultaneously from Reaper under the same timeline selection), then the metadata marker in the original sweep is used to identify the start of a measurement in the recorded sweep. A Matlab routine was created which identified the sample index of such markers in a recording, extracted the corresponding recorded sweeps and then convolved the recordings with a time-reversed and amplitude-compensated version of the original sweep to extract the impulse responses. In the case of MV4, for instance, the first metadata marker is for receiver position PA1 with all sources north-facing. Under this measurement, seven impulse responses were extracted, each IR separated by δ_{int} seconds. Within each IR window, the front of the IR was detected and windowed to allow only 128 samples before the direct sound with any preceeding harmonic distortion components set to zero amplitude.

Impulse responses were then equalised to remove the diffuse field responses of the transducers: the Eigenmike, housing 32 DPA capsules, the KEMAR dummy head, 2 GRAS 40 AH low-noise microphone capsules, and the reference omnidirectional microphone, which is a single AKG CK77 capsule. An equalisation filter response was obtained by measuring the response of each microphone within an array of Genelec 8030 and 8040 loudspeakers in the AudioLab listening room at the University of York. The array is arranged in a 50-point spherical Lebedev formation [21]. Impulse responses were windowed to remove any early reflection components, leaving only the direct sound responses. For each microphone, the minimum phase responses were then averaged and an inverse filter response was calculated using Kirkeby regularisation [22]. Each Maida Vale impulse response measurement was then convolved with the correspoding equalisation filter. The IRs were then exported and named based on the studio, measurement setup, source–receiver combination, source orientation and microphone type.

Further processing was implemented on the RAW 32-channel Eigenmike capsules. The signals were converted to third-order spherical harmonic format using the MH Acoustics Em32-Encoder plugin within Matlab. As the Eigenmike was also recorded in endfire position, the soundfield was pitched down by 90 degrees about the y-axis to compensate.

5. Results

The following subsections present the measured acoustic parameters of each of the studios taken from the recorded RIRs and their similarities and differences followed by the approach taken to render VR environments from the captured data.

5.1. Acoustic Comparison

For each room, all ISO-3382 impulse responses were extracted from their corresponding measurements and their acoustic properties derived. A spatial average of the different acoustic properties was then computed for each room. A comparison of the acoustics of the four measured studios is presented in Figure 8. Here, we present the acoustic definition (D50), clarity (C80), Centre Time (CT), Early Decay Time (EDT) and reverberation time (T30).

D50 gives us a comparison of the energy in the first 50 ms of the impulse response to the total received energy and is related to subjective speech intelligibility [14]. We can readily see that studios MV3, MV4 and MV5 have high levels of definition (80% to 90% on average), but this is much lower for MV2, which is the more reverberant space.

C80 gives us a comparison of the energy before and after the first 80 ms in the impulse response [14]. MV2 gives a clarity level closer to that expected of concert halls than any of the other studios. It is therefore no surprise that this room is favored for rehearsals by the BBC Singers. The other studios have a stronger direct sound response and much higher clarity overall, which makes them more favorable for recording rock or pop music, which again is their main function at the studios. This tendency of MV3, MV4 and MV5 to favor the direct sound is reinforced by the measure of Centre Time (CT), which is the time of the centre of gravity of the squared response, measured in seconds [14]. The frequency dependency of MV3/4/5 in this regard is remarkably similar, demonstrating an incredible level of control throughout these studios. Conversely, the Centre time of MV2 averages close to 80 ms, which again shows remarkable balance in the acoustics, given a T30 level around 0.75 s. It is interesting to note that EDT values for MV2 are greater than the T30 values, indicating the presence of strong early reflections from the wood panelling on the walls of the room as well as the exposed wooden floor. However, the effect of the geometric diffusion throughout the space lessens the impact of later reflections, resulting in a lower T30.

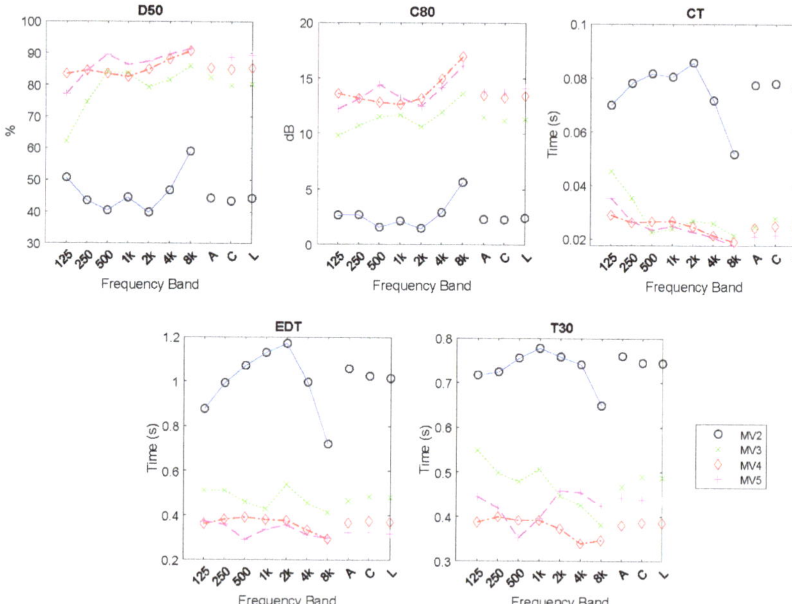

Figure 8. Comparison of acoustic parameters of the four measured live rooms: Top left—Definition D50, Top middle—Clarity C80, Top right—Centre time CT, Bottom left—Early Decay Time EDT, Bottom right—Reverb Time T30. Each image shows responses in octave bands as well as averaged weighted response using A, C, or L (linear) weighting.

The measured reverberation time for MV3 is marginally higher in the lower frequencies at around 0.5 s, but drops off with higher frequency, due in part to the absorptive effect of the curtain material throughout the space. MV4 shows the shortest T30 overall, averaging at 0.4 s. This is largely consistent across the spectrum, with some high-frequency deviation to 0.35 s at 8 kHz. MV5 demonstrates an interesting range of energy decay, (0.45 s at 125 Hz, dropping to 0.35 s at 500 Hz and then rising again) which is not surprising as it is the smallest space overall, with wooden refelective surfaces giving more potential for modal interaction at low to low–mid frequencies. Nonetheless, it is remarkable that MV3, MV4 and MV5 have such distinct similarities in their responses, despite the radical difference in their dimensions.

5.2. Vr Rendering

The Spatial Room Impulse Response (SRIR) measurements extracted from the Eigenmike can be utilised in 3DoF or 6DoF VR rendering [17]. Instrument audio can be rendered in real time through convolution with a desired SRIR decoded to binaural or loudspeaker rendering using an Ambisonic framework.

For applications where movement about the space is required, interpolation across the impulse responses is necessary [23]. This interpolation can be used to increase the density of impulse responses or interpolate the IRs in real-time within the rendering engine as the user moves through the space. Many methods of interpolation have been presented in the literature [24–30], with good results in increasing the measurement density obtained using a non-linear dynamic time-warping approach [31,32].

Another factor for any virtual acoustic rendering to be convincing is the realistic representation of the directional properties of the source audio. Complex sound sources such as drums will not radiate with equal power in all directions. Several approaches to measuring and synthesising source directionality have been proposed in the literature, primarily with regard to wave field synthesis reproduction. The capture of source directivity is traditionally achieved using arrays of microphones surrounding a performer and databases of such measurements are available [33]. Directivity filters can then be applied to a single monophonic recording of a performance to simulate the change in frequency response with source/listener movement. A simple approach has been proposed by Giron [34], where the interference of several monopole virtual sources is used to synthesise the directivity of real sources. However, the resulting frequency-dependent directivity does not behave like that of real world sources. A further approach is that the array of microphones used to capture the directivity measurements can actually be used to capture the performance (in an anechoic chamber) entirely, and virtual loudspeakers can be synthesised at reproduction using monopoles or virtual cardioids [35]. Although, this is not a practical solution to a real performance situation. Another method is the decomposition of the directional response into spherical harmonics, which has been proposed for computational-based auralisation in numerous papers, most notably in Spors [36].

Additionally, it has been found by Jacques et al. [35] that the directivity characteristics of a natural musical instrument do not have to be completely objectively identical to the original response, provided that the end-listener is not familiar with every single aspect of the *particular* instrument's directional response. A close approximation is generally sufficient to create a plausible reproduction.

The use of spherical loudspeaker arrays has also been proposed by Farina [37] and Zotter [38] for modeling the directivity of sound sources. Farina demonstrates how complete directional information can be obtained by measuring sweeps with spherical harmonic emission due to manipulation of the individual loudspeaker feeds in a docdecahedron [37]. This does require customisation of such a loudspeaker array as commercial solutions are not readily available, thereby limiting the uptake of the method. Kessler [11] has broadly outlined a method of approximating source directivity using standard loudspeakers by taking multiple spatial impulse response measurements at the same position, but at different loudspeaker rotations. An approximated directional characteristic can then be derived by simply exciting the room in different directions from the source position, capturing the SRIRs in each source direction and combining the impulse responses in a given ratio for auralisation. Here, we too utilise a similar approach to Kessler's, with the exception that we employ a least-squares optimisation to the measured directional signals and the corresponding SRIRs, such that a plausible directional characteristic is obtained [23].

The broadband directional response of a Genelec 8040 loudspeaker is approximately sub-cardioid, meaning that it has some significant attenuation (around 10 dB) to the rear of the loudspeaker [39]. For each receiver position there are four such sub-cardioid sources, measured in the North, East, South and West directions. We will define the matrix containing the directivity vectors of each of the loudspeakers as $\mathbf{D} = \begin{pmatrix} \mathbf{d}_1 & \mathbf{d}_2 & \mathbf{d}_3 & \mathbf{d}_4 \end{pmatrix}$.

The directivity of our virtual source must match that of real source directivity measurements **p** such that

$$\mathbf{D}\begin{pmatrix} \alpha_1 \\ \alpha_2 \\ \alpha_3 \\ \alpha_4 \end{pmatrix} = \mathbf{p} \tag{1}$$

where $\mathbf{a} = \begin{pmatrix} \alpha_1 & \alpha_2 & \alpha_3 & \alpha_4 \end{pmatrix}^T$, the gain factors to be applied to the loudspeakers such that the desired directional characteristic is obtained. Thus if

$$\mathbf{D}^T\mathbf{D}\begin{pmatrix} \alpha_1 \\ \alpha_2 \\ \alpha_3 \\ \alpha_4 \end{pmatrix} = \mathbf{D}^T\mathbf{p} \tag{2}$$

then we can solve for **a** with

$$\mathbf{a} = \begin{pmatrix} \alpha_1 \\ \alpha_2 \\ \alpha_3 \\ \alpha_4 \end{pmatrix} = (\mathbf{D}^T\mathbf{D})^{-1}\mathbf{D}^T\mathbf{p} \tag{3}$$

For example, imagine we wish to approximate a source with a supercardioid directivity, such as might be the case with a bass drum—high intensity from the front direction with phase reversed output from behind the drum. Starting with sub-cardioid responses, the above optimisation process will then yield $\mathbf{a} = \begin{pmatrix} -0.4211 & 1.0000 & -0.4211 & -0.4211 \end{pmatrix}$, where a negative gain represents a phase inversion. Therefore, at post-production, the final SRIR to be rendered at position i, $h_i(t)$ is the linear summation of the weighted measured impulse responses for that position, given by

$$h_i(t) = \alpha_1\, h_{s1}(t) + \alpha_2\, h_{s2}(t) + \alpha_3\, h_{s3}(t) + \alpha_4\, h_{s4}(t) \tag{4}$$

Note that the true directional excitation of complex sources is also frequency-dependent. Consequently, we propose that the above optimisation be performed in perceptual frequency bands [23].

6. Database of Measurements

The measurements of Studios 4 and 5 are freely available online with the permission of the BBC. The following elements for each recording studio are included:

- 32-Channel Eigenmike capsule impulse responses (RAW);
- 32-Channel Eigenmike capsule impulse responses (Diffuse Field Equalised);
- 16-channel third-order Ambisonic Impulse Responses;
- 2-channel KEMAR impulse Responses (RAW);
- 2-channel KEMAR impulse Responses (Diffuse Field Equalised);
- Mono Reference omnidirectional microphone impulse responses (RAW);
- Mono Reference omnidirectional microphone impulse responses (Diffuse Field Equalised).

Additional TOA impulse responses are presented that simulate instruments at MV4 source positions and demonstrate the source directivity simulation methodology outlined in Section 5.2. These IRs match the MV4 Unity project [17] also included with the database. All data is stored at 48 kHz sample rate, 24 bit resolution. Version 1.0 of the entire database can be downloaded at https://audiolab.york.ac.uk/resources/ (accessed on 10 August 2022). Measurements are available under the Non-Commercial Creative Commons License https://creativecommons.org/licenses/by-nc/3.0/ (accessed on 10 August 2022).

7. Conclusions

We have presented a database of acoustic measurements of the BBC Maida Vale Recordings Studios. The database consists of spatial impulse responses in third-order Ambisonic format as well as measurements from a KEMAR binaural manikin. The acoustic data captured facilitated a comparative study of the studios, that demonstrates the remarkable similarity of acoustic properties of studios MV3, MV4 and MV5 despite the major differences in geometry. MV2, which is equivalent in size to MV3, has a much longer reverberation time, with C80 and D50 values closer to concert hall acoustics than the other spaces, making it an ideal space for choirs or orchestral rehearsal. The short reverberation time and high clarity of the other spaces make them more appropriate for rock and pop recording.

The measurements captured facilitate 3DoF and 6DoF VR rendering and in the case of MV4 also correspond to an interactive Unity project of the space which accompanies the database. The database enables future work in reverberator design using, for example, feedback delay networks to approximate the distribution of SRIRs or alternate methods of spatial impulse response interpolation for 6DoF rendering. Other avenues of exploration include using the SRIRs for shared immersive musical experiences, for instance in networked music performance or fully virtual recording frameworks [17].

Author Contributions: Conceptualisation, G.K. and H.D.; methodology, G.K. and H.D.; measurement, G.K., H.D., P.C., T.R., A.H., J.C., P.T. and B.L. geometric measurement and visualisation, J.C. and D.J.; formal analysis, G.K.; resources, G.K., H.D. and B.L.; data curation, G.K. and B.L. writing—original draft preparation, G.K. and H.D.; writing—review and editing, G.K. and H.D.; project administration, G.K. and H.D.; funding acquisition, G.K. and H.D. All authors have read and agreed to the published version of the manuscript.

Funding: This research was funded by Engineering and Physical Sciences Research Council IAA project "MINERVA: Musical Interactions in Networked Experiences using Real-time Virtual Audio", EP/R51181X/1.

Institutional Review Board Statement: Not applicable.

Informed Consent Statement: Not applicable.

Data Availability Statement: Version 1.0 of the Maida Vale Impulse Response Database can be downloaded from https://audiolab.york.ac.uk/resources/ (accessed on 10 August 2022). Measurements are available under the Non-Commercial Creative Commons License https://creativecommons.org/licenses/by-nc/3.0/ (accessed on 10 August 2022).

Acknowledgments: The authors gratefully acknowledge the assistance of Emma Young, Chris Pike, Jack Reynolds and the BBC R&D team, as well as Andrew Rogers and the staff at BBC Maida Vale Recording Studios. The assistance of Andrew Chadwick in logistical preparation of the measurements is also gratefully appreciated.

Conflicts of Interest: The authors declare no conflict of interest.

Abbreviations

The following abbreviations are used in this manuscript:

ESS	Exponential Sinetone Sweep
DFE	Diffuse Field Equalised
HOA	Higher Order Ambisonic
IR	Impulse Response
KEMAR	Knowles Electronic Manikin for Acoustic Research
OA	Outside performance Area
PA	Performance Area
RIR	Room Impulse Response
SRIR	Spatial Room Impulse Response
VR	Virtual Reality

References

1. Bork, I. Report on the 3rd round robin on room acoustical computer simulation—Part II: Calculations. *Acta Acust. United Acust.* **2005**, *91*, 753–763.
2. Luizard, P.; Otani, M.; Botts, J.; Savioja, L.; Katz, B.F. Comparison of sound field measurements and predictions in coupled volumes between numerical methods and scale model measurements. In Proceedings of the Meetings on Acoustics ICA2013, Montreal, QC, Canada, 2–7 June 2013; Acoustical Society of America: New York, NY, USA, 2013; Volume 19, p. 015114.
3. Postma, B.N.; Katz, B.F. Perceptive and objective evaluation of calibrated room acoustic simulation auralizations. *J. Acoust. Soc. Am.* **2016**, *140*, 4326–4337. [CrossRef] [PubMed]
4. Gomez-Agustina, L.; Barnard, J. Practical and technical suitability perceptions of sound sources and test signals used in room acoustic testing. In Proceedings of the INTER-NOISE and NOISE-CON Congress and Conference Proceedings, Madrid, Spain, 16–19 June 2019; Institute of Noise Control Engineering: Reston, VA, USA, 2019; Volume 259, pp. 7076–7087.
5. BBC. Maida Vale: The Home of the BBC Symphony Orchestra. 2009. Available online: https://www.bbc.com/historyofthebbc/buildings/maida-vale/ (accessed on 12 August 2022).
6. Burton, J. BBC Maida Vale Studios. 2013. Sound on Sound. Available online: https://www.soundonsound.com/music-business/bbc-maida-vale-studios (accessed on 12 August 2022).
7. Borish, J.; Angell, J.B. An Efficient Algorithm for Measuring the Impulse Response Using Pseudorandom Noise. *J. Audio Eng. Soc.* **1983**, *31*, 478–488.
8. Foster, S. Impulse response measurement using Golay codes. In Proceedings of the ICASSP'86. IEEE International Conference on Acoustics, Speech, and Signal Processing, Tokyo, Japan, 7–11 April 1986; Volume 11, pp. 929–932.
9. Papadakis, N.M.; Stavroulakis, G.E. Review of acoustic sources alternatives to a dodecahedron speaker. *Appl. Sci.* **2019**, *9*, 3705. [CrossRef]
10. Farina, A. Simultaneous measurement of impulse response and distortion with a swept-sine technique. In Proceedings of the 108th Convention of the Audio Engineering Society, Paris, France, 19–22 February 2000.
11. Kessler, R. An Optimised Method for Capturing Multidimensional Acoustic Fingerprints. In Proceedings of the 118th Convention of the Audio Engineering Society, Barcelona, Spain, 28–31 May 2005.
12. Armstrong, C.; Thresh, L.; Murphy, D.; Kearney, G. A perceptual evaluation of individual and non-individual HRTFs: A case study of the SADIE II database. *Appl. Sci.* **2018**, *8*, 2029. [CrossRef]
13. Majdak, P.; Balazs, P.; Laback, B. Multiple exponential sweep method for fast measurement of head-related transfer functions. *J. Audio Eng. Soc.* **2007**, *55*, 623–637.
14. *ISO 3382-1:2009*; Acoustics-Measurement of Room Acoustic Parameters—Part 1: Performance Spaces. International Organization for Standardization: Geneva, Switzerland, 2009. Available online: http://www.iso.org/ (accessed on 21 September 2009).
15. Noisternig, M.; Sontacchi, A.; Musil, T.; Holdrich, R. A 3D Ambisonic Based Binaural Sound Reproduction System. In Proceedings of the 24th International Conference of the Audio Engineering Society, St. Petersburg, Russia, 1–3 June 2002.
16. Gorzel, M.; Allen, A.; Kelly, I.; Kammerl, J.; Gungormusler, A.; Yeh, H.; Boland, F. Efficient encoding and decoding of binaural sound with resonance audio. In Proceedings of the Audio Engineering Society Conference: 2019 AES International Conference on Immersive and Interactive Audio, York, UK, 27–29 March 2019; Audio Engineering Society: New York, NY, USA, 2019.
17. Cairns, P.; Hunt, A.; Cooper, J.; Johnston, D.; Lee, B.; Daffern, H.; Kearney, G. Recording Music in the Metaverse: A case study of XR BBC Maida Vale Recording Studios. In Proceedings of the 2022 Audio Engineering Society International Conference on Audio for Virtual and Augmented Reality, Washington, DC, USA, 15–17 August 2022.
18. Moody, A. Adult anthropometric measures, overweight and obesity. *Health Surv. Engl.* **2013**, *1*, 1–39.
19. McKenzie, T.; Murphy, D.; Kearney, G. Assessing the authenticity of the KEMAR mouth simulator as a repeatable speech source. In Proceedings of the Audio Engineering Society Convention 143, New York, NY, USA, 18–21 October 2017; Audio Engineering Society: New York, NY, USA, 2017.
20. Izotope. RX9 User Manual. 2021. Available online: https://www.izotope.com/en/products/rx.html (accessed on 12 August 2022).
21. Lecomte, P.; Gauthier, P.A.; Langrenne, C.; Garcia, A.; Berry, A. On the use of a lebedev grid for ambisonics. In Proceedings of the Audio Engineering Society Convention 139, New York, NY, USA, 29 October–1 November 2015; Audio Engineering Society: New York, NY, USA, 2015.
22. Kirkeby, O.; Nelson, P. *Fast Deconvolution of Multi-Channel Systems Using Regularisation*; ISVR Technical Report No. 255; Southampton, UK, 1996. Available online: https://citeseerx.ist.psu.edu/viewdoc/download?doi=10.1.1.472.184&rep=rep1&type=pdf (accessed on 12 August 2022).
23. Kearney, G. Auditory Scene Synthesis Using Virtual Acoustic Recording and Reproduction. Ph.D. Thesis, Trinity College Dublin, Dublin, Ireland, 2010.
24. Antonello, N.; De Sena, E.; Moonen, M.; Naylor, P.A.; Van Waterschoot, T. Room impulse response interpolation using a sparse spatio-temporal representation of the sound field. *IEEE/ACM Trans. Audio Speech Lang. Process.* **2017**, *25*, 1929–1941. [CrossRef]
25. Matsumoto, M.; Tohyama, M.; Yanagawa, H. A method of interpolating binaural impulse responses for moving sound images. *Acoust. Sci. Technol.* **2003**, *24*, 284–292. [CrossRef]
26. Bruschi, V.; Nobili, S.; Cecchi, S.; Piazza, F. An innovative method for binaural room impulse responses interpolation. In Proceedings of the Audio Engineering Society Convention 148, Online, 2–5 June 2020; Audio Engineering Society: New York, NY, USA, 2020.

27. Mignot, R.; Chardon, G.; Daudet, L. Low frequency interpolation of room impulse responses using compressed sensing. *IEEE/ACM Trans. Audio Speech Lang. Process.* **2013**, *22*, 205–216. [CrossRef]
28. Das, O.; Calamia, P.; Gari, S.V.A. Room impulse response interpolation from a sparse set of measurements using a modal architecture. In Proceedings of the ICASSP 2021–2021 IEEE International Conference on Acoustics, Speech and Signal Processing (ICASSP), Toronto, ON, Canada, 6–11 June 2021; pp. 960–964.
29. Garcia-Gomez, V.; Lopez, J.J. Binaural room impulse responses interpolation for multimedia real-time applications. In Proceedings of the Audio Engineering Society Convention 144, Milan, Italy, 23–26 May 2018; Audio Engineering Society: New York, NY, USA, 2018.
30. Mehrotra, S.; Chen, W.g.; Zhang, Z. Interpolation of combined head and room impulse response for audio spatialization. In Proceedings of the 2011 IEEE 13th International Workshop on Multimedia Signal Processing, Hangzhou, China, 17–19 October 2011; pp. 1–6.
31. Masterson, C.; Kearney, G.; Boland, F. Acoustic impulse response interpolation for multichannel systems using dynamic time warping. In Proceedings of the Audio Engineering Society Conference: 35th International Conference: Audio for Games, London, UK, 11–13 February 2009; Audio Engineering Society: New York, NY, USA, 2009.
32. Kearney, G.; Masterson, C.; Adams, S.; Boland, F. Dynamic time warping for acoustic response interpolation: Possibilities and limitations. In Proceedings of the 2009 17th European Signal Processing Conference, Glasgow, UK, 24–28 August 2009; pp. 705–709.
33. Physikalisch-Technische-Bundesanstalt. Directivities of Musical Instruments. 2009. Available online: http://www.catt.se/udisplay.htm#SD_instruments (accessed on 12 August 2022).
34. Giron, F. Investigations about the Directivity of Sound Sources. Ph.D. Thesis, Ruhr-Universität, Bochum, Germany, 1996.
35. Jacques, R.; Albrecht, B.; Melchior, F.; de Vries, D. An approach for multichannel Recording and Reproduction of Sound Source Directivity. In Proceedings of the 119th convention of the Audio Engineering Society, New York, NY, USA, 7–10 October 2005.
36. Ahrens, J.; Spors, S. Implementation of Directional Sources in Wave Field Synthesis. In Proceedings of the IEEE Workshop on Applications of Signal Processing to Audio and Acoustics, New Paltz, NY, USA, 21–24 October 2007.
37. Farina, A.; Martignon, P.; Capra, A.; Fontana, S. Measuring Impulse Responses Containing Complete Spatial Information. In Proceedings of the 22nd UK Conference of the Audio Engineering Society, London, UK, 25–27 June 2007.
38. Zotter, F.; Höldrich, R. Modeling radiation synthesis with spherical loudspeaker arrays. In Proceedings of the 19th ICA, Madrid, Spain, 2–7 September 2007.
39. Streicher, R.; Everest, F.A. *The New Stereo Soundbook*, 3rd ed.; Audio Engineering Associates: Pasadena, CA, USA, 2006.

Review

Intangible Mosaic of Sacred Soundscapes in Medieval Serbia

Zorana Đorđević [1,*], Dragan Novković [2] and Marija Dragišić [3]

[1] Institute for Multidisciplinary Research, University of Belgrade, 11000 Belgrade, Serbia
[2] School of Electrical and Computer Engineering, Academy of Technical and Art Applied Studies, 11000 Belgrade, Serbia
[3] Institute for the Protection of Cultural Monuments of Serbia—Belgrade, 11000 Belgrade, Serbia
* Correspondence: zoranadjordjevic.arch@gmail.com

Abstract: Religious practice in Serbia has taken place using both indoors and outdoors sacred sites ever since the adoption of Christianity in medieval times. However, previous archaeoacoustic research was focused on historic church acoustics, excluding the open-air soundscapes of sacred sites. The goal of this review paper is to shed light on the varieties of sacred soundscapes that have supported the various needs of Orthodox Christian practice in medieval Serbia. First, in relation to the acoustic requirements of the religious service, we compare the acoustic properties of masonry and wooden churches based on the published archaeoacoustic studies of medieval churches and musicological studies of the medieval art of chanting. Second, we provide an overview of the ethnological and historical studies that address the outdoor sacred soundscapes and investigate the religious sound markers of large percussion instruments, such as bells and semantra, the open-air litany procession that has been practiced during the annual celebration of a patron saint's day in rural areas, and the medieval assemblies that took place on the sacred sites. This paper finally points out that the archaeoacoustic studies of sacred soundscapes should not be limited to church acoustics but also include open-air sacred sites to provide a complete analysis of the aural environment of religious practice and thus contribute to understanding the acoustic intention of medieval builders, as well as the aural experience of both clergy and laity.

Keywords: sacred soundscapes; medieval Serbia; bells; semantron; medieval churches; acoustic measurements; litany; medieval assemblies

Citation: Đorđević, Z.; Novković, D.; Dragišić, M. Intangible Mosaic of Sacred Soundscapes in Medieval Serbia. *Acoustics* **2023**, *4*, 28–45. https://doi.org/10.3390/acoustics5010002

Academic Editors: Margarita Díaz-Andreu and Lidia Alvarez Morales

Received: 3 August 2022
Revised: 30 November 2022
Accepted: 14 December 2022
Published: 27 December 2022

1. Introduction

The studies of aural environments of medieval sacred places are dominantly focused on indoor soundscapes. Although the corpus of studies has gradually increased, there is a striking imbalance between the number conducted for sacred places in the Latin West and the Byzantine East. While the numerous churches in Western and Central Europe are thoroughly studied [1–3], the aural aspects of Byzantine and Balkan medieval churches are addressed in few research projects. Hagia Sophia in Istanbul was of particular interest over last few decades. The CAHRISMA project (Conservation of the Acoustical Heritage by the Revival and Identification of the Sinan's Mosque's Acoustics) considered it together with two other Byzantine churches, testing virtual acoustic models and providing auralization, aiming to visually and sonically revive architectural heritage [4]. The project "Icons of Sound" examined acoustic models, liturgical texts, and melodic structures to illuminate how the sung liturgy and architectural acoustics worked together, creating an aural experience that facilitated mystical transcendence for the congregation [5]. It was essential that everyone could clearly hear both spoken and sung words in the church [6]. As a part of the same project, the acoustics of Hagia Sophia was recreated in a concert hall setting via the live auralization technique, thus, together with a Byzantine chant performance, providing the sonic experience of this early medieval temple for a wide audience [7]. The recent acoustic simulation of Hagia Sophia tested how the acoustic conditions change with occupation and the new finishing materials used in restoration works, emphasizing that the current state of Hagia Sophia's mid- and low-frequency reverberation time (T30) averages around 9 s [8]. The study of another early Byzantine church, San Vitale in Ravenna, emphasized that its long reverberation time of 4.6 s favors musical performance rather than speech [9]. Byzantine churches in Thessaloniki, the second largest city of the Byzantine Empire after Constantinople, were acoustically studied, which included the impulse response measurements of 11 churches [10–12] and the examination of two acoustic models in order to comprehend the changes in church acoustics that occurred from shifts from the basilica-type to cross-plan churches with a central dome [13]. The recent archaeoacoustic project, "Soundscapes of Byzantium", took a step further, investigating the connections of church acoustics, Byzantine chanting tradition, and wall paintings (the representations of hymns and hymnographers in particular) in eight churches built from the 5th to the 14th century in Thessaloniki [14–16]. The acoustic measurements supported what was up to that point considered only as a metaphorical interpretation: the intermingling of human and angelic voices. It was argued that the presence of angelic voices corresponded to the distinct aural phenomena that were observed in Thessaloniki churches, such as melodically synchronous overtones, disconnected flutter echo, and unique reverberation patterns [17]. In comparison with Hagia Sophia in Istanbul, Byzantine churches in Thessaloniki have an exceptionally low reverberation time (lower than 3 s), which makes them more suitable for spoken word [15]. Even lower reverberation times are measured in a recent study of five Byzantine churches in Albania; however, they reported low speech clarity [18]. Similar is found for the Byzantine churches in Cappadocia, for which the virtual acoustical models were tested to explore the effect of architectural changes on church acoustics [19].

The study of the impact of sound outside medieval churches and monasteries on the open-air surrounding landscape has been very limited and focused mostly on bell ringing in towns and villages [20,21] or the audibility range of bells with mapping using the Geographic Information System (GIS) [22,23], or addressed briefly in general [15,24]. However, a recent study of medieval Italian soundscapes explored the reach of bell sounds in rural areas that surround various monastic complexes. Combining catchment, viewshed, and sound-mapping analysis, it was shown that in some cases, sites that were outside the 30-minute travel threshold from the monastery and the visibility area were still within audibility range. This finding emphasizes the communicative capacity of soundscapes in medieval monastic landscapes [25]. In addition, research into open-air medieval assemblies, which often took place at sacred sites, points out that practical and ceremonial needs include the acoustic properties of these open-air medieval places of gathering [26].

This review paper strives to outline the variety of sacred soundscapes that compose the intangible mosaic of acoustic environments nurtured since the medieval times in Serbia. The reason for this is that, as we will show in this paper, the published interdisciplinary archaeoacoustic studies are exclusively focused on church acoustics in Serbia, leaving aside all open-air soundscapes despite their important role in Serbian religious practice. Therefore, we tend to draw attention to the variety of medieval sacred soundscapes and contribute by expanding the interests of archaeoacoustic research to an outdoor aural environment by providing a review of both the studies of medieval church acoustics and the studies that are a bit more distantly related to open-air sacred soundscapes.

The sacred art and architecture of medieval Serbia developed under the influence of both the Byzantine East and Latin West ever since the adoption of Christianity. In addition to churches and their specific indoor sonic environment, medieval sacred soundscapes also include open-air sanctuaries and related religious practices. In this paper, we are particularly focused on the period from the late 12th century when the Nemanjić dynasty came to the throne until the fall of Serbia under Ottoman rule in the mid-15th century, because this period was fundamental for the development of the Serbian Orthodox Church as well as sacred architecture. Rapid developments in church architecture started with the endowments of Stefan Nemanja, the eponymous founder of the Nemanjić dynasty. When he inherited power, his son Stefan the First-Crowned obtained the crown from Roman Pope in 1217, and his other son—later known as Saint Sava—initiated the autocephalous status for the Serbian Orthodox Church that became an independent Archbishopric of Žiča in 1219. Consequently, the close interactions between the medieval Serbian state and the Serbian Orthodox Church led to the spread of Christianity in the region and the foundation of numerous monasteries.

In this paper, we first review the archaeoacoustic studies of indoor sacred soundscapes, focusing on the characteristics of sound content in religious practice and its correlation with medieval church acoustics, including both larger monastic churches built from masonry and smaller local churches built from wood. Then, in the next section, we review the studies that considered the outdoor sacred soundscapes and related practices: 1. the usage of large percussion instruments (bells and semantra); 2. litanies in rural areas; and 3. medieval state and church assemblies. In conclusion, we point out that the archaeoacoustic studies of medieval sacred soundscapes should not be limited to church acoustics but also include open-air soundscapes to provide a complete analysis of the aural environment of the religious practice, thus contributing to understanding the acoustic intention of medieval builders as well as the aural experience of both clergy and laity.

2. Indoor Sacred Soundscapes

2.1. The Aspect of Sound in the Medieval Serbian Churches

Considering the acoustical problems, it is necessary to keep in mind the basic purpose of the examined space, as well as its end users' expectations regarding the sonic environment. The Serbian Orthodox Church adopted the Byzantine comprehension of sacred space as an ideal image of the universe in which sound consequently contributed to the overall religious experience. Saint Sava, the first archbishop of the Serbian Orthodox Church and the youngest son of the Grand Prince Stefan Nemanja, referred to the sound of divine service as the very soul of the sacred place [27]. The divine service, as the most significant religious act, has three main purposes: to spread the Christian faith and lift the faithful's spirit to God; to facilitate and induce praying, expressed in both spoken pious conversation with God and ecclesiastical chanting; and to induce redemption and the unification of man and God through faith, and a Christian life full of virtues, prayer, and secrets [28]. In other words, both spoken and chanted word has been important in Orthodox Christian religious practice.

Along with Christian faith, Byzantine chanting tradition was adopted in medieval Serbia. This monophonic *a cappella* chant is based on recognizable musical formulas, with a gradual melodic flow and using musical intervals smaller than a semitone [29]. Each spoken word is given the quality of a song so that the chanting highlights the essence of the lyrics. Saint John of Damascus systematized church chanting in the Octoechos, the church book for chanters that contains eight voices or modes, each "ruling" the divine service for a week. The adoption of the eight modes of the antique music enabled the expression of a variety of feelings in Byzantine chanting in accordance with the liturgical cycle [30]. The spiritual importance of chanting is also emphasized by medieval hagiographers: Teodosije wrote that while on his deathbed, Stefan Nemanja (later canonized as Saint Simeon) requested to be escorted with monks' chanting, until finally he looked as if he himself "chanted an angel song with angels" [31]. The chanting was expected to be acoustically intensified inside a church by blurring and dissolving sound into an immersive acoustic experience, thus enhancing the stimuli of vision and scent [32].

To achieve the adequate dynamics and variability of the divine service, the chanting of psalms is combined with Bible reading, common prayers, and the sermon. From the ambo—the elevated stand opposite of the Royal Doors in the naos (or today, occasionally from the southern choir)—gospels are read and litanies are spoken. During the divine service, the faithful stand facing the altar, occasionally bowing, kneeling, and lifting their arms and eyes to the sky [28]. As the aim was to convey the teaching, it was of the utmost importance that the speech was clear and intelligible [33]. Since both spoken and chanted word was essential in Orthodox divine service, the church space was expected to not only acoustically facilitate the intelligibility of spoken word but also intensify a spiritual experience through chanted melodies.

2.2. Church Acoustics

2.2.1. Monastic Churches

The acoustics of medieval monastic churches have been studied in the last several decades, including the acoustic vessels embedded in the massive walls and the acoustics of the church's interior space.

Acoustic vessels are found in 15 medieval churches that are now under the jurisdiction of the Serbian Orthodox Church in what today is Serbia. Several of them have been examined theoretically [34], but only the ones from two churches—in the villages Komarane [35] and Trg [36]—have been tested in the laboratory. The resonant frequency of the tested vessels is in the range of a male voice. However, from the archaeoacoustic point of view, it is not only the question of acoustic efficiency that is important but the ideas that lie beneath the building practice of embedding acoustic vessels into church walls [37]. Therefore, it is necessary to include an analysis of all available vessels from medieval Serbia in order to arrive at a statistically justified conclusion to contribute to the existing body of knowledge on acoustic vessel practices in medieval Europe [38–40], possibly leading to a better understanding of the transmission of this practice between East and West.

Previous acoustic measurements of monastic churches built in medieval Serbia included Lazarica church [41] and the monastic churches of Ljubostinja Monastery and Naupara Monastery [42], Manasija Monastery, Ravanica Monastery, and Pavlovac Monastery [33]. All these churches were built in the Morava architectural style (1371–1456), which is considered to be the climax of church building in Serbia. They have a triconch plan combined with a developed or compressed inscribed cross, chanting apses on the northern and southern side, and a central dome (Figure 1). The developed inscribed cross plan has four columns supporting the central dome, which enabled building a larger church, as can be observed in Figure 1. For the purpose of comparison, Manasija church (Figure 1, second from the right) has a volume of about 4000 cubic meters, while the smallest one also built in the Morava style is Pavlovac church (Figure 1, the first one from right), which has a volume of about 400 cubic meters.

Figure 1. Plans and longitudinal sections of the Morava-style medieval churches (range of years of building), from the left to the right: Ravanica (1375–1377), Lazarica (1377–1380), Naupara (1377–1382), Ljubostinja (1381–1388), Manasija (1407–1427), and Pavlovac (1410–1425).

When discussing acoustics, it is important to note the finishing of large surfaces. Medieval monastic churches in Serbia usually have stone floors and fresco painted walls from the floor to the apex of the dome. However, the level of fresco paintings preservation varies from one church to another. Figure 2 shows the case of Ravanica monastic church, in which the upper zone of the wall paintings is significantly damaged and large areas are mortared as a part of conservation works. There are also churches that have no wall paintings at all, such as Lazarica and Pavlovac. The iconostases (the screens separating the naos from the sanctuary apse) are found today made of stone with a height of about 2.5 m (as in medieval times), or made of wood and significantly taller, with a large wooden cross on top. The church furniture is made of wood, and is dominantly located in the apses and under the central dome.

Figure 2. The nave interior of the Ravanica monastic church: view towards the altar (**left** and **middle**) and upwards to the central dome (**right**) (photo: Zorana Đorđević).

Acoustic measurements of impulse responses were carried out for all of the above-mentioned churches using EASERA acoustic software (Version 1.0b (2005), SDA—Software Design Anhert GmbH, Germany). The level of measured background noise in all churches was extremely low, ranging from 21 to 23 dBA, which was expected considering that the locations of the churches are isolated from any external sources of noise, far from urban areas. The only exception is the Lazarica church, which is located in an urban environment, and where the background noise level during measurement was 26 dBA.

Interesting conclusions can be drawn from the comparison of acoustic parameters of churches of different volumes (Table 1), such as Ljubostinja (circa 1500 m^3) and Naupara (circa 500 m^3). Although Ljubostinja church has a volume three times larger than that of Naupara, differences in the average reverberation times between the two churches are not so pronounced. This can be explained by the fact that Ljubostinja has more wooden surfaces within the naos, which is an additional reason for the even higher average absorption coefficient of the Ljubostinja naos compared with the same area in Naupara. The significantly shorter-than-expected reverberation time in Ljubostinja could be explained by the columns in the naos, which are dominant acoustic diffraction elements. Sound diffraction exerts its own impact on shortening reverberation time due to an overall shortening of the average path distance that sound waves pass. Such acoustical situations increase the possibility of significantly different behavior in sound fields (that can result in the shortening of reverberation time), which cannot be observed simply by comparing only the overall volumes of two or more spaces. The similar reverberation times in both churches' naos might indicate similar values of acoustic parameters that describe speech intelligibility; however, this is not the case (Table 1). Namely, Speech Transmission Index (STI) which was derived from the impulse response measurement using EASERA software, has much greater values for Naupara, indicating noticeably higher levels of speech intelligibility for this church. Such a situation can be explained by its smaller interior volume and the reduced path distances that sound waves pass in Naupara, resulting in the greater part of the reflection energy arriving to the listener within the ear's integration time zone (first 50 ms after the arrival of direct sound). The sound energy that reaches listeners within this time zone is considered as useful in the sense of speech intelligibility. More energy comes to a listener after this period, which is characteristic for Ljubostinja, distracting the speech recognition process and resulting in lesser values in speech intelligibility parameters. It is notable that the best speech intelligibility is under the central dome of the Naupara naos, which is close enough to a listener to provide reflection that improves overall speech intelligibility. Finally, acoustic parameters clearly indicate a significant reduction in speech intelligibility in the case of sound field excitation from the altar. Singing in the altar area behind the iconostasis has a specific role in Orthodox divine service, with the main intention of encouraging a sense of the holy mystery, while the comprehensibility of spoken and sung text is not so important [42].

Table 1. Mean values of acoustic parameters for naos based on the in situ acoustic measurements.

Architectural Style		Church (Year of Building Completion)	Volume [m^3]	T_{30}[s] @500 Hz	EDT[s] @500 Hz	STI	Cited From
Raška architectural style		Trg (1382)	400	0.61	0.63	0.738	[36]
Morava architectural style	developed inscribed cross plan	Ravanica (1377)	1800	1.56	2.09	0.499	[33]
		Ljubostinja (1388)	1500	1.79	1.41	0.554	[42]
		Manasija (1418)	4000	3.60	4.22	0.406	[33]
	compressed inscribed cross plan	Lazarica (1381)	720	2.13	2.35	0.473	[41]
		Naupara (1382)	500	1.61	1.19	0.623	[42]
		Pavlovac (1425)	380	2.04	2.09	0.559	[33]
Wooden church		Brzan (19th century)	200	0.84	0.94	0.735	[33]

In the case of Lazarica church, the STI parameter indicates that intelligibility under the dome highly depends on the position of the sound source. When the excitation signal was from the altar, the STI parameter had a value of 0.491 when measured directly under the dome, and when the excitation was from the chanting apse, the STI for the same receiver position was 0.673. Such a situation can be explained with specific relations between chanting apse and the "under the dome" position. A person standing under the dome is in front of the chanting apse, having the highest "direct to reflected" sound ratio, which directly impacts speech intelligibility, while there is a physical barrier (iconostasis) between the sound source in the altar and the same receiver position [41].

In addition to the comparison of acoustic parameters, the subjective assessment of church acoustics provides valuable additional information on the experience of sound in a sacred place. A comparative analysis of acoustic parameters measured in three monastic churches built from masonry—Manasija, Ravanica, and Pavlovac—along with a wooden church in Brzan village was conducted together with a subjective assessment of sound field characteristics with 119 respondents, aged 15 to 61, based on listening to the auralization files of Byzantine chanting and speech. Pavlovac church is perceived as the largest space by 39% of respondents, ahead of the Manasija church, which has a nearly 8-times-larger volume than Pavlovac. Such an unexpected result is due to the acoustic influence of the dome. A listener beneath the dome in Pavlovac church observes acoustic phenomena that are not so audibly conspicuous in higher churches, such as Ravanica and Manasija. The dome directs the sound waves towards its geometric focus, producing a unique play of late delays and consequential changes of the phase differences of the reflected sound waves, finally resulting in a subjective perception of a much larger space than it actually is. The results of acoustic measurements support this as well. The subjective evaluation of speech intelligibility and the STI parameter are matching. They are both inversely proportional to the sizes of the church, so that their values decrease as the volume grows. This is fairly logical situation, since higher amounts of late sound energy coming to the listener, which is a characteristic of larger spaces, is the main reason for decreases in overall speech intelligibility. The results of the subjective evaluation suggest the same: the smallest church (Brzan) is evaluated as "the best", while the biggest one (Manasija) is seen as having "the worst" speech intelligibility. Additionally, different positions within the same church are characterized by different values of STI parameter. Thus, the speech generated inside the naos, in the case of Ravanica and Manasija, is almost completely incomprehensible in the narthex: most of the respondents (85%) were evaluated as "unsatisfied" in those positions regarding speech intelligibility [33].

Changes to the original church interior might dramatically affect the experience of sound. For example, the Trg church deviates from the expected values of the reverberation time. The average RT @ 500 Hz for small churches of overall volume similar to the volume of Trg church (around 400 m^3) is around 1.6 s [43]. However, the average measured values of the T30 parameter in Trg church is 0.60 s at the octave with the central frequency of 500 Hz. This might be due to the presence of wooden furniture but also a thick layer of carpets that authors were not allowed to remove during the acoustic measurements. Such floor carpeting, often justified by the necessary thermal insulation of the church floor, produces a significant deviation from the original acoustic condition of the church. Accordingly, it should be noted that the "original" reverberation time of the church could be even higher than 30% in the higher frequency range but not significant in the lower frequency range [36].

Additionally, Trg church is a good illustration for an acoustic situation where the Early Decay Time (EDT) parameter has consistently greater values than the T30 parameter in the case of sound excitation from the altar area, while in the case of excitation from the naos, the situation is the opposite. Since EDT correlates with reverberance, or as a perceived reverberation during music play or speech (at least as perceived by the audience), the conclusion is the that overall perceived reverberance inside of the church is greater in the case of chanting from altar, then from the naos. In other words, divine service from the naos position has a tendency toward intimacy due to the nearness of the talker and listener, while the performance from the altar causes difficulty in locating a sound source, thus adding to the sense of mystery of the religious service. The same conclusion can be drawn for most of the other analyzed medieval churches (Table 1).

2.2.2. Wooden Churches

Wooden churches have been built in Serbia since the medieval times. The earliest recorded mention of log churches in Serbian lands was in the Life of Saint Sava, written in the 14th century. The hagiographer Teodosije wrote that Saint Sava himself encouraged people to build wooden churches whenever needed [31]. Under the Ottoman rule in Serbia from the 15th to the 19th century, wooden churches retained their simple architecture that allowed them to be easily repaired, disassembled, and transferred to another well-hidden location [36]. Therefore, the shape of wooden churches did not change from the medieval period. That entitles us, in the scope of this paper, to examine the archaeoacoustic studies of log churches built after medieval period, because turbulent historical circumstances, including uprisings and migrations, left us with log churches not older than the 17th century [44]. There are about 40 wooden churches preserved today in Serbia. After the First (1804) and the Second (1815) Uprising, and particularly after the official Ottoman recognition of liberty to fully practice Orthodox religion (Hatisherif from 1830), the revitalization of wooden churches began.

Although traditionally called log churches, these buildings are not made of logs but planks. Usually, the oak tree was used, as it is perceived as the most durable building material and provides good sound insulation, rousing feelings of warmth and intimacy [44].

The architecture is quite simple: single-nave church, consisting of narthex, nave, and apse, and a high roof with eaves. The log church in Brzan village (Figure 3), built in the first half of 19th century, is particularly interesting as it was equipped with acoustic vessels—ceramic jugs pierced on the bottom. These vessels used to hang from the roof beams, between the roof and the false wooden vault made in a trough shape. As previously mentioned, the acoustic vessels were usually embedded in the masonry walls of medieval churches. The case of Brzan church suggests that the same acoustic practice was applied in wooden architecture as well. The vessels could not be embedded in thin wooden walls but they were hung on the roof beams. When Brzan church was reconstructed in 1960s, mortar was peeled from the interior walls, leaving the visible notches on the wooden planks that are usually made to apply the mortar on the wood. The iconostasis, reaching the ceiling, is also made of wood, and the floor is paved using bricks. Pavlović, who led the reconstruction, wrote that the Brzan church was an acoustical space [45].

Figure 3. Wooden church in Brzan village: acoustic vessels hanging on the roof construction as found during the reconstruction works in the 1960s (**left**), church exterior (**middle**), church interior (**right**) (photo: Institute for the Protection of Cultural Monuments of Serbia (**left**), Zorana Đorđević (**middle** and **right**)).

The results of the on-site acoustical impulse response measurement in the Brzan church are presented in Table 2 [33]. The values of the obtained parameters are quite expected for that type of building. Wooden churches are characterized by small volumes and an interior design that is completely different in its acoustic properties from masonry churches. That is the main reason for their very specific acoustics. The main acoustic properties of wooden churches are extremely high intelligibility of speech (STI = 0.735, the borderline case of speech intelligibility that might be qualified as "almost excellent"), as well as the overall feeling of acoustic intimacy and warmth that the present believers have during the service [33]. On the other hand, the acoustic ambience of wooden churches lacks a pronounced sacral moment, which is reflected in longer times of reverberation and various psycho-acoustic phenomena that occur due to the existence of specific architectural elements, especially the dome.

Wooden churches could not be characterized by a "huge" and "divine" sound due to their fairly small reverberation time compared with masonry churches. Yet, the warm and intimate acoustics of this ambience with high intelligibility of speech fully corresponded to the basic task of wooden churches—to provide a suitable aural environment to communicate religious ideas to the congregation.

Table 2. Acoustic parameters measured in the wooden church in Brzan village: Reverberation Time (T_{30}), Early Decay Time (EDT), Articulation Loss of Consonants (AL_{cons}), Bass Ratio (BR), and Average Sound Absorption Coefficient ($\bar{\alpha}$) [33].

T_{30} [s]	EDT [s]	AL_{cons} [%]	STI	BR	$\bar{\alpha}$
0.84	0.94	3.2	0.735	1.02	0.18

3. Outdoor Sacred Soundscapes

While semantra were traditionally used in Byzantium, bells were introduced gradually to the liturgical practice of the Eastern Church. Religious soundscapes in the medieval Serbian state under the Nemanjić dynasty and their successors were actively created by both semantra and bells. The sound of these large percussion instruments could be heard from various distances around a sacred site, depending on terrain configuration, vegetation, and weather conditions. In addition to accommodating these particular sounds, outdoor sacred soundscapes had a role in the religious activity of a village's patron saint's day celebration, which includes a litany procession from one sacred tree to another around a village. Open-air soundscapes of sacred sites were also important for the medieval state and church assemblies that usually took place in the vicinity of a (monastic) church. In this section, we will investigate the sacred soundscapes in medieval Serbia regarding the use of semantra and bells, litany processions, and finally the state and church assemblies.

3.1. Semantra and Bells

As a traditional instrument in Byzantium, the semantron was adopted in medieval Serbia, known under the names of klepalo and bilo [46]. The klepalo is a wooden board played with wooden mallets. As Figure 4 shows, today we find two types of klepalo—a thin and long plank, which is played while a person leans it on their shoulder, or a massive piece of wood hanging on ropes outdoors, usually in the vicinity of the bell tower or church entrance. The bilo is made of bronze plates, which are played with metal hammers while hanging. These three types of semantra were mentioned in the Hilandar Monastery typikon, drafted by Saint Sava in 13th century based on the model of the rule of the Evergetis Monastery in Constantinople. The Hilandar typikon recommends to follow the custom of playing the klepalo and bilo as the first call to divine service, which would then be carried on with bell ringing [27]. As Rodriguez noted, the Evergetis typikon does not mention bells; however, Saint Sava did mention them in the place which the former refers to the bronze semantron (probably bilo). Arguing further that bell ringing was introduced earlier in medieval Serbia than in Byzantium, Rodriguez pointed out that monuments, such as the endowments of St. Sava's father, the Grand Prince Stefan Nemanja, such as the monastic church of Saint Nicholas near Kuršumlija (1159–1166) and Đurđevi Stupovi (1166), were both originally built with bell towers. However, while bell ringing did not assume the role of semantra, these percussion instruments were used in conjunction throughout the medieval period, creating an eclectic religious soundscape [47]. Listing and analyzing the extant bells from medieval Serbia and the wider region, Rodriguez concluded that the pealing of bells had an important role in the local religious soundscape [48].

The sound of bells and semantra have reverberated for long distances, broadcasting an auditory message across rural landscapes—marking the canonical hours, summoning the congregation to pray, alerting to danger, and informing of important events. The number of hits depended on the message that was conveyed [46]. However, the quality of the aural communication varied, depending on both the acoustic environment and sociocultural context [49]. Therefore, extensive archaeoacoustic research is necessary to draw conclusions on the quality of this eclectic religious soundscape.

Figure 4. The setting of large percussion instruments in the Žiča monastery. From left to right: bell, *bilo*, *klepalce* or manual *klepalo*, *klepalo* (photo: Zorana Đorđević).

The meaning of the sound of the bell and semantron has been investigated more thoroughly, pointing out that these sounds were perceived as a sign of divine presence [50]. Symbolic interpretations of semantron in early Byzantine texts compared them to angelic trumpets that will "awaken all peoples on the doomsday" and described them as calling monks and Christ's soldiers to the invisible warfare with demonic forces [51]. The sound of bells has been a part of the manifestation of sacrality ever since they became a central element of Christian identity; however, its social and spiritual meanings varied markedly depending on the context, use, and sound reception [52]. It provided local cohesion, social inclusion, and a sense of a place within the acoustic range [53], provoking the feeling of attachment to locality as an intended emotional response to sound [22], relating people strongly to their surroundings, shaping the identity of places by defining or delimiting their use [54], and producing spiritual community in medieval times [52].

Under the Ottoman rule of the Serbian territories, bells were forbidden and systematically destroyed for religious reasons [47,55]. This resulted in the more frequent use of kelpalo and bilo. Since people tended to save bells as sacred objects by hiding them, throwing them into wells and rivers or burying them in the ground, some bells were successfully preserved and reintroduced after the liberation. Their significance grew beyond religious needs and they became a symbol of the freedom and unity of the Serbian people. However, during the First World War, the Austro-Hungarian authority ordered taking-away all Serbian bells to melt them for arms. For this reason, there are not many medieval bells preserved today [48,56].

3.2. Litany Procession

Zavetina is a day devoted to a patron saint who protects a village. When a misfortune befalls the village—disease, flood or drought—the villagers would take a vow (*zavet* in Serbian) to a saint to celebrate his/her day if in return he/she protects them from the occurring accident. This vow would then pass on to their descendants [57]. An essential role in the celebration of *Zavetina* is placed on *Zapis*—a sacred tree with a cross incurved into its bark. This open-air sacred place is a primitive temple where people prayed and made sacrifices. Each village has several sacred trees. The main *Zapis* is usually located in a village center, while others are located on a village border, often on private properties [58].

The most important part of the *Zavetina* celebration is a litany procession that starts with a gathering of villagers around the main *Zapis*. Then, a priest reads a prayer while villagers carry red flags, crosses, and icons around the sacred tree (Figure 5). After completing three circles around the *Zapis*, the congregation goes around the village borders, visiting all the sacred trees and repeating the same activities of three times circling around each sacred tree while a priest reads a prayer. Then, they would decorate the tree and renew the cross in its bark by recurving it. Consecrating the sacred trees on the village borders, the area inside or along the circle is separated as sacred, leaving as profane everything that remains outside the circle [59]. The intention of drawing this imagined circle around a village is to protect its inhabitants from various misfortunes. The litany procession was welcome to go across the crop fields to influence the abundant harvest. There was also a belief that litany can contribute to healing people, who were therefore brought onto the procession path [60]. The litany completes in front of the main *Zapis* with the common feast, followed by numerous rite activities, including breaking the festive cake, naming the hosts by the priest, and giving toasts [59].

Figure 5. Litany on Saint Marco's Day in the village Lužnice (photo: Marija Dragišić).

Each village in Serbia used to have its own *Zavetina*, which would usually be in the period from Easter to Saint Peter's Day [61]. In this period, from April to July, crops need rain the most. Therefore, during the celebration of *Zavetina*, the villagers prayed and practiced various rite activities for summoning the rain. The congregation in litany sung a prayer to invoke rain. These songs were sung in unison and with strong voice by all the villagers. One of such songs is noted in the village Boljevac, in eastern Serbia [59]:

> I carry a cross, praying to God,
> Lord, Lord, have mercy!
> We pass across the field
> Additionally, clouds pass across the sky,
> Lord, Lord, have mercy!
> We pray to God:
> To give us morning dew,
> Lord, Lord, have mercy!
> To mow our fields,
> Lord, Lord, have mercy!
> To make a sheaf from two ears of wheat,
> Lord, Lord, have mercy!
> A cask of wine from two clusters,
> Lord, Lord, have mercy!

(In Serbian: Крста носим, Бога молим, Господе, господе, помилуј! Ми идемо преко поља, А облаци преко неба, Господе, господе, помилуј! Ми се Богу помолисмо:Да удари етња роса, Господе, господе, помилуј! Да пороси поља наша, Од два класа шиник жита, Од две гице, чабар вина, Господе, господе, помилуј!)

Litany included two types of sound events: priestly prayer and the common singing of the villagers. Taking place outside populated areas—on meadows, fields, and forest slopes—acoustically implies the absence of reflections and good speech intelligibility but also the need to speak in elevated levels so that information would be conveyed to the congregation and encourage the unity of a village community. Usually, each household had at least one participant in the litany to ensure being included in the prayer and having a good harvest [62]. Thus, the number of people depended on the size of a village. In addition to the aim of gathering the community, litanies were also meant to protect the village from an atmospheric disaster. Collective singing was used to propitiate the forces of nature. It was often performed on the move while the litany was passing from one sacred place to another around the village using the shortest route, even if it meant crossing the crop field. In some villages, the litany goes through a river, with the villagers believing that it would summon rain. Consequently, the collective singing could be heard with different intensities on different parts of the route, thus announcing the arrival of the litany to a certain place. At the head of the procession, in some cases, was a man who struck the klepalo [62], thus giving a loud rhythmic component while ensuring that the sound of the litany reached the distant places of the landscape—houses at the far ends of the village, and behind hills, forests, and rivers. The gathering of all the participants of the litany around the sacred tree was followed with prayer and the addressing of the priest. The sacred tree has always been located in a quiet area, which is suitable for people addressing each other with intelligibility and a sense of intimacy.

3.3. Open-Air Assemblies Related to the State and the Church

Sacred places in medieval Serbia were also used for various gatherings. The most important one was the state assembly—the representative body of medieval Serbia. To discuss the acoustics of open-air sacred places where the medieval assemblies were held, it is essential to know the purpose of these assemblies, the approximate number of participants, and the physical characteristics of the location.

The state assembly gathered the high clergy of the Serbian Orthodox Church, including all the episcopes headed by the archbishop (from the mid-14th century, the patriarch) and an equal number of hegumens or elders of Serbia's most renowned medieval monasteries. Towards the end of the 14th century, the assembly consisted of the 29 most prominent representatives of the Serbian Orthodox Church—the archbishop, 14 episcopes, and 14 hegumens. In addition to the clergy, the state assembly included the representatives of the state's civil and military administration, which was, at the most, two or three times the number of the representatives of the church assemblies. If the ruler was absent, he was substituted by his wife. The total number of the assembly participants was limited to about 100 of the most influential persons in the state. The participants were divided into groups and seated around separate tables [63].

The state assemblies were initiated by the ruler, who chose the time and place, usually in the capital or in a monastery [64]. On these occasions, rulers were crowned; heirs to the crown were proclaimed; deceased rulers were canonized; church dignitaries, archbishops, and patriarchs were elected; laws were passed; wars were declared; and decisions on important state issues were made. Prior to making the decision, the ruler was obliged to discuss the matter with the most prominent representatives, after which he would announce his decision to the assembly. For example, on the first state assembly in the town of Ras, which was summoned by the progenitor of the Nemanjić dynasty, Stefan Nemanja, both political and religious debate was on the matter of heresy. Later assemblies also show that accordance on state matters often came through long debates, and the assembly could take several days. On the second assembly held in 1196 outside the Saint Peter's Church in Ras, Nemanja abdicated and appointed his middle son Stefan as heir. To do so, Nemanja needed the support of the assembly, because appointing the younger son as a successor was in breach of tradition and custom laws regarding the rights of the firstborn son and as such needed to be ratified by the state representative body. On this occasion, Nemanja gave a short speech to the assembly. Although the Church did not limit the sovereignty of the ruler, it had the obligation of the ethical supervision but could not deprive the ruler of the throne. On the other hand, the ruler proposed and then the state assembly elected the new archbishop. For that purpose, the assembly would summon at least twice—the first time to elect the archbishop and the second time to attend the church ceremony and the celebration. However, the assembly did not always approve the ruler's suggestion for the archbishop. During the rule of the king Milutin Nemanjić (1282–1321), the assembly was summoned three times in one year to elect a suitable archbishop [63].

In addition to the state assemblies, there were also the church assemblies and the church–people's assemblies. With the foundation of the Serbian autocephalous Archiepiscopy in 1219, Saint Sava became the first archbishop of the Serbian Orthodox Church. He had introduced the church assemblies, which became particularly important after Serbia fell under Ottoman rule in 1459, as they also took over the role of the medieval Serbian state assemblies [65].

The third type was the church–people's assembly, which included high clergy, the royal family, and nobility, but also a significant number of poor people. The Church organized them, supported by the ruler, with a clear aim to enhance the reputation of both the Church and the Serbian State. One of the great church–people's assemblies was organized for the transportation of the relics of Saint Luke the Evangelist in 1453. When the relics, which were believed to have miraculous power, arrived in Serbia from Epirus where they were purchased, they were welcomed by a great solemn procession of the state and church officials, as well as numerous common people. The relics were transported to the capital of Smederevo, where they were then carried through and around the city while the procession attendees chanted church songs. Finally, the relics were placed at the Smederevo Metropolitan Church [63].

In all three types of assemblies described above, people gathered in the immediate vicinity of churches or monasteries. The sacrality of these spaces emphasized the importance of the assembly for the wider community, as well as the mutual dependence of the state and the church. This kind of event had such significance that it was occasionally presented on church frescos (Figure 6). The priority of these assemblies was to convey a message, decision, or regulation; thus, the intelligibility of speech was particularly important. The proximity of monastic buildings or church walls, certainly without any ceiling, created a specific acoustic environment. Reflective surfaces were sound reflectors (usually high stone walls), which significantly enhanced sound by stimulating the reflected sound in the area of temporal integration of the sense of hearing. At the same time, open-air places implied the absence of all those acoustic anomalies and problems that negatively affected speech intelligibility, especially the existence of pronounced late reflections (over 50 ms) and excessive reverberation times. Additionally, it is reasonable to assume that the address of the speakers, who always had a strong authority in relation to the people present, meant a quiet and peaceful environment, without noise and the murmur of voices. Given the absence of any unnatural noise sources in medieval times, this parameter of interfering with speech intelligibility (high levels of ambient noise) was also reduced to a minimum. In that sense, this type of gathering, with the default address of the speaker with an increased level of voice, provided high speech intelligibility, which was certainly the basic acoustic requirement. Additionally, this acoustic ambience can be described as intimate, in the sense that it formed a sound environment of close connection with the speaker, who was in the immediate vicinity of the people present.

Figure 6. Fresco The Assembly of the Saint Simeon the Myrrh-Gusher in the Patriarchy in Peć (photo: William Taylor Hostetter, Institute for the Protection of Cultural Monuments of Serbia).

4. Discussion

The above review of archaeoacosutic studies of medieval churches in Section 2 and the review of studies that corroborate the need for including the open-air soundscapes in archaeoacoustic analysis of sacred sites in Section 3 of this paper, impose some questions that need to be further discussed: 1. Are there acoustic characteristics common to all tested Serbian medieval churches? 2. do they have the optimal reverberation time for the medieval religious practice? and 3. what should be included in the archaeoacoustic open-air soundscape studies? We will further discuss these questions in the same order.

Both spoken and chanted words have been important in religious practice in medieval Serbia. It has been equally important that the acoustical environment supports the intelligibility of speech but also adds to the spiritual experience of the religious chanting. Previous archaeoacoustic research showed that this balance was achieved to a variable extent. To examine it, it is inevitable to include at least acoustic parameters, such as reverberation time, early decay time, and speech transmission index. The acoustic analysis of medieval Serbian churches indicates some specificities: although they differ significantly in terms of volume, objective and subjective analyses indicate an acoustic environment that fully corresponds to the basic purpose of these buildings—a sacred sound environment with preserved speech intelligibility. The analysis of churches built in the Morava architectural style pointed out that even when the churches have significantly different volumes (for example, Ljubostinja has a volume three times larger than that of Naupara), RT differences are not necessarily pronounced. The reason might lie in the macrogeometry (columns supporting the central dome) or microgeometry (wooden furniture, carpets). Intelligibility of speech differs among the researched churches depending on the sound source position. It is significantly reduced when the sound source is in the alter, divided from the naos with the iconostasis. It also decreases as the church volume grows. One of the main architectural features of churches built in the Morava style is the central dome, which has its footprint on the acoustic quality of the space. The dome produces a unique play of late sound delays, finally resulting in a subjective perception of a much larger space than it is, which also sonically adds to the spiritual experience of the religious service. This effect is particularly pronounced in Pavlovac church, the smallest one examined. In the case of wooden churches, this acoustic–spiritual effect is missing; however, the speech intelligibility is very high. However, to provide a higher resolution image of medieval church acoustics in Serbia and the proclaimed goal to obtain a balance between speech intelligibility and spiritual experience, it is important to include churches built in other medieval architectural styles.

When considering a broader picture, it is important to note that Serbian medieval churches, even the largest ones, such as Manasija, are significantly smaller than early Byzantine churches, such as Hagia Sophia in Istanbul, which is 150,000 cubic meters [8] or Hagia Sophia in Thessaloniki, which is 15,250 cubic meters [11]; however, they are comparable to late Byzantine churches, which are reported to range from 250 to 3000 cubic meters [10,18,19,66]. Thus, although the study of reverberation times in Serbian Orthodox churches noted their values without recommending the optimum reverberation time for Byzantine chanting tradition, which had been practiced in medieval Serbia [67], there are several recommendations for an optimal reverberation time in churches that should be discussed regarding Serbian medieval churches. Testing of the subjective preference of reverberation that is considered pleasing in relation to Byzantine chanting showed a clear preference for having a reverberation time of 4 s [32]. The Perez-Minana formula for speech in occidental churches T = 0.08 $\sqrt[3]{V}$ [68] gives significantly lower results compared with those that were measured in Serbian medieval churches. Thus, applying this formula for Manasija church (V = 4000 m^3) gives the reverberation time value of 1.3 s, while the measured RT30 @500 Hz is 3.6 s. Another calculation for the optimal reverberation time for liturgical music at mid-frequency range was recommended by Leo Beranek, according to Adeeb and Sü Gül, and is calculated as RT = 0.55 × log 10 (Vol.) − 0.14 [19]. For the analyzed Serbian churches (400–4000 m^3), that would mean that the RT should be between 1.2 and 1.8 s, which matches three out of eight studied churches (see Table 1).

The variety of sacred soundscapes in medieval Serbia supported the religious needs that go beyond a church space and included the surrounding landscape—in the vicinity of a church or monastery in the case of state/church assemblies or outlining the protective circle of sacred places around a village, thus spiritually uniting the community for a common goal (protecting from disease, summoning rain, ensuring fruitful yield). These soundscapes were permeated with the sound of large percussion instruments—bells and semantra—that conveyed various messages over a great distance, maintaining the connection between the wider community and the monastery but also giving a pulse to monastic life itself. Therefore, to understand the aural environment of medieval rural areas, in addition to church acoustics, it is necessary to include the multidisciplinary archaeoacoustic examination of the open-air soundscapes of sacred sites, starting from mapping the range of bell and semantron sounds, their relation to visual reach, and their capacity for communication. As pointed out in this paper, the existing studies of medieval Serbian soundscapes mainly addressed the sound aspect from a monodisciplinary point of view—anthropology, ethnology, and history. However, a multidisciplinary approach is key to researching past sound environments because it is a complex question that needs to be examined from various sources, including the fields of acoustics, archaeology, the history of art and architecture, anthropology, and musicology.

5. Conclusions

In this paper, we strived to argue the need for expanding the archaeoacoustic research of medieval Serbian sacred sites to open-air surrounding soundscapes, in addition to church acoustics. Therefore, to outline the mosaic of sacred aural environments in medieval Serbia, we reviewed the archaeoacoustic studies of medieval Serbian churches, both monastic and village, and we examined the studies that addressed outdoor sacred soundscapes. This review paper points out that open-air sacred soundscapes have not been acoustically measured in Serbia so far; however, their investigation would remarkably advance our understanding of sonic environments in sacred medieval sites.

Author Contributions: Conceptualization, Z.Đ.; investigation, Z.Đ., D.N. and M.D.; writing—original draft preparation, Z.Đ.; writing—review and editing, Z.Đ., D.N. and M.D. All authors have read and agreed to the published version of the manuscript.

Funding: This research received no external funding.

Institutional Review Board Statement: Not applicable.

Informed Consent Statement: Not applicable.

Data Availability Statement: Not applicable.

Acknowledgments: We are grateful to Aleksandra Subić for the song translation. We would also like to thank the two anonymous reviewers and Guest Editors for their helpful suggestions to improve the manuscript.

Conflicts of Interest: The authors declare no conflict of interest.

References

1. Girón, S.; Álvarez-Morales, L.; Zamarreño, T. Church acoustics: A state-of-the-art review after several decades of research. *J. Sound Vib.* **2017**, *411*, 378–408. [CrossRef]
2. Alonso, A.; Suárez, R.; Sendra, J. The Acoustics of the Choir in Spanish Cathedrals. *Acoustics* **2018**, *1*, 35–46. [CrossRef]
3. Boren, B. Word and Mystery: The Acoustics of Cultural Transmission during the Protestant Reformation. *Front. Psychol.* **2021**, *12*, 564542. [CrossRef] [PubMed]
4. CAHRISMA: The Conservation of Acoustical Heritage by the Revival and Identification of Sinan's Mosques' Acoustics', 2000–2003. Available online: https://cordis.europa.eu/project/id/ICA3-CT-1999-00007 (accessed on 13 December 2022).
5. Pentcheva, B.V. *Hagia Sophia: Sound, Space, and Spirit in Byzantium*; The Pennsylvania State University Press: University Park, PA, USA, 2017; ISBN 978-0-271-07726-0.
6. Woszczyk, W. Acoustics of Hagia Sophia: A scientific approach to the humanities and sacred spaces. In *Aural Architecture in Byzantium: Music, Acoustics, and Ritual*; Pentcheva, B.V., Ed.; Routledge, Taylor & Francis Group: London, UK, 2018; pp. 176–197. ISBN 978-1-4724-8515-1.
7. Pentcheva, B.V. (Ed.) *Aural Architecture in Byzantium: Music, Acoustics, and Ritual*; Routledge, Taylor & Francis Group: London, UK, 2018; ISBN 978-1-4724-8515-1.
8. Sü Gül, Z.S. Acoustical Impact of Architectonics and Material Features in the Lifespan of Two Monumental Sacred Structures. *Acoustics* **2019**, *1*, 493–516. [CrossRef]
9. Knight, D.J. The Archaeoacoustics of a Sixth.Century Christian Structure San Vitale, Ravenna. In *Music & Ritual: Bridging Material & Living Cultures*; Jiménez, R., Till, R., Howell, M., Eds.; Ekho Verlag: Berlin, Germany, 2013; pp. 133–146. ISBN 978-3-944415-13-0.
10. Tzekakis, E.G. Data on the acoustics of the Byzantine churches of Thessaloniki. *Acustica* **1979**, *43*, 275–279.
11. Tzekakis, E.G. The acoustics of the Early-Christian monuments of Thessaloniki. *Arch. Acoust.* **1981**, *6*, 3–12.
12. Karampatzakis, P. *Acoustic Measurements in 11 Byzantine Temples Thessaloniki*; Hellenic Institute of Acoustics (ELINA), Democritus University of Thrace: Xanthi, Greece, 2008.
13. Karampatzakis, P. *Comparison of the Acoustic Parameters of the Monuments of Achiropoiito and Panagia Chalkeon Using Measurements and Mathematical Models*; Hellenic Institute of Acoustics (ELINA): Athens, Greece, 2010.
14. Antonopoulos, S.; Gerstel, S.E.J.; Kyriakakis, C.; Raptis, K.T.; Donahue, J. Soundscapes of Byzantium. *Speculum* **2017**, *92*, S321–S335. [CrossRef]
15. Gerstel, S.E.J.; Kyriakakis, C.; Raptis, K.T.; Antonopoulos, S.; Donahue, J. Soundscapes of Byzantium: The Acheiropoietos Basilica and the Cathedral of Hagia Sophia in Thessaloniki. *Hesperia J. Am. Sch. Class. Stud. Athens* **2018**, *87*, 177. [CrossRef]
16. Gerstel, S.E.J. Monastic Soundscapes in Late Byzantium: The Art and Act of Chanting. In *Resounding Images: Medieval Intersections of Art, Music, and Sound*; Boynton, S., Reilly, D.J., Eds.; Studies in the visual cultures of the Middle Ages; Brepols: Turnhout, Belgium, 2015; pp. 135–152. ISBN 978-2-503-55437-2.
17. Gerstel, S.E.J.; Kyriakakis, C.; Antonopoulos, S.; Raptis, K.T.; Donahue, J. Holy, Holy, Holy: Hearing the Voices of Angels. *Gesta* **2021**, *60*, 31–49. [CrossRef]
18. Sukaj, S.; Bevilacqua, A.; Iannace, G.; Lombardi, I.; Parente, R.; Trematerra, A. Byzantine Churches in Albania: How Geometry and Architectural Composition Influence the Acoustics. *Buildings* **2022**, *12*, 280. [CrossRef]
19. Adeeb, A.H.; Sü Gül, Z. Investigation of a Tuff Stone Church in Cappadocia via Acoustical Reconstruction. *Acoustics* **2022**, *4*, 419–440. [CrossRef]
20. Vissière, L. Le paysage sonore parisien aux XIIIe et XIVe siècles ou la naissance des cris de Paris. *Bull. Société Natl. Antiq. Fr.* **2015**, *2010*, 136–158. [CrossRef]
21. Leroux, L. *Cloches et Société Médiévale: Les Sonneries de Tournai au Moyen Âge*; Tournai—Art et Histoire: Tournai, Belgium, 2011; ISBN 978-2-87419-045-2.
22. Mileson, S. Sound and Landscape. In *Oxford Handbook of Medieval Archaeology*; Gerrard, C., Gutierrez, A., Eds.; Oxford University Press: Oxford, UK, 2016.

23. Mlekuz, D. Listening to the Landscapes: Modelling Soundscapes in GIS. *Internet Archaeol.* **2004**, *16*. [CrossRef]
24. Troelsgard, C. Byzantine chant notation. Written documents in an aural tradition. In *Aural Architecture in Byzantium: Music, Acoustics, and Ritual*; Pentcheva, B.V., Ed.; Routledge, Taylor & Francis Group: London, UK, 2018; pp. 52–77. ISBN 978-1-4724-8515-1.
25. Bertoldi, S.; Castiglia, G.; Castrorao Barba, A.; Menghini, C. Soundscape and catchment analysis for a spatial geography of Medieval monastic estates in southeastern Tuscany (11th–12th centuries). *AeC* **2022**, *33*, 95–114. [CrossRef]
26. Baker, J.; Brookes, S. Identifying outdoor assembly sites in early medieval England. *J. Field Archaeol.* **2015**, *40*, 3–21. [CrossRef]
27. Sava, S. *Sabrana Dela*, 2nd ed.; Srpska Književna Zadruga: Beograd, Serbia, 2018; ISBN 978-86-379-1375-7.
28. Mirković, L. *Pravoslavna Liturgika ili Nauka o Bogosluženju Pravoslavne Istočne Crkve*; Sveti Arhijerejski Sinod Srpske Pravoslavne Crkve: Beograd, Serbia, 1965.
29. Stefanović, D. *Old Serbian Music—Examples of 15th Century Chant*; Institute of Musicology SASA: Belgrade, Serbia, 1975.
30. Popmihajlov, N. Crkvena muzika i bogosluženje. *Istina—Bogoslovski Časopis Pravoslavne Eparhije Dalmatinske* **2004**, *5*, 9–11.
31. Teodosije. *Žitije Svetog Save*; Antologija Srpske Književnosti: Belgrade, Serbia, 2009.
32. Mourjopoulos, J.; Papadakos, C.; Kamaris, G.; Chryssochoidis, G.; Kouroupetoglou, G. Optimal acoustic reverberation evaluation of Byzantine chanting in churches. In Proceedings of the International Computer Music Conference, Athens, Greece, 14–20 September 2014; pp. 14–20.
33. Đorđević, Z.; Novković, D.; Dragišić, M. Acoustics of Orthodox churches from late medieval Serbia: Questioning objective and subjective approach. In *Archaeoacoustics: Scientific Explorations of Sound in Archaeology*; Forthcoming; Acoustical Society of America (ASA) Press: Monroe, MI, USA; Springer: Berlin/Heidelberg, Germany.
34. Kalić, D. Acoustic Resonators in Serbian Medieval Churches. In Proceedings of the Second Joint Meeting of Greek and Yugoslav Acoustical Societies, Athens, Greece; 1984; pp. 91–97.
35. Mijić, M.; Šumarac-Pavlović, D. Analysis of Contribution of Acoustic Resonators Found in Serbian Orthodox Churches. *Build. Acoust.* **2004**, *11*, 197–212. [CrossRef]
36. Đorđević, Z.; Novković, D.; Pantelić, F. The Ceramic Vessels of Trg: Acoustic Wall Construction in a Medieval Serbian Church. *Change Over Time Int. J. Conserv. Built Environ.* **2020**, *9*, 2. [CrossRef]
37. Valière, J.-C.; Bertholon, B.; Đorđević, Z.; Novković, D. Revisiting archaeoacoustic methodology in the studies of acoustic vessels: Application to Brittany and Serbia. In Proceedings of the 2nd Symposium: The Acoustics of Ancient Theatres, Verona, Italy, 6 July 2022.
38. Valière, J.-C.; Bertholon, B. Towards a history of architectural acoustics using archaeological evidence: Recent research contributions to understanding the use of acoustic pots in the quest for sound quality in 11th–17th-century churches in France. In *Worship Sound Spaces: Architecture, Acoustics and Anthropology*; Research in architecture; Guillebaud, C., Lavandier, C., Eds.; Routledge: New York, NY, USA, 2020; ISBN 978-0-429-27978-2.
39. Kanev, N. Resonant Vessels in Russian Churches and Their Study in a Concert Hall. *Acoustics* **2020**, *2*, 399–415. [CrossRef]
40. Valière, J.-C.; Palazzo-Bertholon, B.; Polack, J.-D.; Carvalho, P. Acoustic Pots in Ancient and Medieval Buildings: Literary Analysis of Ancient Texts and Comparison with Recent Observations in French Churches. *Acta Acust. United Acust.* **2013**, *99*, 70–81. [CrossRef]
41. Đorđević, Z.; Novković, D.; Andrić, U. Archaeoacoustic Examination of Lazarica Church. *Acoustics* **2019**, *1*, 423–438. [CrossRef]
42. Đorđević, Z.; Novković, D. Archaeoacoustic Research of Ljubostinja and Naupara Medieval Monastic Churches. *Open Archaeol.* **2019**, *5*, 274–283. [CrossRef]
43. Mijić, M. *Akustika Pravoslavnih Crkava—Sinteza Objektivnog i Subjektivnog u Akustičkim Komunikacijama*; 8th Telecommunications Forum TELFOR: Belgrade, Serbia, 2000.
44. Pavlović, D. Crkve brvnaare—Graditeljstvo. In *Molitva u Gori: Crkve Brvnare u Srbiji*; Republički Zavod Za Zaštitu Spomenika Kulture: Beograd, Serbia, 1994; pp. 21–37.
45. Pavićević-Popović, R. Molitva u gori. In *Molitva u Gori: Crkve Brvnare u Srbiji*; Republički Zavod Za Zaštitu Spomenika Kulture: Beograd, Serbia, 1994; pp. 5–19.
46. Stanojević, S. Bila, klepala i zvona kod nas. *Glas Srpske Kraljevske Akademije* **1933**, *CLIII*, 79–90.
47. Suarez, A.R. When did the Serbs and the Bulgarians adopt bell ringing? *CAS Work. Pap. Ser.* **2018**, *10*, 3–31.
48. Suarez, A.R. Bells and bell ringing in medieval Serbia and Bulgaria. *Etudes Balkaniques* **2021**, *LVII*, 453–504.
49. Kang, J.; Aletta, F.; Gjestland, T.T.; Brown, L.A.; Botteldooren, D.; Schulte-Fortkamp, B.; Lercher, P.; van Kamp, I.; Genuit, K.; Fiebig, A.; et al. Ten questions on the soundscapes of the built environment. *Build. Environ.* **2016**, *108*, 284–294. [CrossRef]
50. Neri, E. Les cloches: Construction, sens, perception d'un son. *Cahiers Civilisation Medievale Xe-XIIe Siecle* **2012**, *55*, 473–496.
51. Miljković, B. Semantra and bells in Byzantium. *Zbornik Radova Vizantološkog Instituta* **2018**, *55*, 271–303. [CrossRef]
52. Arnold, J.H.; Goodson, C. Resounding Community: The History and Meaning of Medieval Church Bells. *Viator* **2012**, *43*, 99–130. [CrossRef]
53. Corbin, A. *Village Bells: Sound and Meaning in the Nineteenth-Century French Countryside*; European perspectives; Columbia University Press: New York, NY, USA, 1998; ISBN 978-0-231-10450-0.
54. Coronado Schwindt, G. The Social Construction of the Soundscape of the Castilian Cities (15th and 16th Centuries). *Acoustics* **2021**, *3*, 60–77. [CrossRef]
55. Pavićević-Popović, R. Zvono—s osvrtom na Žička zvona. *Povelja* **2007**, *37*, 153–159.

56. Aksić, N. Druge zove, sebe ne čuje—zvono u srpskoj narodnoj tradiciji [Tolling for others while not hearing itself—Bell in Serbian folk tradition]. *Philol. Median.* **2014**, *6*, 151–165.
57. Kulišić, Š.; Petrović, P.Ž.; Pantelić, N. *Srpski Mitološki Rečnik*; Nolit: Beograd, Serbia, 1970.
58. Dragišić, M. Zapis as an object of protection of non-portable cultural properties. *Communications* **2009**, *XLI*, 227–244.
59. Bandić, D. Narodna Religija Srba u 100 Pojmova. In *Biblioteka Odrednice*, 2nd ed.; Nolit: Beograd, Serbia, 2004; ISBN 978-86-19-02328-3.
60. Milićević, M.Đ. *Život Srba Seljaka*; Prosveta: Belgrade, Serbia, 1984.
61. Stefanović Karadžić, V. *Srpski Rječnik*; Štamparija Kraljevine Jugoslavije: Beograd, Serbia, 1935.
62. Todorović, I. *Ritual Uma: Značenje i Struktura Litijskog Ophoda*; Posebna Izdanja/Etnografski Institut. Srpska Akademija Nauka i Umetnosti; Etnografski Institut, Srpska Akademija Nauka i Umetnosti: Beograd, Serbia, 2005; ISBN 978-86-7587-031-9.
63. Blagojević, M. Serbian Assembly and the Aassembly of the Nemanjić and Lazarević Fatherland. *Glas. CDX L'academie Serbe Sci. Arts Cl. Sci. Hist.* **2008**, *14*, 1–39.
64. Ćirković, S.M.; Mihaljčić, R. *Leksikon Srpskog Srednjeg Veka*; Knowledge: Beograd, Serbia, 1999; ISBN 978-86-83233-01-4.
65. Vlahović, P. Saborovanje. In *Narodna Kultura Srba u XIX i XX Veku: Vodič Kroz Stalnu Postavku*; Stojaković, V., Ed.; Etnografski muzej u Beogradu: Beograd, Serbia, 2003; pp. 11–19. ISBN 978-86-7891-013-5.
66. Gerstel, S.E.J.; Kyriakakis, C. Revealing the acoustic mysteries of Byzantine churches. *Faith Form* **2016**, *49*, 15–17.
67. Šumarac Pavlović, D.; Mijić, M. Preferred Reverberation Time of serbian Orthodox Churches. In Proceedings of the 17th International Congress on Acoustics, Rome, Italy, 2–7 September 2001.
68. Fernandez, M.; Recuero, M. Data Base Design for Acoustics: The Case of Churches. *Build. Acoust.* **2005**, *12*, 31–40. [CrossRef]